feral future

To my neighbour Jill Skattebol, and to all the other doughty weeders working tirelessly around Australia to protect precious bushland remnants from advancing walls of weeds.

Praise for Feral Future

'An important work . . . easy to read, fascinating, often eye-popping.'

Australian

'A gripping work that deserves a wide readership.'

Age

'A lucid and harrowing account of the background and implications of bioinvasion.'

Sydney Morning Herald

'A new and sure-to-be-important book.'

Canberra Times

feral future

The Untold Story of Australia's Exotic Invaders

TIM LOW

The University of Chicago Press
Chicago & London

Tim Low is a biologist, writer and photographer best known for his books, which include the prize-winning best-sellers *Bush Tucker* and *Bush Medicine*, and for his long-running column in *Nature Australia* magazine. Apart from writing and photographing wildlife, he conducts fauna and flora surveys, and assesses bushland for cities and shires, often advising about weed problems. He has represented the conservation movement on government weed committees and has spoken at international conferences on exotic pests.

Published by arrangement with Penguin Books Australia Ltd.
The University of Chicago Press, Chicago 60637
First published by Penguin Books Australia Ltd. 1999;
second edition 2001

University of Chicago Press edition 2002
Printed in the United States of America

11 10 09 08 07 06 05 04 03 02   1 2 3 4 5
ISBN: 0-226-49419-5 (cloth)

Library of Congress Cataloging-in-Publication Data

Low, Tim, 1956–
Feral future : the untold story of Australia's exotic invaders / Tim Low.
p. cm.
Originally published: Ringwood, Vic. : Viking, 1999.
Includes bibliographical references (p.).
ISBN 0-226-49419-5 (cloth : alk. paper)
1. Biological invasions–Australia. 2. Exotic plants–Australia.
3. Plant introduction–Australia. 4. Pest introduction–Australia.
5. Animal introduction–Australia. I. Title.

QH353 .L68 2002
577'.18–dc21   2002067534

♾ The paper used in this publication meets the minimum requirements of the American National Standard for Information Sciences–Permanence of Paper for Printed Library Materials, ANSI Z39.48-1992.

PREFACE, 2002

In 1996 a member of the Western Australian parliament, Richard Evans, called for the eradication of all cats from Australia. He wanted diseases brought in to kill pet cats and their feral counterparts. 'The only good cat is a flat one', he declared. Five years earlier conservationist John Wamsley had worn a cat-skin hat to a public ceremony. He boasted of killing his first cat when he was ten. Wild-roaming cats are reviled for the toll they take on wildlife (Smith 1999).

Australians take their alien invaders seriously. Australia was never invaded by foreign armies, it was devastated by bioinvasion instead. Almost every Australian knows the old tales about Biblical-style rabbit and mouse plagues and about farmers abandoning their land to prickly pear. To be Australian is to accept an ironic legacy – a land overrun by foxes, toads, rabbits, blackberries and more.

Ecologically, Australia has suffered more from bioinvasion than any other large land. It's an oft-quoted fact that the land 'down under' boasts the worst mammal extinction record of historic times. Almost nowhere in temperate Australia can an intact mammal fauna still be found. Eighteen mammal species have vanished altogether and seven more survive precariously on small islands (Strahan 1995). Foxes and cats attract the most blame for this situation, although the truth is more complex. Ten frogs have recently disappeared, with most experts blaming an exotic fungus. Another fungal disease, *Phytophthora cinnamomi* from Asia, kills more than 2,000 native plant species, some of them endangered. No other large region has suffered this much from bioinvasion, except perhaps Africa, where fish vanished by the score after Nile perch were freed in Lake Victoria.

Australia has proved especially vulnerable because it is an island, a remote mass (7,682,000 square kilometres) that cleaved away from Gondwana – the southern supercontinent – 45 million years ago. The mammal fauna was marsupial-dominated when Englishmen came in the late eighteenth and nineteenth centuries with dreams of improving upon this picture. In came pigs, goats, horses, deer, donkeys, foxes, rabbits, sparrows, starlings, trout, salmon, perch and thousands of plants. Ancient ecosystems found themselves invaded by new carnivores, by hoofed beasts, toads, salmonid fish, pine trees and cacti. The clashes were dramatic. Colonial naturalists watched aghast as native mammals fell before foxes; just as biologists today watch rare fish and frogs succumb to trout.

The fall of Australia's 'primitive' marsupials before Europe's 'modern' fauna (p. 139) reinforced nineteenth-century ideals about European racial superiority. Australia was deemed archaic, a continent forgotten by time. Englishmen were usurping the 'primitive' Aborigines and the same logic dictated that European animals would replace Australian ones.

This picture of Australia as an evolutionary backwater has lingered on into modern times. So keen was I to repudiate it that I devoted several chapters to chronicling Australian invasions of other lands (part IV). Most Australians know nothing about the dramatic conquests achieved by our plants in Florida and South Africa. My tales about Australians invading Britain (chapter 19, Colonial Revenge) are meant to stimulate nationalistic juices. Australia suffered so ignominiously from invasion not because it was backward but because the newcomers were so diverse and so unlike anything encountered before. Foxes hunted differently, trout outgrew native fish, and new diseases struck naïve victims.

Notions of Australian inferiority, tied to deference toward British and then American culture, pervaded evolutionary thinking as well.

Biologists until recently assumed that most of Australia's birds evolved from northern hemisphere ancestors spreading south through Indonesia. Recent DNA and fossil evidence imply otherwise. Australia was probably the ancestral home of all the world's songbirds, order Passeriformes (Boles 1995, 1999), and it certainly gave rise to the corvoid lineage that includes all the world's crows, shrikes and magpies (Sibley & Ahlquist 1986).

Why Australia lost so many mammals (mostly from the dry Outback) remains a topic for hot debate. According to the mosaic burning hypothesis, Aborigines managed their lands by maintaining a mosaic of habitats, burning land in patches to ensure a diverse resource base. Wallabies could hide in unburnt thickets and feed nearby on shoots sprouting from the ash. When Aborigines lost their lands the vegetation thickened up until hot fires swept through and bequeathed a monotonous landscape that exposed animals to attacks by native and exotic predators. An opposing theory argues that small native mammals (including certain rodents as well as marsupials) could not withstand the combined onslaught of overgrazing by rabbits, sheep and cattle, and hunting by foxes and cats. Both theories have their adherents (Bowman 1998) but the evidence implicating foxes is especially compelling, as chapter 26 explains.

Because Australia was settled partly by convicts and partly by Irish nationalists, Australians today display a healthy disrespect for authority, coupled with a readiness for self-criticism. We can derive pleasure from ridiculing those among our forebears who imported foxes, rabbits, starlings and toads to 'improve' Australia (chapters 5 and 7). Given this attitude and the position won by alien invaders in Australian folklore, one might question why I felt a need to write this book. Australians, after all, are more pest-aware than almost any nationality (only New Zealand can compare). Our quarantine laws are among the strictest anywhere. We see the difference when we go through foreign airports. Australia's quarantine service, helped along by our island-nation status, has kept out rabies, foot-and-mouth disease, screw-worm and many other agricultural threats.

Australians place an unusually strong emphasis on what is native and what is not. Native animals and plants are widely cherished while exotic pests are despised. Cane toads are so loathed in Queensland that the city of Brisbane, the state capital, organised a series of toadbuster nights urging residents to kill them. Cane toads are hardly an economic problem, but they are deemed a threat to wildlife and they look repulsive. I attended one party at which children were awarded prizes for catching the most toads (which went into the freezer). A passionate love or hate relationship, bound to the idea of native verses exotic, exists more strongly in Australia than anywhere else (except perhaps New Zealand). It shows up in gardening tastes – most Australian nurseries devote a section to native plants, and whole natives-only nurseries are commonplace.

This situation contrasts with Europe, where a long history of introductions has blurred any sense of the differentiation between indigenous and introduced. Research in Britain has shown that rabbits, hares, house mice, rats, sparrows and Scotch thistles entered during Roman or medieval times, but few people know this. The status of many plants remains in question. This kind of uncertainty is rare in Australia, with its distinctive flora and fauna, and brief history of European occupation dating back to 1788. In chapter 35 I do mention species of uncertain status, but Australia has far fewer of these to wonder about than Europe, Asia or Africa.

I wrote this book because Australians have become complacent about future pest threats. A belief prevails that we have learned from past errors, that Australia is now well protected from further invasions. I argue the opposite: that Australia remains remarkably vulnerable, that new pests continue to invade at an unacceptable rate, and often under foolish circumstances. My approach in *Feral Future* is to highlight the hordes of alien pests in Australia that few people know about – starfish, earthworms, ants, seaweeds and the like – to point out that some of them arrived only recently, and to warn that others will invade soon. There was so much to say that I glossed over some of our worst rogues – foxes, rabbits, and prickly pear – because their stories are so well known. Had this book been aimed mainly at non-Australians the rabbit would have earned its own chapter (like the toad'). Rabbits have done more harm here, economically and ecologically, than almost any other pest, but Australians already know this. The deserts of drifting sand loosed by rabbits are ably discussed in Eric Rolls' classic *They All Ran Wild* (1966), and more recently in B.J. Coman's book, *Tooth and Nail* (1999). Rabbits were the 'chainsaws of the outback', a 'grey blanket' that smothered the land, wringing ruin.

Soon after *Feral Future* was first published in late 1999 I found my alarmist approach vindicated. In February 2001, red imported fire ants (*Solenopsis invicta*) were found in gardens in Brisbane, not far from my home. If these ants advance across the land they will rewrite the ecological equation. Native to South America, they have become simply the worst pest in the southern United States. Each year, eleven million Americans are stung and US $1.3 billion are spent on control. People in nursing homes have died after ants crawled into their beds. Bradley Vinson, a Texan professor, explained how life changes: 'in infested areas, people do not have picnics on the ground, go barefoot, sit or lay on the ground, or stand in one place without constantly looking at the ground' (Vinson 1997). You can't cross fields where nests are dense.

The ants threaten the outdoor 'Aussie' way of life. Already there are streets in Brisbane where children play on roads because lawns are dangerous. 'You only have to see people's backyards where there are nests every square metre to understand the seriousness of it all', says Brisbane fire ant operations director Keith McCubbin. 'I have seen backyards where the kids can't play, where the dog won't go and where they can't even mow their lawn'. One fire-ant sting isn't too bad (like a brush with nettles) but they attack in unison, bringing agony, raising blisters. They kill foals, fawns, calves, ground-dwelling birds, lizards – even fish. Attracted to electric fields, they crawl inside light switches and ruin computers and air conditioners. Houses burn down. They belong in an *X-Files* episode.

The first edition of *Feral Future* did not mention fire ants, for although I had seen their nests in Georgia they were considered unlikely to invade Australia. They actually entered Brisbane while I was writing this book. The size of established nests implies entry sometime between 1995 and 1998.

Brisbane has two large infestations, extending over more than 700 properties. Brisbane journalists made much of the discovery, running dramatic 'Killer Ant' articles, but the media further south has made light of it all, despite climatic modelling to show that fire ants, if not checked, will invade most of the Australian continent, avoiding only the drier outback regions and loftier mountains. Fire ants have already reached Melbourne in a potted plant brought from Brisbane, although those few were soon quelled. Colonies travel about in potted plants, loads of soil, and other garden supplies, and queens hitch rides on trucks.

Australia is taking the threat very seriously. A massive US $64 million five-year eradication campaign is underway. The federal government and every Australian state has contributed. American experts were flown in to offer advice. Baits are dropped regularly around tens of thousands of Brisbane homes in what amounts to one of the largest eradication efforts ever attempted anywhere in the world. The baiting is going well, with death rates of nests higher than expected. I now chair the government's Fire Ant Environmental Advisory Group.

How seriously wildlife here suffers will depend upon how successfully the ants invade intact habitats. In North America fire ants are confined largely to gardens, parks, farms and other disturbed settings. They infest campgrounds and walking trails but seldom penetrate natural habitats. That held true until the 1970s when a new form of fire ant appeared in America, a polygyne form in which each nest holds many queens instead of one (Wojcik et al. 2001). Polygyne worker ants roam from one nest to another with impunity, forming vast super-colonies instead of maintaining exclusive territories. Ant densities can reach 30 times that of the native ant fauna they displace and ten times that of the monogyne (one queen) form. In heavily infested areas 99 per cent of ants are fire ants. The polygyne ants often form new nests by budding – a group of ants that excavates a new nest within striking distance of the old. In this way they can work their way from clearings into natural forests. These ants are expected to exact a dramatic toll on wildlife, including endangered species. Australia has the world's richest fauna of lizards, many of which eat ants. A preliminary study conducted in Brisbane found a complete absence of lizards from one fire-ant infested site (Nattrass & Vanderwoude 2001). In America the ant-eating Texas horned lizard (*Phrynosoma cornutum*) disappears from fire ant sites.

Brisbane has two main fire ant infestations, and one is monogyne, the other polygyne. DNA work shows that the latter could only have come from Argentina, whereas the monogyne ants came directly from Brazil or indirectly via North America. How they arrived is anyone's guess. The Wacol ants presumably travelled among industrial goods, perhaps in stacked terracotta pots. The ants around the port are probably descended from a queen that flew off a ship. It could have travelled from North America in earth attached to second-hand farm or industrial machinery. Huge loads of soil reach Australia on equipment, as chapter 13 explains. Whatever the pathway, we are left with the chilling thought that Brisbane was invaded twice by these ants in a few years. We can expect more and more incursions in the years to come.

Only two months after fire ants were found in Brisbane, another exotic ant turned up – the yellow crazy ant (*Anoplolepis gracilipes*). Many decades ago these ants invaded Christmas Island, an Australian territory south of Java. Fifteen million red land crabs (*Gecarcoidea natalis*) have been killed there by the ants in recent years and ten bird species were left critically endangered. These ants appeared in north Queensland in April 2001, and an eradication campaign is now underway. They appeared at the Port of Brisbane soon after, but were soon extinguished. Australia will also be invaded soon by the little fire ant (*Wasmannia auropunctata*), or electric ant, a mean-stinging Latin American species that now thrives in plantations and forests on many Pacific islands east of Australia (Fabres & Brown 1978). Bill Crowe, a quarantine entomologist, says these ants are regularly intercepted at airports. 'I'm fairly confident there's been a huge increase in the amount of ants coming in on cargo', he told me. If I were writing *Feral Future* today I would probably devote a whole chapter to travelling ants. They are part of the high price we are paying for growing world trade. According to invasion biologist Dennis O'Dowd from Melbourne (originally from the U.S.), 'Tramp ants don't turn up by accident. They are an inevitable consequence of international trade. If you're moving millions of cargo containers around the world, it's going to happen'.

Fire ants are one example of a pest conquering two continents in parallel. A letter came to me recently from Daniel Hilburn of the Department of Agriculture in Oregon stating: 'I have worked in regulatory plant protection for twenty years in various parts of the United States and Bermuda. A surprising number of the invasive species you mention in your book are familiar to me'. He presumably had in mind such plants as privet (*Ligustrum sinense*), Brazilian pepper (*Schinus terebinthifolia*), English ivy (*Hedera helix*), Japanese honeysuckle (*Lonicera japonica*) – worse in the U.S. than here, buffel grass (*Cenchrus ciliaris*), periwinkle (*Vinca major*), pampas grass (*Cortaderia selloana*), broom (*Cytisus scoparius*) and water hyacinth (*Eichhornia crassipes*). The chytrid fungus killing our frogs has also invaded California, although it is less harmful there, and cane toads have colonised southern Florida.

U.S. readers should spare a thought for all the pests America has given us. They include mosquito fish, prickly pear (*Opuntia stricta*), mesquite (*Prosopis*), pond apple (*Annona glabra*), Monterey pine (*Pinus radiata*), slash pine (*P. elliottii*), ponderosa pine (*P. ponderosa*), groundsel bush (*Baccharis halimifolia*), century plant (*Agave americana*), American sea rocket (*Cakile edentula*), evening primroses (*Oenothera* species), and even coast redwood (*Sequoia sempervirens*). Black widow spiders turn up from time to time in cargo, but they have not yet established.

Australian exports to America include eucalyptus, wattles, saltbushes, rainforest trees, wallabies, lizards, lice, weevils, wasps, bugs, worms and even tobacco blue mould, a major crop disease (chapters 18, 21 and 22). These species count as unwelcome invaders, but Australia also gave California a ladybug that saved her citrus groves from destruction by an Australian sap-sucking bug (chapter 21), in one of the world's first biocontrol campaigns.

Cultural differences between the two countries have helped shape their pest plights. Americans place so much emphasis on liberty and personal rights that American governments are often loathe to ban even risky imports. I am thinking in particular of all the foreign reptiles sold in American pet stores, and the internet sites that offer cobras, poison arrow frogs and African scorpions. Florida and Hawaii are now blighted with all manner of escaped (or released) reptiles and frogs, a situation with no counterpart in Australia. Our pet shops aren't allowed to import reptiles at all. They are bland and relatively benign places (although see chapter 9) peddling puppies, kittens, mice, birds, fish, and little else.

Foreign readers may be unfamiliar with some of our animal names. Wombats are big burrowing marsupials with short legs. Bandicoots are much smaller marsupials resembling very large rats. Cassowaries are enormous flightless birds, related to emus but confined to tropical rainforests and second in weight only to the African ostrich. Goannas (*Varanus*) are very big lizards resembling the iguanas of Latin America (although unrelated). The CSIRO (Commonwealth Scientific and Industrial Research Organisation) is Australia's main scientific institution.

Since I wrote *Feral Future* many new issues have come to hand. Ants I have already mentioned. Foxes, previously confined to mainland Australia, turned up recently in Tasmania, perhaps introduced illegally by would-be hunters. A massive eradication campaign is underway. Cane toads reached Kakadu National Park in 2001, many years ahead of schedule. In chapter 7 I urge readers to see the splendour of Kakadu before toads move in, but already it's too late.

Monkeys may also pose a problem for Australia one day. Long-tailed macaques, brought to Irian Jaya (the Indonesian half of the New Guinea island) by American soldiers during World War II (to test wild foods), escaped into the rainforests around Jayapura. If they aren't killed off soon, they will spread through the New Guinea rainforests and wreak ruin by devouring bird-of-paradise eggs and fruits and seeds of rare trees. One day in the distant future they will reach the coast opposite Australia and become something new for Australian Quarantine to worry about. The fear will be that young monkeys will be carried illegally across the strait as pets.

I predicted that bumblebees would worsen our weed problems (chapter 40), and it's coming true. Tree lupins (*Lupinus aboreus*), major invaders in New Zealand (where bumblebees are rampant) but modest weeds here, are now showing 'a dramatic increase in seed set' in Tasmania, says entomologist David Goulson, who warns of 'a serious threat' (Goulson 2000). Evidence implicating chytrid fungus in frog extinctions keeps accumulating. A survey of museum specimens found the fungus first appearing on a frog caught near Brisbane in 1978.

Australians keep making inroads abroad. Eastern banjo frogs (*Limnodynastes dumerilii*) turned up recently in New Zealand, deliberately released it seems. Our deadly redback spiders (close relatives of the black widow) have invaded Belgium, and we have exported a centipede (*Lamyctes emarginatus*) to far off Greenland, and ants to the Galápagos Islands. The Australian spider beetle (*Ptinus tectus*), found naturally in bird's nests and bat colonies, has become a pest of stored food worldwide.

I am more pessimistic now than when I wrote *Feral Future*. In 2000 I was flown to a major government meeting on the notorious pasture grass, hymenachne. It proved a farce. The few farmers present (growers of pasture grass themselves) watered down every statement about control. Declaration as a Weed of National Significance (WONS) won't stop graziers planting this rampant weed. The government accepts this. Later I found a major new infestation in central Queensland.

Colin Wilson of the Parks and Wildlife Commission of the Northern Territory depressed me further with his own hymenachne woes. After it became a WONS in 1999 his department twice offered to help the Department of Primary Industry and Fisheries (DPIF) remove the small areas (fewer than a hundred hectares) in cultivation. DPIF never replied and Colin says that hymenachne is now 'totally out of control and beyond hope of eradication' in the Northern Territory. It is spreading rampantly through national parks. All of this since my book came out. Colin tells me gamba grass (growing four metres tall) is looming as the worst weed of northern woodlands because trees die when it burns. DPIF finally stopped promoting it and now is pushing a grass (*Digitaria milanjiara*) that failed to pass the weed risk assessment until it was run through again with 'new' information. The federal government has pulled back from promoting new grasses but DPIF is powering ahead. Pasture grasses are an absolute curse for Australia – presenting a (so far) irreconcilable conflict of interest.

But enough gloom. I've had great responses to *Feral Future*. I've been flown all over Australia to speak at conferences, festivals and other events. The book has won acceptance as a realistic assessment of the situation. The Australian Quarantine and Inspection Service is happy with me, despite my criticisms of some of their work – I have this in writing. Their funding received a massive boost after the recent foot-and-mouth scare. Various government departments have asked me to sit on committees to assess pest issues. TV New Zealand flew me to Auckland to appear in a 'Feral Future' special. Best of all, the University of Chicago Press has seen fit to publish this book for an international audience.

I was also flown to Cape Town twice to speak at international pest fests involving the Global Invasive Species Program (GISP). The U.S. Undersecretary of Agriculture, Mike Dunn, sounded threatening when he warned us to solve global pest problems without impeding free trade. 'We can't let pseudo-science restrict our trade'. We went and saw squatters hacking and poisoning weedy Australian wattles. I met the American biologist in GISP responsible for promoting global awareness about pests. I asked if he had a media or educational background and he said, oh no, he only volunteered because no one else wanted the job.

I am delighted that a new national conservation group has formed in Australia, the Invasive Species Council (ISC), to counter the pest threat. The coordinator, Paola Parigi, has sought advice from the Nature Conservancy and the American Lands' Alliance. Said Barry Traill, the Melbourne campaigner who initiated the group after talking to me, 'We urgently need an organisation that's whole and sole focus is to protect us from further incursions and to stop existing pests spreading more and more'. The ISC is the first green group in the world to focus solely on exotic pests – animal, plant and disease. I wish them every success.

BIBLIOGRAPHY

These references to topics discussed in this preface supplement those appearing at the end of the book.

Boles, W.E. (1995) The world's earliest songbird (Aves: Passeriformes). *Nature*. 374: 21–22.

Bowman, D.M.J.S. (1998) The impact of Aboriginal landscape burning on the Australian biota. *New Phytology*. 140: 385–410.

Coman, B.J. (1999) *Tooth and Nail: The Story of the Rabbit in Australia*. Text, Melbourne.

Fabres, G. & Brown, W.L. (1978) The recent introduction of the pest ant *Wasmannia auropunctata* into New Caledonia. *Journal of the Australian Entomological Society*. 17: 139–42.

Goulson, D. (2000) Do exotic bees pose an environmental threat? *News Bulletin/ Entomological Society of Queensland*. 28(3): 54–57.

Nattrass, N. and Vanderwoude, C. (2001) A preliminary investigation of the ecological effects of red imported fire ants (*Solenopsis invicta*) in Brisbane, Australia. *Ecological Management & Restoration* 2(3): 220–23.

Sibley, C.G. & Ahlquist, J.E. (1986) Reconstructing bird phylogeny by comparing DNA's. *Scientific American*. 254(2): 68–78.

Smith, N. (1999) The howl and the pussy: feral cats and wild dogs in the Australian imagination. *The Australian Journal of Anthropology*. 10(3): 288–305.

Vinson, S.B. (1997) Invasion of the red imported fire ant. (Hymenoptera: Formicidae): spread, biology, and impact. *American Entomologist*. 43: 23–39.

Wojcik, D.P., Allen, C.R., Brenner, R.J., Forys, E.A., Jouvenaz, D.P. & Lutz, R.S. (2001) Red imported fire ants: impact on biodiversity. *American Entomologist* 47(1): 16–21.

PREFACE

The idea for this book took hold of me one day and wouldn't let go. I was borne off on a strange exotic journey, and life hasn't been the same since.

Wherever I travel in Australia I'm amazed by the hold that exotic species have gained over this land. Visiting Cape York a decade ago I expected to find unsullied wilderness; instead I found rainforests harbouring herds of feral cattle, grubbing pigs, flushes of Latin American carpet grass and slopes choked with hyptis. Cattle had kicked over a termite mound in which paradise kingfishers – one of Australia's most spectacular rainforest birds – were nesting. It's little better anywhere else. Exotic pests are stealing into all of our national parks and wilderness zones. As an environmental consultant I work mainly in southern Queensland surveying remnant bushland, and here I am struck by the variety and number of garden plants escaping into forests. Jacarandas and umbrella trees look winsome in suburban streets, but when you see them over and over again invading forests you begin to wonder what is going on. It disturbs me that most naturalists, ecologists, conservationists and land managers can't even recognise many of our more serious weeds. A scan of the scientific literature shows that all over the world, ecosystems are in disarray – thousands of exotic species, ranging from mites and mosses to elephants and trees, are invading new lands. Many exotics are benefiting from international trade and travel by stowing away with cargo. It is a major world phenomenon, but not one you hear discussed much in the media or tackled by conservation groups or politicians. So I wrote this book.

It is aimed at the general reader, and all the journal citations are hidden at the back. Almost every fact-based statement can be backed up by a journal article or government report, and most are listed in the Source Notes. Scientific names have been pared back to a minimum. Most of the vertebrate scientific names appear in the appendices and nowhere else. Plant, fungus and invertebrate scientific names are provided the first time a species is mentioned in an important context, and wherever else I have thought they would prove useful. Species cited by common name only are also listed with their scientific names in appendix VI. I have not included scientific names for well-known foods. Weed scientific names mainly follow Lazarides, Cowley and Hohnen (1997) and bird common names nearly always follow Christides and Boles (1994).

Chapter 26 (on cats and foxes) is loosely adapted from an article I wrote in 1996 for *Nature Australia* magazine (25[4]: 80).

Those who would know more about exotic species should read Eric Rolls' classic account *They All Ran Wild*, re-released by the University of Queensland Press in 1998, the recent American book *Life Out of Bounds* (1998) by Chris Bright, and the pivotal government report *Plant Invasions: The incidence of environmental weeds in Australia* by Humphries, Groves and Mitchell (1991). Other milestone reports on weeds include Groves (1998), Csurhes and Edwards (1998), *The National Weeds Strategy* (1997), Lonsdale (1994) and Carr (1993). For a long list of articles on environmental weeds, see Swarbrick and Timmins (1997).

Many people helped and I am very grateful. Jem Bates, my editor, showed great dedication and fortitude extending well beyond the bounds of duty. Jenny Horwood showed me Adelaide's olives, Linda Kelly got me to Florida, and Geoff Carr opened up his files on weeds and sent reports. For encouragement and help I thank Trish Worthington, Shona Sharman and Maryanne Bache. I thank

Bryan Hacker for inviting me to critique pasture plants and the many AQIS officers who were helpful. I am also indebted to Robert Whyte, Dane Panetta, Owen Foley, Doug Laing, Eril McNamara, Bernard Low, Randall and Shelley Stocker, Greg Holcomb, Mark Read and Emily Slobodvian. I cannot list all of the people who responded to requests for information, but some of the most helpful were Dane Panetta of the Allan Fletcher Research Station (weeds), Steve Csurhes of the Queensland Department of Natural Resources (weeds), Steve Goosem of the Wet Tropics Management Authority, Peter Beers of AQIS (fish policy), Keith McDonald of the Queensland Department of Environment (dead frogs), Barbara Waterhouse of AQIS (NAQS), Tim Heard of the CSIRO (biocontrol), Greg Keighery of the WA Department of Conservation and Land Management (phytophthora), Penny Lockwood of AQIS (ballast), Jack Craw of the Northland Regional Council (New Zealand) and Leigh Winsor of James Cook University (flatworms). The following chapters were fact-checked by experts in government: 10, 11, 14, 15, 31, 32, 37. At Penguin Books I would like to thank Madonna Duffy, designer Nikki Townsend and publisher Clare Forster.

As a biologist writing about pests my greatest strengths are vertebrates and weeds and I am most likely to have stumbled while describing invertebrates and disease. Throughout the project I was amazed by how much there was to say, and by how much had to be left unsaid.

Tim Low
March 1999

CONTENTS

Preface, 2002 v

Preface xvii

Introduction xxv

PART I: SETTING THE STAGE 1

1 'Cursed is the Ground' – Pests in Ancient Times 3
2 First In – Invasions in Old Australia 11
3 'A Seasonable Gift' – Explorers Seeding the Wilds 17
4 Ecological Insurrection – Settling the Empty Land 24
5 'To Our Heart's Content' – The Mad Dreams of the Acclimatisers 30

PART II: CARELESS AMBITIONS 39

6 By Design – Planned Introductions 41
7 Ode to the Toad – The Cane Toad Conquest 46
8 'Peopling a Barren River' – A Fishy Business 55
9 Wet Pets – Aquarium Escapes 62
10 When Beauty is the Beast – Garden Plants Run Amok 72
11 Seeking Greener Pastures – CSIRO Welcomes Weeds 81

PART III: INVASION BY STEALTH 93

12 'Every Creeping Thing' – Species on the Move 95
13 Soil Travellers – Worms, Seeds, Spores 103

14 Ballast Blues – Something in the Water 108
15 Where Have All the Flowers Gone? – Phytophthora's Curse 116
16 The Sick and Dying – Exotic Diseases Strike 122
17 The Price of Free Trade – Opening the Door to Invasions 129

PART IV: AUSTRALIANS AS PESTS 137

18 A Source of Perverse Pride – Australianising the World 139
19 Colonial Revenge – British Wallabies and Budgerigars 144
20 Ecologically Entwined – Australians in New Zealand 150
21 Colouring the Landscape – Our Animals Abroad 157
22 Inheriting a Degraded World – Exporting Our Flora 165
23 It's Civil War – Natives can be Pests Too 172

PART V: A ROGUES' GALLERY 177

24 The Shuffled Pack – An Alien Who's Who 179
25 Seizing the Advantage – Exotic Roads to Success 184
26 A Bad Rap – Cats: Scoundrels or Scapegoats? 190
27 Where the Deer and the Antelope Roam – Hoofed Introductions 195
28 The Ultimate Pest – Our Destructive Ways 200

PART VI: WHERE ARE WE HEADED? 205

29 Expanding and Infilling – Pests Old and New Tighten Their Grip 207
30 Sleepers Wake – The Pests that Bide Their Time 215
31 Knocking at the Door – The Next Wave of Invaders 222
32 Whither the Wet Tropics? – A Hot Wet Case Study 229
33 The Homogocene – Visiting the Future 237
34 The New Architects – Redesigning the Land 245
35 Cryptogenic World – Native or Not? 250
36 It Happens Naturally – Invasion as Natural Process 261

PART VII: THINKING AND ACTING 267

37 Seeking Magic Bullets – Biocontrol Often Misses the Mark 269
38 The Quarantine Quandary – AQIS Wields a Small Sword 279
39 Are We Blind? – Barriers to Enlightenment 290

40 Wild Organisms – Understanding Plants 298
41 What to Do? – Embracing a New Ethos 307
42 Life Goes On 314

Appendices

I Australia's Worst Environmental Weeds 318
II Weeds of National Significance (WONS) 319
III Introduced Fauna in Australia 320
IV Australian Animals Abroad 324
V Some Recent Quarantine Highlights 326
VI Animals and Plants – A Checklist of Scientific Names 329

Glossary 332

Source Notes 333

Bibliography 349

Index 376

INTRODUCTION

> 'If a man will begin with certainties, he shall end in doubts, but if he will be content to begin with doubts, he shall end in certainties.'
>
> **Francis Bacon, 1606**

This is a book about exotic pests. The themes are simple: people and their products are crisscrossing the world as never before, and on the new global highways so created, plants and animals are travelling too. On top of this, domesticated plants and animals are escaping our control on an unprecedented scale. A globalisation of ecology is under way with profound implications for us all.

It's happening everywhere. Exotic plants and animals are turning up in Antarctica and on the remotest islands on earth. Australian redback spiders have made their way to Tristan da Cunha, a remote speck lost in the middle of the South Atlantic, and they are invading homes in New Zealand and Japan as well. Australia now has rabbits scampering about on Uluru (Ayers Rock), European weeds on top of Mt Kosciuszko and Amazonian earthworms burrowing beneath her rainforests. More than 140 exotic species haunt the murky waters of Melbourne's Port Phillip Bay alone.

World ecology is now locked onto the same trajectory as popular culture. Just as American pop music and blue jeans, burgers and Coke, have displaced indigenous cultures and foods in every land, so too are vigorous exotic invaders overwhelming native species and natural habitats. Some biologists warn of a 'McDonaldization' of world ecology. The earth is hurtling towards one world culture and (maybe) one world ecosystem.

Australia is a key player in all this. We are known around the world for the appalling pests we have imported over the years, and we export them too. Australian shrubs and trees rank as some of the worst weeds anywhere and our sap-sucking bugs are hated pests in citrus groves worldwide. A shrubby tree native to one small district in New South Wales, the cootamundra wattle (*Acacia baileyana*), has invaded woodlands in Africa, Europe, America, New Zealand and Australia. An obscure flatworm (*Parakontikia ventrolineata*) from southern Queensland has made its way into the gardens and greenhouses of three other continents. Britain is now home to our barnacles, beetles, wallabies, earthworms, mushrooms, tree ferns and daisies. Even our small neighbour New Zealand has emerged as a major exporter of pests, her worms killing off European earthworms and eels and her snails now infesting the Thames.

Australians, by and large, are asleep to all this. We have grown complacent about our pests. We've heard it all before, we think, and in any case we've wised up since the bad old days when we offered a welcome to sparrows, foxes and toads. But really we haven't changed much at all. Dangerous pests are still brought into the country for foolish reasons. A major theme of this book is that we do not learn from history, that we continue to emulate the mistakes of the past. In 1990 a serious American weed, kochia (*Kochia scoparia* var. *scoparia*), was imported for land reclamation, and by 1995 half a million dollars had been spent on quelling it. The pasture grass hymenachne (*Hymenachne amplexicaulis*), released in north Queensland in 1988 against strong objections, took just eleven years to become one of Australia's top twenty weeds. Singapore daisy (*Wedelia trilobata*), a garden creeper imported in the mid-1970s, now ranks as one of the four worst invaders along the southern Queensland coast.

Our collective ignorance about pests is remarkable. Most Australians have never even heard of our worst invaders. Even among conservationists few can name harungana, hymenachne, pond apple, tilapia and green crabs, although these pests may pose a graver threat to wilderness than any tourist resort or mine. Australians *have* heard of other pests – honeybees, for example, and trout, olives, ivy and arum lilies – but have yet to cast them in the role of villains. If I end up by saying very little about certain pests (rabbits, mice, starlings), it's not that I think them unimportant but there just isn't the room.

In the course of my work as a biologist and conservationist I have become increasingly concerned that these matters are not receiving due attention. Exotic invasions, because they are irreversible and cumulative, deserve to be taken more seriously. More than 2700 weeds have become established in Australia so far, at a cost to the economy of more than $3 billion, and each year another ten take root. Weeds now make up 16 per cent of Australia's wild plant species. Two hundred or more foreign animals and algae infest our seas. The number of exotic insects, fungi and micro-organisms is anyone's guess. These figures will keep growing, and while some ecological problems (pollution, greenhouse gases, deforestation) are partly reversible, our pests are here to stay. Over the next hundred years, as world trade and travel continue to grow, the environmental fallout may become intolerable.

When most people think of 'pests' they think of farm or household pests, not environmental threats. When the latter do attract interest, a handful of rogues – foxes, cats, toads, a few weeds – steal the limelight. Scant heed is paid to the bulk of the invaders – to the insects, diseases and weeds stealing into our forests, to the foreign fish fouling our rivers and the strange things skulking in our seas. I have tried to redress this imbalance by focusing squarely upon environmental pests, especially upon the lesser known of the major pests.

In recent years many scientists have come to realise that our pest problems are spiralling out of control. From the Commonwealth government have come milestone reports on weeds, ballast pests, fish diseases, and more weeds. *Plant Invasions* by Stella Humphries, Richard Groves and David Mitchell has proved the most influential. Produced by the Australian National Parks and Wildlife Service in 1991, its authors warned that our weed problems are much worse than we think, that most weeds are multiplying, that control problems are 'formidable'. They were the first to pull together a list of Australia's worst environmental weeds, one I refer to throughout this book.[1] The list (see appendix I) makes interesting reading. The eighteen plants are unknown to most people, yet with the exception of Japanese kelp, *all of them* were brought to Australia deliberately, usually as garden and pasture plants. (Mission grass [*Pennisetum polystachion*] was trialled unsuccessfully for pasture then went on to fuel ferocious bushfires in the Northern Territory.) Humphries and colleagues were among the first to suggest that Australia's *carte blanche* approach to plant imports might be irresponsible. The same could be said for aquarium fish, fish product and flower imports. Most of our pest disasters have been self-inflicted.

The National Weeds Strategy (1997) is another blunt-talking, landmark report. 'Weeds are among the most serious threats to Australia's primary productivity and natural environment', it declares, noting 'evidence of an increasing rate of weed encroachment through or towards almost all ecosystems of immediate economic, social or conservation value within Australia'. No beating about the bush here.

The word 'weed', used in this context, needs explaining. The *Strategy* says a weed is 'a plant which has, or has potential to have, a detrimental effect on economic, social or conservation values'. I prefer the simplest of all definitions: it is 'a plant in the wrong place'.[2] Any foreign plant growing and breeding in a forest is an environmental weed. A camphor laurel tree (*Cinnamomum camphora*) by a stream, for instance, has pushed aside some native tree and now exerts a sway over the ecosystem. A forest, after all, is not just a random assemblage of plants but an interacting system. Camphor laurel leaves are concentrated packages of camphor – a powerful insecticide – and when enough of them fall into pools, fish and tadpoles die. Camphor laurel is one of many weeds that secrete soil poisons that stunt nearby plants. Nothing much grows under mature camphor laurel trees.

Most people think of 'weeds' as little things, like the dandelions disfiguring their lawn, but environmental weeds can be enormous. Most of them are trees, shrubs, vines, grasses, lilies and waterweeds. The worst of them – the 'ecological engineers' – can form whole forests, shrublands or grasslands. Some of them – foxgloves and freesias, for instance – are very pretty and began their careers as garden plants. Many of the weeds we see in bushland today are not common and probably not very harmful. But some of them are multiplying fast, and those dismissed as trivial today, because they only infest a few road verges, may well turn into triffids tomorrow. The history of white settlement in Australia is shorter than the lifetime of many trees; we are still in the first stages of invasion.

Armies of foreign animals are invading Australia as well, and like the weeds, they come in all shapes and sizes, from millipedes and sponges to swans and camels. They also exert an influence on the environment, by preying on native plants and animals, taking over their homes and stealing their foods. Honeybees, for instance, deprive birds of nectar and possums and parrots of nestholes. Foreign oysters evict native shellfish from rocks.

Unfortunately, we can't do much about the pests already here; most of them are unstoppable. No matter how much we tried we could never destroy every spore, egg or seed. Farmers can spray their crops to counter foreign bugs and weeds, but we can't go around fumigating forests. We have a much better chance of keeping new pests out. The important point to realise is that most new weeds will invade, not directly from abroad, but indirectly from our gardens and farms – they will be escaping garden, timber, pasture and crop plants, with a few aquarium weeds thrown in. Many of our new animal pests will be aquarium fish dumped into streams, and these escapes too are preventable.

The problems today are momentous, but biological invasion is not a recent phenomenon; in fact it harks back to ancient times. The first part of *Feral Future*, Setting the Stage, surveys the history of exotic invasions. Part II, Careless Ambitions, looks at the groups of pests brought to Australia deliberately – as pets or pasture plants, for our gardens, for sport or for pest control – while Part III, Invasion by Stealth, assesses unintentional introductions. Part IV, Australians as Pests, looks at all the pests we have sent to other lands (conservationists in Africa, America and New Zealand have good cause to be indignant). Part V, A Rogues' Gallery, asks the question, 'Why are exotic invaders more successful than the natives they displace?' then introduces some of Australia's worst villains. Part VI, Where Are We Headed?, looks at future trends (the Homogocene cometh), and the final part, Thinking and Acting, addresses the implications of all these changes, suggesting courses of action.

The most important conclusion that emerges from all of this: we need to think more seriously about all the domesticated plants (and animals) under our control; living things are a lot more wilful than we realise. Most domesticated species remain wild at heart, animated by the drive to survive and multiply. At every opportunity they will escape our dominion and return to nature. Our nurseries, gardens and pet shops are, viewed properly, carelessly conceived holding pens for wild creations that regularly escape. Darwin Botanic Gardens, from which mimosa bush, leucaena and cutch tree escaped, is likely in time to damage Kakadu more than any uranium mine. Gardening has done far more harm to Australia's environment than mining.

Ignore the warnings and we will end up as pest-infested as some of the worst places overseas – southern Florida, for instance. A couple of years ago on the edge of a Florida swamp I caught a clear glimpse of the future. Indian mynas and South American monk parakeets were frolicking among wild Australian paperbarks.

Here were four continents in collision, a hybrid world not yet imagined, the new ecology – our feral future.

1. During the final stages of editing this book, in June 1999, the federal government released a very important new list, Weeds of National Significance (see appendix II), listing Australia's twenty worst weeds (ranking environmental and economic weeds together). Because of deadline constraints it has not been possible for me to draw much upon this list, although I have amended a couple of chapters to make the point that hymenachne, cabomba and willows are now ranked among Australia's twenty worst weeds.
2. Other definitions are provided in the Glossary at the back of the book.

Arafura Sea
Torres Strait
Cape York
Darwin
Kakadu National Park
Gulf of Carpentaria
Indian Ocean
Kimberley
Coral Sea
Port Douglas
Cairns
Derby
NORTHERN TERRITORY
Karumba
Townsville
Mount Isa
Mackay
QUEENSLAND
Alice Springs
Rockhampton
WESTERN AUSTRALIA
Ayers Rock
Charleville
Brisbane
SOUTH AUSTRALIA
Gold Coast
Geraldton
Nullabor Plain
Kalgoorlie
NEW SOUTH WALES
Perth
Port Augusta
Port Macquarie
Newcastle
Adelaide
Sydney
Albany
Kangaroo Island
Canberra
VICTORIA
Melbourne
Indian Ocean
TASMANIA
Hobart
Australia
National Capital
Perth City
International Boundary
State / Territory Boundary
VICTORIA State / Territory Name
500 km
0
500 Miles

I

SETTING THE STAGE

Exotic pests have been around for a very long time. Archaeologists unearth some surprising evidence. Seed-sowing explorers and madcap acclimatisers advance Australia's exotic invasion during the nineteenth century.

1

'Cursed is the Ground'

Pests in Ancient Times

'It is an ancient story, yet it is ever new.'
Heinrich Heine

In the Book of Genesis, God vents his wrath upon Adam for tasting the forbidden fruit: 'cursed is the ground because of you; in toil you shall eat of it all the days of your life; thorns and thistles it shall bring forth to you'. Thorns and thistles appear in the Bible so frequently as to suggest that vast tracts of Palestine may have been overgrazed by goats. (Thorns indicate overgrazing.) The wise men of the time blamed the thorns upon sin, not overstocking, and since that time Western societies have put moral problems ahead of ecological ones, with appalling consequences for the land.

Weeds were evidently a problem for the ancients. The Hebrew prophet Hosea warned his people that 'judgement springs up like weeds in the furrows of the field'. Matthew (13:36) recorded Christ's Parable of the Weeds, in which Jesus says, 'Just as the weeds are gathered and burned with fire, so will it be at the close of the age'.[1] Theophrastus, Aristotle's devoted disciple, told of how dodder ('vetch-strangler') choked crops, and the Roman agronomists Cato and Varro both wrote about pests and how to defeat them. Pliny the Elder, in his *Natural History* (AD 77), described armies of fieldmice and rabbits, plagues of cockroaches and locusts, gnats on figs and blight, mildew and caterpillars attacking the vine. Pest control in those distant times ranged from the chemical to the magical. 'Many people kill ants and also moles with the dregs of olive oil,' Pliny explained, 'and to protect the tops of the trees against caterpillars and pests productive of decay they advise touching them with the gall of a green lizard, but as a protection against caterpillars in particular they say that a woman just beginning her monthly courses should walk round each of the trees with bare feet and her girdle undone.'

Weeds accompanied agriculture on its march across the world. They sprouted wherever Neolithic or Bronze Age settlers cleared forests of elms. Pollen records show that wild radish (*Raphanus raphanistrum*), charlock (*Sinapis arvensis*) and prickly sowthistle (*Sonchus asper*) spread across Europe as farms advanced, exploding in numbers as forests around them fell. These weeds later travelled to Australia as contaminants of pot plants and crop seed.

The Roman Empire, with its vast trade routes, unrivalled road networks and large farms, offered splendid opportunities for dispersal. Many of England's finest wildflowers and weeds arrived with the Romans. Fennel (*Foeniculum vulgare*), hemlock (*Conium maculatum*) and Scotch thistle (*Onopordium acanthium*) appeared in English pollen deposits from this time. Scotch thistle, an exotic species in Scotland, probably originated in the Mediterranean. These Roman imports would colonise Australia two thousand years later, brought here as garden plants. Archaeologists in England have unearthed deposits of Roman grain infested with weed seeds and beetles. Some of these weeds do not grow in England today, which suggests the grain had been brought from Mediterranean lands, perhaps to brew beer. At a Roman rubbish pit in Warwickshire archaeologists unearthed a bounty of insects, including a wood-boring longicorn beetle unknown in Britain today. Perhaps it was imported as a grub inside a Roman chair.

Exotic species spread with other imperial migrations, although we know much less about these. Who can say what European insects invaded India with Alexander, or which marine creatures crossed the seas on Phoenician ships? The Mongols must have carried burrs on their horses, and livestock parasites would have travelled south with the Bantu migrations in Africa. The conquest of the Americas by European species is better documented, and we know that cattle and pigs were roaming free in North America by 1560, and horses by the eighteenth century. Many horses won their freedom after Apache and Comanche Indians raided Spanish settlements.

So when did it all begin? We don't know, but it's exciting to discover that the oldest evidence for a feral invasion anywhere in the world comes from Australia's doorstep – from New Ireland in eastern New Guinea. Here, in caves, lie hundreds of bones from discarded human meals. One limestone shelter, Balof 2, is littered with remains of cuscuses (*Phalanger orientalis*) and pademelons (*Thylogale brownii*). But dig down further into the past through a metre or two of refuse and the pademelon remains disappear (at about 7000 years ago), then, at about 14 000 years ago, evidence of cuscuses also ceases. This sequence implies that neither pademelons nor cuscuses were original inhabitants of New Ireland, that each was brought to the island at a different time in the past. Other caves tell similar stories. Thousands of years ago Melanesian boatmen must have taken trussed wallabies and cuscuses from New Britain further west. The animals escaped or were set free to supply future meals. Either way, we are left with the remarkable conclusion that the first exotic animals in the world may well have been marsupials.

Melanesians, it seems, liked to carry animals when they travelled. Most island groups around New Guinea now support colonies of exotic rainforest mammals, mainly cuscuses and rodents, but also tree kangaroos (on Umboi) and apparently sugar gliders as well.

(Melanesians still shift mammals around, the Rofaifo of New Guinea recently freeing echidnas on Mount Erimbari as a new source of food.) Cuscuses (*P. orientalis*) were taken west as far as Timor (four or five thousand years ago) and east to the Solomon Islands, a major expansion of range. Remarkably, one island – Seram, west of New Guinea – even has cassowaries that appear to be exotic. Great mystery surrounds one of the transportees, the delightful admiralty cuscus (*Spilocuscus kraemeri*), found on Manus, a remote island north of New Guinea. Early archaeological sites yield no trace of this creature, suggesting that it too was introduced, perhaps a thousand or two years ago – but from where? It's not known anywhere else.

The early history of feral species around the world, a much-neglected subject, is full of such riddles. Here I have drawn up a chronology of some of the milestone events (though many of the dates are necessarily very approximate).

8000 to 18 000 BC: Cuscuses are introduced into New Ireland. They are later distributed widely through the islands of the New Guinea region, where they affect the ecology to this day.

10 000 BC: Wolves are tamed in western Asia, leading to the development, by genetic selection, of the dog, a valued hunting companion because of its speed and acute sense of smell. (The wolf is one of the few animals to live in cooperative hierarchical societies, producing pups unusually amenable to training, unlike the offspring of cats, which grow into solitary hunters.)

7000 BC: Goats, sheep, wheat and barley have been domesticated in the Near East. The first weeds of agriculture no doubt appear at this time, and the sparrow begins its long association with people, learning to nest under roofs and steal grain.

5000 BC: Pademelons are taken to New Ireland. In China, rice is brought under cultivation in the lower Yangtze Delta, and the first weeds of rice probably appear. The domestication of pigs and water buffalo may have begun. Rice, pigs and water buffalo later become exotic species in northern Australia (rice along the Ord River).

3000 BC: Grain infested with flour beetles is placed in Egyptian tombs.

3000 to 2000 BC: In one of the most eclectic introductions of exotic mammals anywhere, pigs, dogs, monkeys, goats and cuscuses are brought to Timor.

1500 to 2000 BC: Indonesians import the dingo into Australia, along with its parasitic worms, which go on to infect kangaroos and wallabies.

1600 BC: Egyptians have domesticated the cat. It is welcomed into homes as a mouse-killer, a protector of grain. Later it is revered as divine; at Beni Hasan in Middle Egypt, 300 000 cats are embalmed, some of them to be sold off during the nineteenth century as fertiliser in England. The cat goes on to become a feral pest all over the world.

1400 BC: The Lapita people (from present-day Indonesia) start to spread through the Pacific region, taking pigs, Pacific rats and plant foods, and probably distributing the mourning gecko and certain snails and insects. Aided by their rats and pigs, they exterminate vast numbers of island birds.

1000 BC: Sparrows and starlings are living at the Bronze Age settlement of Apalle, in central Sweden. The sparrows have spread well north of their original home in the Middle East, and later they colonise England, from which they are brought to Australia.

31 BC to AD 476: During the Roman era many plants and animals are distributed across Europe, North Africa and the Middle East. Mice are so prolific that Pliny claims a single mother can give birth to a 120 offspring; they are thought to become pregnant by licking salt, or each other. The feral porcupines found in southern Europe today may date from this time. Nettleleaf goosefoot (*Chenopodium murale*), vervain (*Verbena officinalis*) and other weeds appear in England.

From about the tenth century the Asian house rat is transported by islanders throughout Micronesia, and Pacific rats reach Norfolk Island, presumably carried there by Polynesians, who also bring banana plants.

During the twelfth century rabbits reach England, imported from continental Europe as a luxury food; by the 1250s they have become a localised nuisance. At first they live mainly along the coasts, where the sandy soils assist burrowing, but during the nineteenth century they explode into farming districts throughout the land.

By the thirteenth century black rats reach plague numbers in western Europe, anticipating disaster for the human population.

During the fourteenth century bubonic plague, carried by infected fleas of the black rat, devastates Europe. It probably arrives with Genoan trading ships from the Black Sea port of Caffa, reaching England and Italy in 1348, Germany, Poland and Scandinavia in 1349–50. In its first three years the Black Death fells more than one-third of the population of Europe. No one knows the death toll in Asia. Plague outbreaks recur in Europe at ten-year intervals throughout the Middle Ages. It reappears in London in 1665 as the Great Plague.

Early in the fifteenth century, Chou Ting-wang compiles the *Chiu-Huang Pen-ts'ao*, a treatise on plants fit for food in times of famine. Many farm weeds are mentioned, including dandelions (*Taraxacum officinale*), sowthistle (*Sonchus oleraceus*), amaranths (*Amaranthus* species), plantains (*Plantago* species), fat hen (*Chenopodium album*) and shepherd's purse (*Capsella bursa-pastoris*). All of these weeds later find their way to Australia.

During the sixteenth century the dodo, a symbol of extinction today, is wiped out by two exotic pests – pigs and monkeys – brought to its island home of Mauritius by European seafarers.

In 1788 the First Fleet anchors in Botany Bay and disgorges a host of exotic species. Australia will never be the same again.

The historical record suggests many of the pests Australia gained from England that we think of today as British are just as foreign to England as they are to Australia. This is true of our rats, rabbits, sparrows, cockroaches and large numbers of weeds (hemlock, Scotch thistle and hedge mustard [*Sisymbrium officinale*] among others). It is also true of the pheasant and fallow deer and probably of the starling and common garden snail (*Helix aspersa*).

Black rats and brown rats originated in Asia, and the bubonic plague that so ravaged medieval Europe travelled west with exotic black rats and their exotic fleas. The hotbed of rat evolution is South-East Asia, on Australia's doorstep, and rats (*Rattus* species) reached our shores by rafting southwards on debris a good half a million years before they marched into the slums of London. Here they evolved into seven native species. One of them, the canefield rat, is now a multi-million-dollar pest of sugarcane crops. The native bush rat, which I often encounter during fauna surveys of rainforests and woodlands, looks remarkably like the hated black rat (but with a shorter tail). England has no native *Rattus* at all, and ecologically the black rat is less out of place in Australia than it is in Europe.

Rabbits are native to the dusty plains of Spain and northern Africa, which helps explain why they do so well in outback Australia. Two thousand years ago, according to Pliny, they brought famine to the Balearic Islands east of Spain by 'ravaging' the crops, compelling the islanders to petition the Emperor Augustus for military aid against the pest. Spain, Pliny noted, was teeming with rabbits, their fertility 'beyond counting'. England's earliest rabbit remains come from a kitchen midden at a castle in Essex dating from the late twelfth or early thirteenth century. Henry III presented rabbits from his royal park at Guildford as a gift to favoured visitors in 1235, and they soon became a popular luxury item at feasts and banquets. Inevitably they escaped into the wild, and by 1257 the Burgess of Dunster in Somerset was complaining of their destructiveness. The rabbits in England (and Australia) are thus descended from domestic livestock gone wild. If those men who gave Australia the rabbit in 1830 had only paid more heed to its past.

Did a wrathful God throw an ecological curse over the land? One could easily take this to be true. For ever since biblical times thorns and thistles, rabbits and rats, and numberless other pests have done their best to blight our lives. The following chapters have more to say about the turbulent history of our early exotic invaders.

1. Although the word 'weed' appears in Bible translations there was probably no collective noun for weeds in ancient times, only names for individual species, which cannot be identified with certainty today. Hosea's weed may have been a wormwood (*Artemisia*) and Matthew's has been identified as either darnel (*Lolium temulentum*) or tares (*Vicia sativa*).

2

First In

Invasions in Old Australia

> 'I love to sail forbidden seas, and land on barbarous coasts . . .'
>
> **Herman Melville, *Moby Dick,* 1851**

Some years ago in Launceston this startling newspaper headline caught my eye: 'Tassie Tiger tracks found?' Tasmanian tiger footprints had been discovered by a couple who claimed to have seen two tigers ten years earlier. Such sightings are common in Tasmania but never come with the hard proof needed to verify that these animals still exist. In fact, this dog-like striped marsupial – the thylacine – is undoubtedly extinct, killed off by settlers defending their sheep, although an epidemic may have helped carry it off. I have leafed through the bulging file of tiger sightings held by Tasmania's national parks office and come away more sceptical than ever. Many of the reports are from mainland Australia, from places not at all remote. Wishful thinking obviously illuminates most of the sightings, which are often made in dim light late at night.

It is ironic that most of these reports originate on the mainland, because tigers did live there just a few thousand years ago, as shown by Kakadu cave paintings and by a mummified carcass found in a Nullarbor cave. Tigers on the mainland could not, it seems, contend with the dingo, a carnivore that competed for the same foods. Tigers probably found it harder and harder to feed their young, and slowly they dwindled away while the dingo's fortunes rose. Dingoes may have brought in diseases too. Rising seas walled off Tasmania before dingoes could get there, and the island became a tiger sanctuary, a last outpost – until the settlers came.

Textbooks tell us that settlers brutally exterminated the thylacine, but most of the blame, I say, should rest with the dingo. The tiger, after all, had vanished from most of Australia well before any white people chanced along. Tasmania was not an ideal sanctuary (the bush mostly too wet and thick) and its tiger population was small, perhaps a couple of thousand, no more. The thylacine was a vulnerable species even before the settlers came bearing guns and snares. We can be thankful those settlers did not exterminate the Tasmanian devil as well, for it too left the mainland after dingoes arrived. Some devil remains in Western Australia date from less than 500 years ago. The Tasmanian native-hen, a flightless bird that once roamed as far north as Central Queensland, was probably another mainland dingo victim.

Exotic animals have been a force in Australia for a very long time. Feral dogs have been slaughtering our wildlife – marsupials, birds and reptiles – for thousands of years. Where did these dogs come from? Biologist Laurie Corbett looked closely at primitive breeds overseas (collecting skulls from a dog meat market in Thailand) and found 'dingoes' in remote villages all over Asia. New Guinea's 'singing' dogs, the basenji of the Congo, the extinct kuri of New Zealand, perhaps even the Carolina dogs of Native Americans, are all dingoes, more or less. In his book *The Dingo in Australia and Asia*, Corbett mapped the dingo's distribution as a solid block running north to China and Japan and west to Egypt, with outliers in Madagascar and west Africa.

Dingoes were bred from wolves in Asia between 6000 and 10 000 years ago. The dingo's scientific name, *Canis lupus dingo*, celebrates the fact that Australia is home to a subspecies of wolf (*Canis lupus*). We might well say that Australia is inhabited by wild wolves – and by millions of tame wolves as well, since pet dogs share the same ancestry. Wolves were not brought to Australia by Aborigines, who came at least 40 000 years ago, but by later visitors, Indonesians presumably, visiting our northern shores some 4000 years ago. Corbett believes dingoes were taken on voyages as a living meat larder. In many Asia-Pacific countries dog meat remains a delicacy.

These wolves carried worms. Two parasites, the nematode *Cyathospirura seurati* and the tapeworm *Echinococcus granulosus*, apparently entered Australia inside dingoes' bellies. The tapeworm today infests kangaroos, wallabies, tree kangaroos and pademelons, entering dingoes through the food chain. The worms seldom kill their hosts, although an endangered bridled nailtail wallaby and a vulnerable Proserpine rock-wallaby died recently – an insidious spin-off of the dingo's introduction. Aborigines themselves must have brought a parasite load. According to Jacques-Julien de La Billardière, in the 1790s Aborigines on Tasmania's remote southern shores were infested with 'vermin', perhaps head lice. These insects must have reached Australia more than 3500 years ago, before Tasmania was cut off as an island.

If ancient seafarers brought us the dingo and its parasites, what else were they carrying? Did they bring in yams, nuts and fruits, whose seeds later sprouted? Were seeds and insects hidden in their canoes, and were barnacles and seaweeds fastened to their hulls? And what of the Aborigines: did they bring in other exotic species? Although it is difficult to know for sure, the answer to all of these questions is probably yes. In Australia today scores of exotic species may be pre-European introductions.

The Pacific rats on Norfolk Island are one obvious example. These rats, native to South-East Asia, occur today on islands throughout the Pacific to which they were taken by Polynesian seafarers, either as unintentional cargo or as quick snacks (rat meat is very tasty, as I can attest). Norfolk Island was uninhabited when Captain Cook passed by in 1774, but we know from archaeological remains that people and rats lived there a thousand or so years ago. The people left (probably after exhausting their resource base) but the rats stayed on, along with one of the Polynesians' crops – the banana. (English colonists found bananas ['plantains'] growing on the island in 1788.) Pacific rats have been found on three other Australian possessions – Christmas Island in the Indian Ocean, Mer (Murray Island) in Torres Strait and Adele Island north of Derby in Western Australia – providing evidence of other long-forgotten voyages.

Certain plants in northern Australia also serve as archaeological markers. The best known are the tamarind trees (*Tamarindus indica*) that tower over beaches in the Northern Territory. They point out the camps of Macassan voyagers who for two hundred years, beginning in the eighteenth century, sailed south from Macassar in southern Sulawesi (Celebes) to harvest bêches-de-mer – edible sea slugs. After oven drying in Australia the slugs were exported from Macassar to China as part of a vast trade network predating Captain Cook. While camped on our northern shores the Macassans seasoned their meals with tamarind pulp and then discarded – or planted – the seeds. Aborigines today harvest the fruits of the wild trees, which they know by local names, such as *angkayuwaya*. Ivy gourd (*Coccinia grandis*), an Asian vegetable found trailing around the old camps, was another probable Macassan introduction. Other exotic plants may have entered as seeds mixed in with the rice that the Macassans traded for Aboriginal turtleshell.

The wild coconuts on Torres Strait beaches are certainly ancient introductions, and so too, I believe, are the Asian five-leaf yam (*Dioscorea pentaphylla*) and wild ginger (*Zingiber zerumbet*) found sprouting in the rainforests. The yams, first discovered on Thursday Island in 1897, were later thought to be extinct in Australia, until

I stumbled upon a few of them growing in jungle on Thursday Island and on nearby Hammond Island. Further south, on Cape York, a colony of true ginger (*Zingiber officinale*) survives in the rainforests of Lockerbie Scrub. Botanist Frederick Bailey recorded this plant when he visited remote Somerset Station run by rugged pioneer Frank Jardine. Bailey was handed some pieces of ginger dug from the jungle gloom; back in Brisbane he had them planted and lo, common ginger emerged. This ginger wasn't seen again for ninety years until I tracked it down growing near Cape York in 1988. I was thrilled to be the first person this century to see 'Australian' ginger growing wild, and I wonder to this day who brought it to Cape York (Macassans, Javanese, Lapitans, Melanesians?) and how long ago. The plants grow inside rainforest only twenty kilometres from Australia's northern tip, and in Jardine's day there were no other settlements for hundreds of miles. The small rhizomes taste just like cultivated ginger.

Jequirity bean (*Abrus precatorius*), a creeper with shiny red and black seeds, may be another ancient introduction. In 1898 anthropologist William Haddon was touring Torres Strait when he obtained on Mer a prized cassowary bone dagger, traded down from New Guinea, with jequirity beans adorning its handle. The hard-coated seeds survive many years as colourful ornaments and they may well have entered Australia as beads thousands of years ago. Other plants to ponder include taro (*Alocasia macrorrhizos*), round yam (*Dioscorea bulbifera*) and weeds such as sida-retusa (*Sida rhombifolia*), urena burr (*Urena lobata*) and Chinese burr (*Triumfetta rhomboidea*). In 1844 a seventeen-year-old Scots woman, Barbara Thompson, was shipwrecked in Torres Strait, the sole survivor of an ill-fated expedition led by her husband. During the five years she lived among the islanders, she learned of the trade networks linking Cape York Aborigines with the tribes of New Guinea. Many weed seeds, not to mention insects, mites, worms and snails, must have crossed the strait with consignments of tobacco, coconuts, bananas, clay, ochre and dugout canoes.

The truth of these movements would be difficult to establish, but some of the speculation is fascinating. Dr Chad Hewitt of the CSIRO believes we may have east African organisms in our waters dating back to ancient trading empires that linked Asia and Africa hundreds of years ago, and at other times, Asia and Australia. Another intriguing possibility is that Macassan fishermen brought us feral cats. Many desert Aborigines regard the cat as a native animal, and early travellers were surprised by how many cats they saw in tropical Australia. Paul Wagner, a PhD student at Macquarie University, is assessing genetic evidence suggesting that cats may have entered northern Australia some time prior to European settlement.

Even if this proves untrue, we can still say with confidence that exotic animals and plants have been flowing into Australia for thousands of years. But now that the volume of trade has multiplied, and our canoes have grown into vast cargo ships, the trickle of exotic immigrants has become a dangerous flood.

3

'A Seasonable Gift'

Explorers Seeding the Wilds

> 'I hoped that this benefaction might prove one of no small value . . .'
>
> **George Grey, explorer, on planting seeds, 1838**

The first Europeans to introduce exotic species into Australia were the early explorers. Deep in the wilderness they would pause in their travels to free chickens and pigs and sow vegetables and fruits. These men felt duty-bound to prepare the virgin land for the settlers and sailors who would follow in their wake. They dreamed of pioneers arriving in Australia to be greeted by fields of wild cabbages and wheat and forests yielding cherries, apples, chicken and pork. Flinders sowed lemons and leeks in South Australia, Vancouver planted grapes, oranges and almonds in Western Australia, rye and wattles were planted in the deserts, coffee in the Wet Tropics, and many other foods – a cornucopia of vegetables and fruits – on islands in the south.

Captain Cook was the first European to free food in Australia. On Cook's second great Pacific voyage, Tobias Furneaux, who commanded the expedition's second ship, the *Adventure*, released pigs on Bruny Island off Tasmania, and in 1776 Cook freed two pigs in Tasmania. 'I ordered the two pigs, being a boar and sow, to be carried about a mile within the woods, at the head of the bay', his journal recounts. 'I saw them left there by the side of a fresh-water brook. A young bull and a cow, and some sheep and goats, were also, at first, intended to have been left by me as an additional present to Van Diemen's Land. But I soon laid aside all thoughts of this, from a persuasion that the natives, incapable of entering into my views of improving their country, would destroy them. If they should meet with the pigs, I have no doubt this will be their fate. But as that race of animals soon becomes wild, and is fond of the thickest cover of the woods, there is great probability of their being preserved.' They did not survive.

In New Zealand, to help the Maori, whom he esteemed more highly than the indigenous Australians, Cook released geese, chickens, rabbits, pigs and goats and planted out two gardens, sowing radishes, turnips, onions, peas, beans, potatoes and purslane (pigweed *Portulaca oleracea*). The potatoes were adopted by the Maori, and during the early nineteenth century many tons of 'Maori potatoes', descended from Cook's plants, went to Sydney to feed hungry Australian colonists.

The first among the explorers to sow seeds in Australia was the ill-fated Captain William Bligh, en route to Tahiti in 1788 in search of breadfruit. On Bruny Island he planted out fruit trees brought specially from Cape Town. He doubted they would survive fires and Aborigines, and he was right. Four years later, after the infamous mutiny, while once again pursuing his search for breadfruit, he went back to Bruny Island. A single apple tree remained. Bligh's gardeners planted a larger plot with fruit trees and seeds of oaks, firs, strawberries, apricots, peaches and rosemary. He also released a cock and hens. (These chickens did not survive, but feral chooks can be seen today on islands in the Great Barrier Reef, some of them descended from stock freed last century by Japanese guano traders.)

That same year of 1792, Frenchman Joseph-Antoine D'Entrecasteaux also set out gardens while exploring southern Tasmania. His botanist, La Billardière, recorded: 'We had a great variety of different kinds of European grain on board, which might be advantageously propagated at this extremity of New Holland.' In February 1793 they returned to Bruny Island, and while ashore La Billardière convinced a young Aboriginal woman to trade her kangaroo skin for pantaloons, which he helped her to don. He then set off for the spot where they had planted vegetables the year before. 'We saw, with regret', wrote La Billardière, 'that there remained only a small number of cabbages, a few potatoes, some radishes, cresses, wild succory, and sorrel, all in a bad condition: no doubt they would have succeeded better, had they been sown nearer to a rivulet, which we observed to the west. I expected to have found some cresses at least sown on its banks; that I did not must surely have been owing to forgetfullness on the part of the gardener.'

With La Billardière was a local Aborigine, who was much impressed by this bold attempt at agriculture. 'He examined attentively the plants in the garden, and pointed them out with his finger, appearing to distinguish them perfectly from their indigenous vegetables.' Later that month, still on the island, La Billardière stumbled upon Bligh's second garden, now a year old. Near the shore were cresses, acorns, celery, three young figs, two pomegranates, a sapling apple and thriving quince. D'Entrecasteaux then freed two goats. 'Near the north shore of this bay', wrote La Billardière, 'we landed a young he-goat, and a she-goat big with kid, putting up all our prayers that the savages might allow them to propagate their breed in this island. Perhaps they may multiply in it to such a degree, as to occasion a total change in the manner of life of the inhabitants, who may then become a pastoral people, quit without regret the borders of the sea, and taste the pleasures of not being obliged to dive in search of food, at the risk of being devoured by sharks.'

Matthew Flinders, the great English-born navigator, was another early explorer/gardener. In 1802 his gardener deposited into the unpromising soil of Kangaroo Island the doomed seeds of oranges, lemons, cherries, rock melons and leeks. The island soils are notoriously infertile and these plants had no hope of surviving. Nine months later Frenchman Nicolas Baudin donated to Kangaroo Island a rooster and hens, a boar and a sow, 'for the benefit of future navigators'. His gardener planted fruits and vegetables in Western Australia. Flinders' early survey work was later completed by Lieutenant Phillip King, whose accompanying botanist, Allan Cunningham, sowed seeds in many places during his travels. In this era botanists and other naturalists were often attached to voyages of exploration to bring back live and preserved plants and animals to feed Europe's curiosity about natural history in remote lands.

In the 1830s George Grey probed vast tracts of north-western Australia, a dismally barren land. A devoutly religious man, Grey felt a strong calling to improve the land. On 1 January 1838, near the mouth of the Prince Regent River, he sowed coconuts, breadfruit and other foods as a new year's gift to Australia. 'I had resolved to do homage to the country by a seasonable gift,' he wrote, 'and, therefore, rising with the earliest dawn, spent the whole day in planting, in various positions, seeds of the most useful fruits and vegetables. Those we had already planted were doing well, and I hoped that this benefaction might prove one of no small value – perhaps to civilized man, or, at least, to the natives of the vicinity.'

When he staggered back to base camp a few months later, after horrendous ordeals in the interior, his plants were doing well and his mood softened; he became sentimental:

> I considered what a blessing to the country these plants must eventually prove, if they should continue to thrive as they had yet done, and as I called to mind how much forethought and care their transport to their present position had occasioned, I would very gladly have passed a year or two of my life watching over them, and seeing them attain to a useful maturity. One large pumpkin plant, in particular, claimed my notice. The

> tropical warmth and rains, and the virgin soil in which it grew, had imparted to it a rich luxurience: it did not creep along the ground, but its long shoots were spreading upwards amongst the trees. The young cocoa-nuts grew humbly amidst the wild plants and reeds, – their worth unknown. Most of these plants I had placed in the ground myself, and had watched their early progress: – now they must be left to their fate.

Other explorers marking out the interior of the country were similarly motivated. Ernest Giles was trekking north of Uluru in 1872 when he found a fine oasis, fringed by fig trees bearing delicious fruit. 'I sowed a quantity of vegetable seeds here,' he wrote, 'also seeds of the Tasmanian blue gum tree, some wattles and clover, rye and prairie grass'. How Giles came to be wandering through the desert with Tasmanian blue gum, wattle and pasture seeds he never explained. He was the only explorer to plant natives and pasture plants, for reasons that remain obscure.

Botanist Daniel Bunce, travelling with Ludwig Leichhardt in 1846–7, was near the Condamine River when he recorded: 'I here commenced a practice which I afterwards followed at every convenient opportunity where the soil and situation was suitable, of sowing seeds of the most useful fruit and vegetables.' On one occasion Bunce sowed cress to furnish fresh greens for the seriously ill men of the expedition. When he returned to harvest them some days later Leichhardt (who was healthy) had eaten the lot. Patrick White alludes to this incident in his epic novel *Voss*.

Eighteenth- and nineteenth-century explorers of Australia sowed at least thirty-seven kinds of plants in the 'wilderness'.[1] This idea of civilising with seeds survived into the twentieth century. Queensland government botanist Walter Hill planted coffee, coconuts and other crops while exploring north Queensland in 1905. Coconuts were often sown on tropical beaches, and sisal (*Agave sisalana*) was planted as a fibre source for sail-making; on some tropical isles it is now a nasty pest, forming prickly impenetrable thickets behind idyllic beaches.

If these explorers could only see Australia today! Coffee and tea now grow wild in our rainforests; watermelons and dates grace the deserts; figs, olives and parsley flourish in our southern woodlands; and other parts of Australia carry crops of wild celery, beets, chillies, persimmons, mulberries, hops and much more, while millions of pigs and goats grow plump in our forests, and even a few wild chickens and cattle thrive. Yet hardly any of this bounty is harvested, save by birds and rabbits. Most wild food plants are worthless wildlings growing far from markets – they are weeds, nothing more. Who today would want to drink a brew made from the wild coffee beans planted in the rainforest by Walter Hill so long ago? Coffee has instead become a threat to the rainforest, listed by the Wet Tropics Management Authority as a 'priority weed'. Cassowaries and bats spread its seeds from old plantations in the hills. Around Lake Barrine the Department of Environment eradicated thousands of seedlings in 1992.

Today, olives, melons (*Citrullus lanatus*), blackberries, fennel and Chinese apples (*Zizyphus mauritiana*, the source of Chinese dates) are declared noxious weeds. Olive trees around Adelaide have converted whole hillsides into vast gloomy thickets, their seeds spread about by foxes and starlings; and blackberries have become one of Australia's worst pests, costing farmers more than $40 million a year to control. In all, more than a hundred foods and culinary herbs earn a place in the CSIRO's *Handbook of Australian Weeds*. In Western Australia, according to the Department of Conservation, weeds infesting national parks and reserves include figs, dates, grapes, mangoes, watermelons, carrots, turnips, Indian mustard, peppermint, fennel and wild oats. A recent federal government report, *Potential Environmental Weeds of Australia* (1997), even raises concerns about liquorice, pistachio, cashew, peppermint and chamomile. All of these plants are invading forests or woodlands, replacing native species and exerting an unwelcome impact on the local ecology.

Feral foods threaten the survival of native plants. The Phillip

Island hibiscus (*Hibiscus insularis*) was all but eaten out by rabbits and goats landed there long ago as a food supply. Upon their removal, feral olive trees threatened to smother the last two remaining clumps. The Phillip Island glory pea (*Streblorrhiza speciosa*) did disappear in this way – it is now extinct. Some feral foods have interesting stories to tell. In Millstream-Chichester National Park in Western Australia, ancient date palms growing along streams mark out trails plied by camel drivers long ago. During bushfires, their dried fronds burn so fiercely that they kill nearby plants; they are considered a serious invader.

We can absolve the explorers from blame for most of these problems, for very few of their plantings survived. The feral foods we see today have spread instead from farms and gardens and from seeds discarded by travellers. Some of them date back to the first white settlement, others to recent experiments by permaculturists and new crop farmers. Tomorrow's weeds will include genetically engineered crops – foodplants designed to withstand pests and herbicides. Scientists have already found that transgenic oilseed rape (*Brassica napus*), given genes to resist herbicides, will cross with wild radishes (*Raphanus raphanistrum*) to spawn spray-resistant weeds.

1. Peaches, plums, apricots, cherries, apples, quinces, pomegranates, grapes, figs, oranges, lemons, bananas, rock melons, strawberries, breadfruit, coconuts, almonds, pumpkins, potatoes, cabbages, celery, radishes, leeks, watercress, chicory, sorrel, coffee, wheat, rye, rosemary, oaks, firs, acorns, blue gums, wattles, clover and prairie grass.

4

Ecological Insurrection

Settling the Empty Land

> 'It is curious how we have introduced all our weeds, vices, and prejudices into Australia, and turned the natives (even the fish) out of it.'
>
> **Marianne North, 1880**

It was cloudy and hot on 28 January 1788 when the first cloven hoofs pressed down on mainland Australian soil. Two days earlier, convicts and marines had begun clearing brush for the site of Sydney Town. Then came the landing of mares and stallions, cows and bulls, goats and hogs, and 'a good stock of poultry'. More livestock was unloaded the next day, and rats and cockroaches probably slipped ashore as well, not to mention grain-eating insects and lice and other vermin carried by both convicts and stock. Weevils had spoiled much of the wheat and barley brought out from England and Cape Town – Watkin Tench recorded that 'every grain of rice was a moving body from the inhabitants lodged within it'. On board the *Golden Grove*, rats ate half a barrel of flour. An inventory three months later listed 7 cattle, 29 sheep, 19 goats,

7 horses, 74 pigs, 5 rabbits, 209 chickens, 35 ducks, 29 geese and 18 turkeys.

Unfettered by fences, the cattle, sheep and pigs roamed free. Pigs became one of our first exotic pests, stealing into tents and huts and devouring precious rations. The order was soon given that trespassing hogs could be killed. In June, four months after the First Fleet's arrival, the cattle ran off, and were not seen again for seven years, by which time the herd, living at Camden, west of Sydney, had multiplied to sixty. These cows were perhaps the first European animals to go truly feral – an unlikely achievement for an animal that rarely runs wild today.

As the convicts planted out their crops, weeds burst forth from the fields. They came as seeds mixed in with the grain, in pots bearing fruit trees (from Brazil and South Africa), in hay brought from Cape Town, and perhaps mixed in the mud on soldiers' boots. The pot plants no doubt carried earthworms, insects and mites.

When Matthew Flinders visited Sydney between 1802 and 1804, his botanist, Robert Brown, was so intrigued by all the weeds he saw that he drew up a list of them – twenty or so, in all. They included plantain (*Plantago major*), nettle (*Urtica urens*), scarlet pimpernel (*Anagallis arvensis*), shivery grass (*Briza minor*), petty spurge (*Euphorbia peplus*), mouse-eared chickweed (*Cerastium glomeratum*) and catchfly (*Silene gallica*). These and other weeds probably came from England, but there was also swinecress (*Coronopus didymus*) and Cape gooseberry (*Physalis peruviana*) from South America and swan plant (*Gomphocarpus fruticosus*) from South Africa. The bitter leaves of the gooseberry were used in place of hops for brewing beer.

Imported grain carried disease too; Australia's first wheat crops were struck by 'blight', probably black stem rust (*Puccinia graminis* f. sp. *tritici*). In 1803 it ravaged 265 acres just weeks before harvest. Governor King later complained: 'Our last year's crop was much injured by rusts and smuts.' Human diseases struck as well. A smallpox epidemic ravaged the local Aborigines, killing about half the Sydney population, spreading west into the outback and south to Port Phillip Bay. Captain John Hunter witnessed the carnage around Sydney: 'it was truly shocking to go round the coves of this harbour, which were formerly so much frequented by the natives, where in the caves of the rocks, which used to shelter whole families in bad weather, were now to be seen men women and children lying dead'.

By 1803 the young colony boasted 2450 cattle and 11 275 sheep. Surprisingly, chital deer from India were also present in Australia by this early date. Dr John Harris kept a herd fenced at Ultimo. By 1806 twenty had escaped – and another feral animal was born. Harris placed a warning in the *Sydney Gazette*: 'All persons are hereby strictly cautioned against shooting or in any other way maiming or hurting any of the said deer on pain of prosecution; and owners of dogs are at the same time required to take notice that should any of the said deer be harassed or torn, the Dog will be destroyed, the owner prosecuted and a reward of five guineas paid to any informer.' During the 1820s Australia's fledgling aristocracy hunted these deer with foxhounds, but eventually the chital died out.

As settlement unfurled across the land, the ranks of pests swelled. By 1827 rabbits in Tasmania were said to be 'running about on some large estates by thousands'. Pigs allowed to roam free around settlements often went bush. Honeybees, imported in 1822, soon swarmed into the wilderness. Weeds travelled with the stock. Explorer Major Thomas Mitchell was surprised to find horehound (*Marrubium vulgare*), a medicinal herb from Europe, growing on remote sheep stations in 1845: 'I was assured by Mr Parkinson, a gentleman in charge of these stations, that this plant springs up at all sheep and cattle stations throughout the colony, a remarkable fact.' Sheep transported its burrs in their fleeces, and it soon became a serious weed. On Maria Island I have seen it sprouting around the walls of an ancient hut, a memorial to its long-forgotten tenants.

Katherine McKell, a pioneer at Bolwarrah station in Victoria, remembered the first thistle to flower in her district during the 1860s or thereabouts:

> One day we came on a Scottish thistle, growing beside a log, not far from the stable sheds – a chance seed from the horse-fodder, of course ... This was carefully rolled in a piece of newspaper and put under a stone. In a few days after it was in a beautifully pressed condition, and was shown round with great pride. No one thought then that, some twenty years later, the thistle from Scotland would have spread in the new land and become a nuisance, requiring a special Act in some shires and districts to enforce eradication on private properties.

By the 1850s armies of weeds were claiming lands that settlers had rashly cleared. An ecological insurrection was under way. The English botanist William Harvey, touring Tasmania in 1855, witnessed eco-chaos:

> *Horehound* is everywhere by the roadsides, and *Chamomile* covers the fields and paddocks; in many places to the exclusion of Grasses. *Thistles* are fast going ahead, all through Van Diemen's Land, and no one seems to trouble himself with them, although I have seen, I suppose, hundreds of acres given over to them, and growing so thick in some places that I have walked over my shoes in the bed of thistle-down which had blown from the withered stems. *Sweet-briar*, originally introduced as a hedge plant, is completely naturalized, and in places forms impenetrable thickets. It annually produces millions of hips, and, if let alone, will soon become as great a pest as the thistles.

Reverend J.E. Tenison-Woods was another botanist who bore witness to the invasions. He warned about thistles (covering acres at a time), about 'pernicious' sida-retusa (*Sida rhombifolia*), and about Bathurst burr (*Xanthium spinosum*), 'one of the most formidable weeds' found covering many hundreds of acres. Men were known to spread this weed out of spite. In 1862 it was observed that 'in districts where the burr grows, if a shepherd owes his master a grudge . . . he seeks satisfaction by driving his flock through a bed of the Bathurst Burr, knowing, by doing so, that he inflicts a very serious injury upon him'.

Another weed that provoked alarm was capeweed (*Arctotheca calendulacea*). 'In the early part of spring its flowers cover the whole hill-sides and every vacant spot, with a pale yellow hue that can be seen for miles', Tenison-Woods explained. 'It grows in knee-deep masses by the wayside, it spreads on the paths that are constantly trodden over, and even it is found in ruts and on ridges in the middle of the country roads . . . Its spread over the whole of the districts around Adelaide has something alarming about it, especially as it is known to have been introduced from the Cape of Good Hope within the last 40 years.'

Remarkably, a measure of ecological balance was restored by a native butterfly, the painted lady (*Vanessa kershawi*). Its caterpillars took a liking to the leaves, but so bountiful was the weed that the butterfly itself exploded into pest proportions. In 1889 an astonishing butterfly plague struck Victoria. As biologist Frederick McCoy described the event:

> This Butterfly appeared in extraordinary numbers for two or three weeks, accompanied by a day-flying Moth [the bogong moth, *Agrotis infusa*], almost darkening the sky with their general flight towards the south-east, covering the gear and decks of ships many miles out to sea, and filling the air on land from the northern part of the colony down south to Melbourne. They poured into Gippsland in such quantities as to spread consternation amongst the settlers, who inundated me with letters inquiring whether their crops or orchards or vineyards would be destroyed by the larvae expected to follow. I was glad to be able to assure them that the only likely damage would be to hated weeds. The newspapers mentioned the stoppage of trains in the tunnel on the Castlemaine Railway, from the masses of bodies of these insects crushed, lubricating the wheels

> to such an extent that they could not bite the rails as they turned, and came to a standstill until sufficient supplies of sand could be sent.

Today capeweed is the most prolific daisy in southern Australia, outnumbering hundreds of native species. In Western Australia, around Margaret River, I have strolled across vast sheep fields that glowed golden from the millions of sunny blooms. No grass remained.

The invasion of Australia by Europeans, their stock, weeds and pests proved so comprehensive that Marianne North, travelling in Tasmania in 1881, complained: 'The country was not in the least attractive to me; it was far too English, with hedges of sweet-brier, hawthorn, and blackberry, nettles, docks, thistles, dandelions: all the native flowers (if there were any) were burnt out. One lovely flower I heard of and was taken a long drive to see. It was – a mullein!' (another English weed). By 1909 Victoria's government botanist, Alfred Ewart, was so appalled by all the weeds he was calling for controls. 'It is not too much to say that no new plant should be introduced to this State, and not even in a private garden, if there is any chance of its spreading, unless an official report on its capacities for good and evil has been obtained, and unless the report is a favourable one.' Sound advice, but Australia did not introduce a proper system until – wait for it – 1997! By then, more than a thousand new weeds had invaded.

5

'To Our Heart's Content'

The Mad Dreams of the Acclimatisers

> 'Another object of this Society is that of stocking our waste waters, woods, and plains with choice animals, making that which was dull and lifeless become animated by creatures in the full enjoyment of existence, and Lands before useless, become fertile with rare and valuable trees and plants, teeming with excellent fruits, variety of foliage, and gay and brilliant flowers.'
>
> **Dr George Bennett at an acclimatisation meeting, Sydney, 1864**

In 1862 the Governor of Victoria, Sir Henry Barkly, declared himself in favour of introducing feral monkeys to the state. Addressing the very first acclimatisation meeting held in Australia, he proposed, to enthusiastic calls of 'hear hear' from the floor, that primates be set free in Victoria's eucalypt forests. He was speaking in support of newspaper editor Edward Wilson, who wanted the monkeys, Barkly said, 'for the amusement of the wayfarer whom their gambols would delight as he lay under some gum tree in the forest on a sultry day'.

It nearly happened. A batch of monkeys obtained for a local zoo had proved exceedingly mischievous, troublesome to look after and altogether very expensive. The zoological society wanted them put down; Barkly and Wilson had other ideas. But by the time of the society's second annual meeting Victoria had a new Governor, Sir Charles Darling, who announced he did not like monkeys. He declared himself, however, to 'have no objection whatever to boa-constrictors', which would, he thought, prove 'exceedingly useful in devouring snakes of a highly venomous character'. Australia remained both monkey and boa free, mainly, it seems, because consecutive governors held differing tastes in exotic fauna.

Acclimatisation was one of the most foolish and dangerous ideas ever to infect the thinking of nineteenth-century men. Many great minds all over the world were captured. Professors, politicians, governors, doctors, churchmen, even kings were inspired by grand plans to redistribute animals and plants around the globe for mankind's betterment and pleasure. Had all their dreams been realised, Australia today would be home to wild giraffes, mongooses, hedgehogs, hummingbirds, vultures, flamingoes, oak forests and much else besides. As it was these men gave us sparrows, starlings, carp and many other mischievous things. Some species were sought out for domestication, others for release into the wild. It was an uncontrollable experiment in ecology by Victorian gentlemen of unshakable self-confidence, who never paused to wonder about all the weeds so alarming Tenison-Woods and Harvey, who never considered that their activities ought to be controlled, as botanist Alfred Ewart would later suggest for plants.

The French thought of it first. Anatomist Isidore Geoffroy Saint-Hillaire explained at the world's first acclimatisation meeting, held in Paris in 1854: 'The prospect was nothing less than to people our fields, our forests, and our rivers with new guests; to increase and vary our food resources, and to create other economical or additional products.' Saint-Hillaire's words struck a resonant chord with the intellectuals of his time, and within six years the French society boasted 2000 members worldwide, including the Emperors of France and Brazil, the Sovereign Pontiff and the King of Siam. Medals were offered for domestication of alpacas, kangaroos and emus. The fervor soon crossed the English Channel and a British acclimatisation society was founded in 1860, followed by societies in Victoria and New South Wales in 1861, South Australia and Queensland the following year, Tasmania in 1895 and Western Australia in 1896. Local groups formed in country towns, in Launceston for instance, and over in New Zealand, a hotbed of activism.

Australian colonists, pining for English birds in a strange, often hostile land, were ripe for the idea. Edward Wilson had talked in 1857 about importing songbirds, 'thereby relieving the comparative silence of our woods and gardens'. (Hunters, in fact, were responsible for the silence.) 'The corn field and grove we have already spreading around us', he said. 'Why should we delay the attempt to furnish them with their most agreeable inhabitants?' Wilson obtained five nightingales from a Melbourne dealer (who also held linnets, blackbirds, chaffinches, thrushes, goldfinches, skylarks, woodlarks and robins – a remarkable selection for the time). He freed them in Melbourne's Botanic Gardens, but they fared poorly; some were so battered by the journey from England they could scarcely fly. Wilson urged more experiments of this kind, and by 1860 some starlings, blackbirds and sparrows had been freed.

Acclimatisation, under other names, had been practised well before this of course, first by the explorers who freed pigs and planted seeds during their expeditions. But the societies lent new authority to the activity, along with a lofty name (taken from the French) and a greatly expanded vision. While earlier attempts focused upon English-bred plants and animals, the acclimatisers were true internationalists with the whole world in their sights. Theirs was an almost religious mission to extend God's work of creation by redistributing his creatures. They wanted South American alpacas in our mountain chains, Himalayan pheasants in Gippsland, ostriches and antelope in the outback, and vegetables growing wild along the seashores. Translocation was to work both ways. English acclimatisers wanted wombats, wonga pigeons and Murray cod, New Zealanders imported possums, parrots and prawns, and India received wallabies, cockatoos and swans. Edward Wilson in London even suggested that wombat meat might better suit the English middle classes than mutton or pork.

It is difficult today to imagine the world these men envisioned. Their speeches defy belief. What was George Francis, director of Adelaide's Botanic Gardens, thinking when he urged society to 'put in action a power that hereafter might stock our vast wilds with as many animals as Africa can boast'. He wanted giraffes, kudu, springboks, deer, flamingoes, Numidian cranes, African crowned cranes, secretary birds, Brazilian macaws, argus pheasants, hoopoes and Cornwall crabs. African buffalo and American bison he was less sure about, but he was keen on the smaller creatures: 'We might have spoken of fish and of insects, of sea and land eels, of crustacea.' He wanted prickly acacias in the deserts to feed camels and giant Brazilian grasses as food for stock. 'In conclusion,' he enthused when closing his speech to Adelaide's first acclimatisation meeting in 1862, 'let me express a hope that future generations may hunt their deer, their giraffes, their antelopes, and their ostriches, without either lions, tigers, or gorillas, that they may have, American like, flocks of pigeons, without destructive herds of buffalo; that they may have domestic fowls and game birds surpassing those common to us now; that they may have swifter bullocks, longer and finer fleeced sheep than at present; with dromedaries traversing the desert; and that desert itself, notwithstanding its drought, teeming with animal and human food.'

Just as singular was the speech by Professor Frederick McCoy to the first annual meeting of the Victorian society, in 1862, during which he admitted to a passion for ruminants:

> While nature has so abundantly furnished forth the natural larder of every other similarly situated country on the face of

> the earth with a great variety, and a profusion of individuals of ruminants good for food, *not one single creature of the kind inhabits Australia*! If Australia had been colonised by any of the lazy nations of the earth, this nakedness of the land would have been indeed an oppressive misfortune, but Englishmen love a good piece of voluntary hard work, and you will all, I am sure, rejoice with me that this great piece of nature's work has been left for us to do; that this large continent extending from the 10th to the 40th parallel of latitude, capable of supporting 100 out of the 180 species of known ruminating animals, may by us be filled with such a selection from them as may present the highest points of excellence, omitting the inferior kinds, and that the opportunity is afforded us of not only making the most valuable and beneficial application of acclimatisation principles possible anywhere, but that from the almost total absence in Australia of beasts of prey which everywhere else keep down multiplication of such stock, we are visibly promised a blessing on our labour, beyond what could be found elsewhere; the good we do will live after us, and the work of our hands will thrive and prosper to our heart's content, and so become a lasting benefit to the millions of men who will in the fullness of time inhabit this land.

Fortunately, Australia's acclimatisers, like the explorers before them, were not as successful as they would have liked. Most of their dreams foundered for want of funding or good planning. Animals died in their cramped pens on the slow voyage out to Australia or succumbed to illness upon landing. Usually only a handful of creatures was released into habitats that did not suit, never to be seen again. The list of failures included English glow-worms freed in Melbourne's Botanic Gardens in 1860, medicinal squill planted on beaches, and releases of squirrels, English eels, canaries, bullfinches, bramblings and Algerian sand grouse. But some birds prospered. Sparrows were freed in several places, including fourteen in Ararat, west of Melbourne. 'The cage door was opened on the verandah of the Bull and Mouth Hotel, and the little strangers immediately made for the eaves of the houses', the *Ararat and Pleasant Creek Advertiser* recorded. Many people had congregated, 'and for hours afterwards, wherever one or two could be seen, knots of persons gathered to watch their movements ... We trust that parents will warn their children against throwing stones at the sparrows.'

Acclimatisers dreamed about plants as well as animals. Their botanical champion was the formidable Baron Ferdinand Jacob Heinrich Von Mueller, Victorian government botanist, Director of Melbourne's Botanic Gardens and arguably Australia's greatest botanist. Von Mueller gave many ponderous but enthusiastic speeches, and penned many pompous articles, extolling the virtues of introduced plants. Inordinately fond of Victoria's forests, he dreamed of the day when hosts of exotic species would enrich them. Speaking in 1870 at the Industrial Museum, he fondly depicted a typical forest glen in eastern Victoria, then asked:

> But can this grand picture of nature not be further embellished? Might not the true Tulip-tree, and the large Magnolias of the Mississippi and Himalaya, tower far over the Fern-trees of these valleys, and widely overshadow our arborescent Labiatae? Might not the Andine Wax Palm, the Wettinias, the Gingerbread Palm ... the Date, the Chinese Fan Palms, and Rhapis flabelliformis, be associated with our Palm in a glorious picture? Or, turning to still more utilitarian objects, would not the Cork-tree, the Red Cedar, the Camphor-tree, the Walnuts and Hickories of North America, grow in these rich humid dales, with very much greater celerity than even with all our tending in less genial spots? Could not, of four hundred coniferous trees, and three hundred sorts of Oaks, nearly every one be naturalized in these ranges, and thus deals, select tanning material, cork, pitch, turpentine, and many other products be gained far more readily there than elsewhere in Victoria, from sources rendered our

> own? . . . I do not hesitate in affirming that out of ten thousand kinds of trees, which probably constitute the forests of the globe, at least three thousand would live and thrive in these mountains of ours; many of them destined to live through centuries, perhaps, not a few through twice a thousand years, as great historic monuments.

And so on. Von Mueller wanted the whole continent transformed. Cabbages and asparagus would grow wild along our seashores, he hoped, while strawberries and cranberries would creep through the mountains. Pumpkins and melons would adorn the deserts and truffles would stud the forests. Of the blackberry he was especially fond, scattering its seeds through the wilderness during botanical sorties. 'He used always to carry a packet of blackberry seeds with him,' Mrs Fraser of Wesburn remembered, 'and whenever he boiled his billy, he scattered a few round the ashes of his fire. He said the poor people in time to come would bless him for his thoughtfulness. I wish he could come and see them now.' The blackberry became Victoria's worst weed.

To Von Mueller there were no weeds; every plant had value. The dandelion (*Taraxacum officinale*), for example, bore medicinal roots and leaves, and a prize was awarded for its successful introduction. In his 1870 address he declared, 'All organic structures, however, whether giants or pigmies, whether showy or inconspicuous, have their allotted functions to fulfill in nature, are destined to contribute to our wants, are endowed with their special properties, are heralding the greatness of the creator.' Years after blackberries were declared noxious weeds he continued to sow their seeds.

Von Mueller's book *Select Extra-Tropical Plants Readily Eligible for Industrial Culture* showcased what are today many of Australia's worst weeds. Prickly pears are there, along with boxthorn, gorse, mesquite, castor oil plant, English broom and Yorkshire fog. Today it seems outrageous that Von Mueller could recommend devil's fig (*Solanum torvum*), a prickly weed, simply because its berries 'seem also wholesome'. In his account of African boxthorn (*Lycium ferocissimum*) we can see, tangled together, the features that endeared it to the poor farmer needing a cheap fence and those that made it the pest it is today. 'It is evergreen,' he wrote, 'fiercely spiny, easily raised from seeds, readily transplanted, quick in growth, stands clipping well, seeds freely, is strong enough to resist cattle and close enough to keep off fowls. 1½ lbs. of seeds at a cost of 30 shillings suffices for a mile of hedging.' Some leaseholders in western Victoria were forced by law to sow boxthorn hedges.

Read today, the speeches of the acclimatisers seem extraordinarily confident if not downright arrogant. When the South Australian Acclimatisation Society was berated for freeing sparrows (a crime of which they were innocent), the president reminded members that the society only introduced *beneficial* birds, then went on to announce, in the very next breath, the release of forty-five starlings, fifteen blackbirds, thirty-six skylarks and thirty goldfinches. George Francis was supremely confident in his Adelaide speech: 'An acclimatisation society skims the cream from the more scientific bodies; it would disregard equally the weeds of the botanist and the reptiles of the zoologist. A society like this requires no conservatories, for it wants only those plants which are adapted to the climate; it gets rid at once of the annoyance, the danger, and the expense of the whole of the birds and beasts of prey; it refuses all that is noxious, and aims to procure only that which is useful.'

The acclimatisation movement was spawned during the Industrial Revolution, an era of rapid scientific discovery, when vast factories animated by complex machines demonstrated mankind's ability to manipulate the world for material advancement. The movement's mistake was to believe that animals and plants could be controlled as easily as metals and fibres – an error its successors still make today. While the early enthusiasm for the weird and wonderful soon waned, the acclimatisers did not fade from the scene. English gentlemen to their fingertips, they concentrated their energies on the sporting interests of the gentry, pursuing the introduction of salmon, trout, pheasant and partridge. Some of them busied themselves with state zoos (begun as holding pens for acclimatised animals), while the Queensland society specialised in cultivable plants.

The golden age of acclimatisation left us a legacy that includes the starling, blackbird, song thrush, bulbul, skylark, pheasant, goldfinch, greenfinch, red deer, hog deer, roach, tench, carp, brown trout and rainbow trout. To be fair, acclimatisers also imported, in Queensland at least, useful varieties of crops such as coffee and mangoes. And not all the exotic species established in that era were released by these groups; acclimatisers cannot be blamed for rabbits and foxes, and others must share responsibility for blackberries, sparrows and Indian mynas.

Amazingly, acclimatisation societies survive here to this day. The most venerable, the Ballarat Fish Acclimatisation Society, dates back to the 1870s. Its treasurer and secretary, Tony Gutteridge, told me they run a hatchery that supplies brown and rainbow trout to farm dams. Up until the 1950s they freed stock into rivers and streams, but when Victoria opened a government hatchery they were ordered to close down, then later allowed to continue their activity by selling to farmers. According to Gutteridge, the Ballarat society boasts twenty-five members, for whom membership bestows considerable local prestige. Other societies operate in New South Wales, where they distribute trout reared in government hatcheries. The Monaro Acclimatisation Society, formed in the 1930s, is the state's oldest, with twenty-three branches. Today's societies are fish stockers rather than real acclimatisers – they do not bring new animals into the country. Trout stocking is also practised by government departments such as the Inland Fisheries Commission in Tasmania.

The real acclimatisers today are the pasture scientists in the CSIRO and state agriculture departments. They carry on the work begun by the Queensland acclimatisers, who introduced many grasses and plant legumes during the colonial era. Regrettably, today's acclimatisers are causing as much harm as their nineteenth-century forebears, a point enlarged upon in chapter 11.

II

CARELESS AMBITIONS

Most of Australia's pests were imported deliberately, often for the silliest of reasons. Alas, the mistakes of the past are still made today. We act for short-term gain, oblivious to future consequences.

6

By Design

Planned Introductions

'We learn from history that we do not learn from history.'
G.W.F. Hegel

What makes our pest problems unique is the way in which so many of them were deliberately created. No one ever set out to pollute our rivers or erode our soils, but men knowingly, wilfully and repeatedly released toads, carp, goats, blackbirds and hordes of other rogues into the bush and rivers of Australia. Most of our worst vertebrate pests were purposefully set free by foolish men convinced they were doing the right thing. This holds true of our very worst mammals (rabbit, fox, pig), birds (starling, myna, sparrow) and fish (mosquito fish, tilapia, carp, trout) and of our one alien amphibian, the cane toad. As for our worst weeds, they were not set free in quite the same way, but most of them were deliberately imported, often for reasons that now seem foolish, then grown where they could easily escape, with predictable results. All of the statistics bear this out. Of our eighteen worst environmental weeds, seventeen were intentionally imported – mesquite (*Prosopis juliflora*) for example, promoted in the *Queensland Agricultural Journal* in 1914 as 'one of the most valuable trees that farmers could plant'. In Victoria, two-thirds of weed species were purposely introduced; and in the Wet Tropics, fifty-one of the fifty-three environmental weeds, mostly garden and pasture plants, were deliberate imports.

So a typical pest was brought in intentionally, often by misguided men who championed its dubious virtues, and we live today with the consequences of this past folly. Only among our exotic invertebrates (insects, mites, snails, worms), fungi and diseases do chance introductions predominate. To be fair, a great many exotic species were imported for sound reasons. Foreign plants and animals feed and clothe us, and many of us could not get by easily without pigs, horses, coffee and carrots. But then again, an amazing number of pests were brought in for little good reason, creating the remarkable legacy we live with today.

Some of our thistles, for instance – Scotch thistle (*Onopordium acanthium*), artichoke thistle (*Cynara cardunculus*) and golden thistle (*Scolymus hispanicus*) – came in as garden plants. Since this may seem unlikely I will quote from that scholarly tome *Noxious Weeds of Australia* (1992): 'In the early days, artichoke thistle was highly regarded as an ornamental plant and it is known that the heads were sold in Melbourne to people of Scottish descent as New Year decorations. It could have been introduced to Australia for this purpose but this is not known. The earliest Australian record is from Adelaide in 1839, and it was available from a Tasmanian nursery as a culinary plant in 1845.' Scotch thistle, declared noxious by 1856, was apparently grown as a heraldic symbol.

A desire to embellish our gardens also gave us hemlock (*Conium maculatum*), Paterson's curse (*Echium plantagineum*), crofton weed (*Ageratina adenophora*), mistflower (*Ageratina riparia*), groundsel bush (*Baccharis halimifolia*) and soursob (*Oxalis pes-caprae*) – a rogues' gallery if ever there was one.[1] Even castor oil plant (*Ricinis communis*) and Mexican poppy (*Argemone ochroleuca*) found a place in colonial gardens. None of these plants is favoured in gardens today, and some are so unattractive that the original enthusiasm to grow them seems surprising. But think of any reason why someone might bring in a plant – as food, medicine, ornament, aquarium plant, live fence or lawn grass; for pasture, timber, fibre, birdseed, land reclamation, acclimatisation or botanic garden display – and there are weeds here today to celebrate the fact. There's even a noxious weed, wild teasel (*Dipsacus fullonum*), that was probably imported for teasing wool; it still grows beside at least one old Tasmanian wool factory.

Each of these groups includes serious pests. Our medicinal plants include four invaders – pennyroyal, horehound, castor oil plant and St John's wort – so malevolent they are declared noxious in some states. St John's wort (*Hypericum perforatum*), a new panacea among naturopaths, infests more than 360 000 hectares of Victoria and New South Wales. When horehound (*Marrubium vulgare*) began spreading in the mid-nineteenth century Baron Von Mueller declared that it 'can nowhere be unwelcome, as it has important medicinal properties . . .' Today it has invaded 6 million hectares of Victoria alone, where it taints mutton and fouls fleeces. Most farmers don't know that canary grass (*Phalaris minor*) was brought in for birdseed, that Scotch thistle and hemlock started out as garden plants, or that spiny emex (*Emex australis*), a disastrous crop pest also known as 'doublegee', was introduced as a vegetable. Spiny emex seeds were brought in 1830 from Cape Town to Western Australia, where a gardener, D. Wonsborough, planted them out, then watched in dismay as a horrendous pest emerged: 'The weed spread very rapidly, and at every turn interfered with our work. The seeds even fastened to the staves of casks that had to be rolled up the banks of the river, no wharfs or similar accommodation being known then, and pricked the hands and arms of the workmen; to bales and other goods it stuck still worse, and in the field it was a constant nuisance.' It now infests a million hectares of pasture and half this area in crops in Western Australia alone.

But the past is soon forgotten, so we fail to grasp that our weed problems are largely self-inflicted, which helps explain why we keep repeating our mistakes.

If plants have been imported into Australia for every conceivable reason, so too have animals – as modes of transport (camels, horses, donkeys), food (pigs, dingoes, Pacific rats, oysters, bees), pets (cats, sliders, birds), for pest control (mosquito fish, mynas, ferrets, mongooses, toads, insects), sport (hares, foxes, trout), hat decoration (ostriches), landscape adornment (squirrels, sparrows, deer, glow-worms) and crop pollination (bumblebees, leafcutting bees). But it is one thing to import an animal, quite another to set it free. One might think that the one seldom leads to the other – that aquarium fish stay in tanks and rarely end up in streams – but this just isn't true. People dump their pets. Animals escape. Wild ferrets have been sighted around Perth and axolotls (aquatic salamanders kept as pets) have been found in wetlands around Canberra. All kinds of aviary birds show up in the wild from time to time. In the 1970s a reward of $20 000 was offered for the capture of two mountain lions that were roaming wild and slaughtering sheep near Duranillin in Western Australia. They had apparently escaped from a circus, but the Agriculture Protection Board was never able to prove their existence, though they tried.

One of the most bizarre arguments for importing an exotic species is propounded in Tim Flannery's remarkable book *The Future Eaters.* Flannery advocates a return of large predators to improve ecological stability, and since giant goannas starred in Australia's past, he proposes we import the Komodo dragon, the world's biggest and deadliest lizard, from Indonesia. 'Such an introduction should of course be approached with extreme caution,' he warns, without a hint of irony. 'A long period of experimentation and data collection would be necessary. Initial experiments may involve the introduction of single sex individuals into fenced areas, followed if appropriate, by single sex releases into unfenced suitable areas.' There might be drawbacks. 'The problem of predation by large varanids upon humans should not be understated,' he admits, 'but the risks posed are probably small when compared with the risk of falling prey to the salt-water crocodile (*Crocodylus porosus*). Australians are willing to extend protection to the "saltie" and are rapidly adapting their behaviour so as to minimise that risk. There is no reason to imagine that they could not also adapt to the smaller risk posed by large goannas.' To many people this might sound too much like Jurassic Park.

The soundest reason for introducing new species these days is for biological control. The new bugs and diseases brought in increase the number of exotic species in the country but, when successful, decrease their total mass and impact by controlling pests. More becomes less.

The following chapters look at the most important ways in which pests are deliberately introduced – as garden plants, pasture plants, and as fish imported for food, aquaria and mosquito control.

1. The following weeds also began their careers as garden plants: ragwort (*Senecio jacobaea*), quaking grass (*Briza maxima*), calotrope (*Calotropis procera*), green cestrum (*Cestrum parqui*), white moth plant (*Araujia hortorum*) and bladder dock (*Acetosa vesicaria*).

7

Ode to the Toad

The Cane Toad Conquest

> 'This introduction into Queensland was made only after a careful analysis of the pros and cons, and, according to the behaviour of the toad up to the present, there appears to be no reason for the assumption that we have made an error in our judgment.'
>
> **Reginald Mungomery, 1936**

In 1932 Raquel Dexter stood before a sugarcane conference in Puerto Rico and gave a talk that changed the world. She introduced delegates to the 'giant toad' (*Bufo marinus*), a huge South American amphibian that was devouring hordes of pests in the cane fields of Puerto Rico. Dexter's talk proved so persuasive that delegates from Hawaii took home batches of toads. The cane toad's conquest of the world was under way.

The Queensland government had a man at the talk, Arthur Bell, who monitored progress in Hawaii. Within a year or two there was a burgeoning toad population there, and in 1935 Reginald Mungomery, of Meringa Sugar Experiment Station near Cairns, flew to

Hawaii to bring some back. His batch of 102 toads was collected on a single historic night – 1 June 1935 – half of them from suburban gardens in Honolulu, the others from a sugar station nearby. Mungomery bagged his first fifty toads in an hour. They were already swarming over Honolulu in plague densities just three years after their release in the nearby hills.

On arrival in north Queensland the toads were pampered like princes. Mungomery dared not free them immediately lest they scatter too widely or fall foul of snakes and birds (he need not have worried). An elaborate breeding pond was constructed within a large octagonal cage. Adorning the pond were hyacinth and water-lilies fringed by taro and ferns. Photographs of this regal facility were proudly displayed at the sugarcane congress in Brisbane in 1935. (Use of hyacinth in the pond was apt, since this plant is also a major world pest.) Within six days the toads had repaid this hospitality by squirting out strings of eggs. Mungomery's joy can be imagined. His toads had multiplied a month to the day after their capture in Hawaii, and only eight days after reaching Australia! By the following year 41 800 baby toads had been liberated around Cairns, Gordonvale, Innisfail and Tully. The breeding pond was awash with eggs, carrying up to 200 000 at a time. Surpluses were poured into nearby lagoons. Farmers as far away as Bundaberg were clamouring for toads. Mungomery knew he was launching an invasion. He wrote in the *Cane Growers' Quarterly Bulletin*, in words that now damn him, 'this latest importation of the toad into Queensland will mark another step in the gradual conquest of the warmer regions of the earth by this remarkable animal . . .'

The toads had been enlisted in the fight against one serious pest – the greyback cane beetle (*Lepidoderma albohirtum*). Larvae of this native insect bored into cane stems, wreaking havoc in the crop. Many kinds of control had been tried and had failed, including soil fumigation, bounties on beetles and grubs, even destruction of rainforests to remove beetle harbourage. Mungomery hoped – although he was never entirely sure – that the 'Giant American

Toad' would prove the answer. In Puerto Rico the toad seemed to control a local beetle, but that species returned to the ground each day, bringing it into regular reach of toads. Greybacks were different. They stayed up in trees, the females returning to earth just once, to lay eggs. Mungomery knew this.

In 1935, speaking with fellow researcher J.H. Buzzacott, he told an audience of experts that 'the recent importation of the Giant American Toad into Queensland will *possibly* provide an animal whose habits fufill *most* of the desiderata required of an ideal predator' (my emphases). They went on to say: 'If they do not accomplish all that is desired of them in connection with the control of the greyback beetle, their importation will still have been justified as there can be no doubt as to their value as general insect feeders. In the meantime we look forward to the future with considerable interest, and hope that the Giant Toad will prove the same valuable acquisition to our fauna as it has proved elsewhere.' In fact, toads are not considered a 'valuable acquisition' anywhere.

Release of this grotesque amphibian did not go unchallenged. There were fears that chorusing toads would ruin the peace, prey upon bees, cause cars to lose traction and skid off roads, and generally become pests. Nature groups worried that toads would devour native fauna, especially frogs and baby ground birds. The federal government vacillated – slapping on a ban in late 1935 then lifting it the following year, after Mungomery dissected some toads and showed that insects dominated their diet.

Despite his own doubts Mungomery was always dismissive of his critics. In 1936 he wrote: 'It seems almost inevitable that all attempts to progress should meet with totally unwarranted opposition, in this case resulting largely through an improper appreciation of the habits of the toad.' The year before he had written in the *Cane Growers' Quarterly Bulletin*: 'To others who scent a "nigger in the woodpile" and suggest the possibility that the toad will, in turn, itself become a pest, we can point to the fact that nearly 100 years have elapsed since it was first introduced into Barbados, and there it has no black marks against its character. Experience with it in other West Indian Islands, and in Hawaii, certainly points to the fact that no serious harm is likely to eventuate through its introduction into Queensland.' To enlist Hawaii in this argument was premature since toads had been freed there only three years earlier.

The only drawback Mungomery would admit to was that toads did eat bees: 'they sometimes loiter near hives of honey bees and jump up and catch the heavily laden bees returning to the hives'. But a solution was at hand – hives could be raised above the ground, Mungomery explained, or ringed by wire netting. 'As most of the bee-hives in Queensland are usually placed more than 2 feet above the ground this introduction of *Bufo* should not prove any serious hardship to Queensland beekeepers.' Beekeepers disagreed. One riled apiarist, slicing open a toad, found 300 of his prized bees. It was not true that hives were normally raised. Beekeepers were only forced to do this after the toads invaded. Apiarists shifting around hives were much inconvenienced, the hives on stakes proving clumsy to move. Labour and transport costs rose. So too did the price of honey. To help one sugar-producing industry another, far less powerful one had been handicapped.

Most of the vices of the toad were known to Mungomery from the start: that it grew to an enormous size, fed voraciously, consumed almost anything, bred prolifically, lived at high densities and spread from cane fields to gardens. (He speculated, 'it seems most reasonable to expect that the toad population, when it has reached saturation point in the sugar districts, will be dense'.) These attributes were virtues, Mungomery thought, his only fear being that toads might not be hardy enough to survive in Australia. Once Queenslanders knew more about these creatures they would welcome them into their gardens. After all, in Hawaii 'flattering reports' had come in from housewives after cockroach numbers dropped. In his 1936 talk he was bold enough to suggest that 'more intimate acquaintance with the habits of the toad will serve to popularise even further this immigrant, whose powers of reproduction are really amazing, and, within the next few years, the toad must assuredly become familiar to most residents of North Queensland'. True enough.

It soon became apparent that toads were not controlling the beetles. The 40th Annual Report of the Queensland Bureau of Sugar Experiment Stations, in 1940, commented: 'It would appear that populations of the giant toad have now reached saturation point in many northern areas, and although some growers affirm that at least in localised areas the toad is coping with the grey-back beetle, in general it appears evident, both on the score of observations in the field and by virtue of the fact that grub damage has increased this year in areas where the local [toad] population is high, that the number of beetles destroyed by the toad is relatively small when compared with the number that are nightly on the wing.'

By 1949, fourteen years after unleashing his protégé, Mungomery was still battling to defeat the beetle. Remarkably, by now he had been promoted to officer in charge of the Entomology and Pathology section at the Bureau of Sugar Experiment Stations. His new solution was Gammexane insecticide. In an article for the *Queensland Journal of Agricultural Science* to promote it, Mungomery included a section on 'Early Control Measures' that is most illuminating. Read it and you will find mention of soil fumigation, beetle bounties, beetle-resistant cane, and the late planting of cane, but no allusion whatever to the toad. On that subject Mungomery had, it seems, been shamed into silence.

Even in Puerto Rico, beetle control proved illusory – Dexter had misled the world. After ten or fifteen years toad numbers fell away and beetle numbers rose. In 1943 and 1948 there were big outbreaks, and new biocontrol agents were sought. The early drop in numbers was perhaps due to a run of wet years that drowned their larvae in the soil. From 1931 to 1936 Puerto Rico was exceptionally wet and her dry seasons unusually severe. It may be true to say that Australia ended up with toads because a lot of rain fell in Puerto Rico in 1931.

We were not, alas, the only country to embrace this rogue because of Dexter's praise. Hawaiian toads were welcomed into

Taiwan, Fiji, the Philippines, New Guinea, the Solomons, Guam and the Palau Islands. Earlier introductions took toads to islands throughout the Caribbean, and later to Florida, Japan and (unsuccessfully) to Egypt. Hawaii's toads have themselves had a colourful past. They can claim descent from 149 toads brought from Puerto Rico, which in 1920 obtained toads from Barbados, which in turn had imported them some time before 1844 from Cayenne in Guyana. All of Australia's toads can thus trace their ancestry back through Hawaii, Puerto Rico and Barbados to Guyana in South America.

The toad is now the most common vertebrate across much of Queensland. As many as 5000 per hectare may crowd around waterholes. It occupies eucalypt forests, rainforests, swamps, mudflats, beaches, farms and gardens, even offshore islands. From Queensland's cane farms it has hopped into the semi-arid outback, into the woodlands of the Northern Territory and south into New South Wales (reaching Port Macquarie at last call). It can breed in brackish water (up to 15 per cent seawater), lay 50 000 eggs in a clutch, and mature in less than a year. Remarkably drought tolerant, it can lose half its body water and survive. In the Northern Territory the toad front is advancing thirty kilometres a year and will soon reach Kakadu. Toads eat mainly ants, beetles and termites, but they sometimes tackle frogs, small mammals, snakes, other toads, centipedes and bowls of pet food. Consummate pests, they thrive wherever people lay bare the land. Because they are poisonous they can forage in the open, where prey is easiest to see – on roads and lawns, and in overgrazed paddocks and burned woodlands. They hop into forests along roads and tracks, breeding in roadside ditches and borrow pits. On acreage blocks they use farm dams and paddocks. Often they perch under nightlights to harvest dazzled insects. Many are crushed on the roads; as far back as 1955 Ormsby observed, 'The giant toad's chief enemy is motor cars'.

Their very poisonous skin and glands yield digitalis-like steroids, and dogs have died fifteen minutes after mouthing a toad (most learn to stay away). But some predators do make meals of toads.

The keelback (*Tropidonophis mairii*), a harmless snake, swallows young toads and 'toadpoles' with impunity, and in some toad-infested swamps it thrives, although probably more by eating frogs than toads. As a boy I caught dozens of these snakes in toad-infested paddocks around Brisbane. Other animals, including water rats, black rats, crows, kites and koels, have learned to tear open the poisonous packaging to extract the wholesome contents. (I know one naturalist who once risked his life by eating skinned and fried toad legs.) Scavengers often harvest toads flattened on roads. But some animals that take on toads suffer horrible fates. When toad fronts reach new areas, goannas, frog-eating snakes and quolls (tiger cats) disappear. Goannas slowly rebuild their numbers by learning to avoid this prey, but snakes, and perhaps quolls, do not. In Queensland, the death adder (*Acanthophis antarcticus*) has suffered enormously – it is now officially listed as rare, probably because of the toad. The spotted black snake (*Pseudechis guttatus*) is in trouble, and the red-bellied black snake (*P. porphyriacus*) has also declined dramatically while remaining common in toad-free zones.

Toads also take over the hiding places of native animals. While searching for reptiles in the bush I have sometimes turned over dozens of rocks and logs and found little else but toads – a depressing experience. Toads are also accused of eating and competing with native frogs, although the few studies so far do not support this; indeed, toads may help frogs by eliminating their predators. They do interfere in dung control by sitting beside cowpats and snapping up approaching dung beetles, taking as many as eighty in a night.

One reason why toads fare so well is that they remain almost parasite free. The blood protozoa, gall bladder sporozoans and helminth worms that afflict them in Latin America have not reached Australia. Biologists hope that certain viruses imported by the CSIRO may provide the control so badly needed. Toads will never be eliminated but we may be able to reduce their numbers. We know some harmful organism is already out there, a disease probably, because toads always drop in size and numbers some years after first invading a district.

In October 1936 the *Australian Sugar Journal* observed: 'In now permitting the liberation of Bufo in the cane areas of Queensland, we may be confident that no element of casualness has been involved in the decision, but that every biological aspect known to science has been carefully considered.' In fact, the toad's introduction was unaccompanied by any research, and proved an act of monumental folly and high arrogance. But we may console ourselves with the knowledge that it might have turned out much worse. Back in 1897 a proposal was made to import European common toads (*Bufo bufo*) and natterjack toads (*Bufo calamita*), along with hedgehogs, moles and shrews, to control farm pests in Western Australia. A few years later A.J. Boyd made a serious plea to introduce moles to solve the cane grub problem. 'The mole can be procured from Japan or India,' he wrote in the *Queensland Agricultural Journal* in 1902. 'Let a hundred or so be imported, and place them on a piece of cane country infested with grubs . . . as the moles increased, so would the grubs decrease, and it is quite within the bounds of probability that, with the introduction of an animal which would hunt them underground, the cane grub would become a thing of the past.'

In 1921 there were further calls for moles, along with shrews ('most useful grub-destroyers in other countries'), then in 1935 – the very year Mungomery brought cane toads to Queensland – some European toads were imported for trials against grass grubs (*Oncopera* species). Quarantine tests showed that while the toads devoured all stages of the insect, they could not get at the grubs in their burrows. Common sense prevailed and Australia was saved the indignity of the European toad. Had similar trials been conducted on cane toads they would never have been set free. European toads, had they been released, would have complemented the cane toad's distribution by invading southern Australia.

Many pests attack cane, and over the years many predators have been imported to control them. In the 1880s locusts struck, and Indian mynas were released to save the day. Instead they became pests – decades before the toad. Soon after, rats were attacking cane and the sugar companies imported Indian mongooses to control them. Fortunately the few that were set free did not survive. In Hawaii, Fiji and the Caribbean, other cane growing areas with rats, introduced mongooses are hated more than toads for the toll they take on birds and reptiles. We were very lucky to escape this pest. Then in 1919, when canefield rats were again wreaking havoc, farmers sought to import American owls. Queensland government entomologist Henry Tryon reported: 'In order to control the rat plague, so serious in our northern sugar-cane plantations – those of the Mossman area especially – a proposal has been made to take steps to cause Australian species of Owls to congregate there, as well as to import rodent-capturing species from the United States, in order to supplement the efforts of our own birds of destruction.' Tryon quashed this foolish idea.

So maybe this story has a happy ending after all. Australia got cane toads and mynas but escaped the ministrations of moles, shrews, Asian mongooses, European toads and American owls.

8

'Peopling a Barren River'

A Fishy Business

> 'He who achieves the more difficult task of . . . peopling a barren river with a noble species of fish, should not pass unnoticed by his contemporaries.'
>
> **Arthur Nicols, 1882**

Australia, driest of lands, is remarkably barren of freshwater fish. Fewer than 200 species grace our waters, compared to 400 in Thailand, 760 in North America, and more than 2000 in South America. In Africa a single lake, Lake Malawi, carries more than twice as many species as all our rivers, streams and lakes combined. No wonder the acclimatisers sought to 'improve' upon this picture. Professor McCoy could report to the Victorian society, at its very first annual meeting in 1862, on the introduction of English bream, dace, tench, loach, roach and carp – the roach, tench and carp survived and thrived. Eels and native fish were set free as well, along with oysters and crays. There was even a release of marine English flatfishes into Gunnamatta Bay. According to science writer Peter Macinnis, 'Far more time, money and effort was spent on the acclimatisation of fish in Australia than on all the other plants and animals put together.'

Most of this work went into perch, trout and salmon. Perch did well in southern Australia. Often called 'redfin', they are easy to take, often schooling in sluggish waters, and sometimes reach ten kilograms or more. In Western Australia, Walter Kingsmill explained in 1918, 'These fish are delivered free of all charge to the nearest railway station or siding to the home of any settler who is willing to take charge of them, release them in the waters in which they are to live, and return the cans in which they have made their journey to the station for the carriage to the depot at Chidlow's Well.'

Salmon and trout proved impossible to import from England, time and again succumbing to the heat of the tropics during the three-month-long journey. Undeterred, the acclimatisers turned attention to their eggs. Shipments were brought out in ice, or in gravel beds washed by cold water, still without success. Finally they tried packing eggs in cold moss – and in 1864 the clipper *Norfolk*, guarded on its last leg by the armed sloop *Victoria*, entered Hobart in triumph bearing thousands of Atlantic salmon and brown trout ova. (It took another thirty years to obtain rainbow trout eggs.) This success was trumpeted as a mighty achievement, one glorified in such books as *The Acclimatisation of the Salmonidae at the Antipodes*, a 238-page work published in London in 1882, in which author Arthur Nicols enthused that 'long before the end of this century when probably the ploughshare will have invaded the haunts of the red deer, and a growing population shall have driven the salmon in disgust from most of our rivers ... the sportsman will take his rifle and rod, and seek among the fern-covered ranges of the Australian Alps and the deep tarns and pools of Tasmania and New Zealand, the noble quarry which has found a congenial home at the Antipodes'. Salmon and trout held great symbolic importance as the sportfish of the English gentry.

To help the trout, fishermen demanded that fish-eating fowl, especially shags (cormorants), be culled. The Western Australian Acclimatisation Committee even sponsored destruction of shag colonies. Persecution of these birds persisted into the 1930s even after eminent biologist Dom Serventy found little trace of commercial fish in their bellies. The problem today is not native birds exterminating trout, but trout exterminating native fish. In Tasmania and Victoria this is emerging as a major issue.

Salmon and trout evolved in the cold waters of the northern hemisphere, across a vast realm where ancient glaciers carved out hundreds of lakes, and melting snows fed roaring rivers. Australia's meagre water systems were not conducive to the creation of large cold-water fish, and Tasmania's waters are dominated instead by little fish called galaxias, or native minnows (*Galaxias* species). Brown trout, in particular, have become arch-foes of these fish. Of the fifteen species found in Tasmania, most of them smaller than pencil length, six disappear when trout are introduced, and another two slide into decline. Some species now survive only in remote headwaters or lakes beyond the reach of trout. Four species are listed as endangered.

After Lake Pedder was drowned amid controversy in 1972 as part of a hydroelectric scheme, the Pedder galaxias (*Galaxias pedderensis*), a fish unique to the lake, thrived. Vast schools shoaled in the swollen waters. But the dam linked Pedder to the headwaters of the Huon and Serpentine rivers and trout gained entry to the once isolated waters. Pedder galaxias plummeted in numbers as the trout multiplied. The Inland Fisheries Commission made matters worse by actually stocking trout into the lake; the guts of captured trout were found to be full of galaxias. Today the Pedder galaxias is Australia's most endangered fish. During recent surveys only five could be found, in two small feeder streams. This species, very difficult to breed in captivity, only survives because a few dozen were taken to trout-free Lake Oberon in South West National Park. Future generations will probably call them the 'Oberon galaxias'.

Another fish faring very badly is the swan galaxias

(*G. fontanus*), discovered in 1978 in a few streams in eastern Tasmania (it was probably dispersed more widely before trout came along). Of the five known populations, two vanished recently when trout and perch moved in. To save this rarity the Inland Fisheries Commission has founded nine new colonies in trout-free streams.

Two examples are enough, but I could mention many other trout-endangered fish in the south – among them the barred galaxias (*G. fuscus*) of mountain streams in Victoria, the Clarence galaxias (*G. johnstoni*) of central Tasmania, the mainland trout cod (*Maccullochella macquariensis*), and the vulnerable saddled galaxias (*G. tanycephalus*) and pygmy perch (*Nannoperca variegata*). The tragedy for these fish is that brown trout, and sometimes perch and salmon as well, either eat them or take their food. Some are now confined to headwaters high in the hills, protected by trout-proof barriers. And there is more bad news. Frogs, along with crayfish and other invertebrates, are also disappearing from trout streams. Trout streams are wet deserts. The ecology of Australia's colder waters has been ruined forever, at public expense – all to keep recreational fishermen happy.

Early this century, life in Australian cities was made vexatious by the clouds of mosquitoes that tormented residents at dusk. Remedies proposed to counter the nuisance were many and ingenious and included the stocking of metropolitan waters with insect-eating fish. Experts were quick to point out that most streams already carried native fish with an appetite for mosquito larvae. The firetail gudgeon (*Hypseleotris galii*), said one scientist, considered 'mosquito larvae with the taste of an epicure' and the rainbowfish (*Melanotaenia duboulayi*) 'pursues its bond with such unrelenting fury as would suggest, in manner, the case of the wolf ravishing the fold'. But native fish were impeded in their duty, experts warned, because the larvae hid among weeds, or dwelt in waters too polluted for fish, or in fish-free ponds, puddles, tanks, wells, barrels and buckets. An inspection of 2000 Adelaide premises in 1922 found that 1300 of them contained mosquito breeding places. Native fish were sometimes bred up in hatcheries to stock city waters. In the 1920s the Brisbane City Council ran a vast hatchery for rainbowfish in South Brisbane lavishly stocked with waterweeds, and a hatchery was planned for Adelaide to stock Lake Torrens with rainbowfish, olive perchlets and firetail.

It was unfortunate that health departments waging the mosquito war did not stick with native fish and heed Adelaide medical officer T. Borthwick, who concluded in 1923 that 'exotic fish found to be effective in other countries do not thrive under our climatic conditions; and that it will be necessary to fall back on local fish which can be proved to be larvae-destroying'.

He was wrong, though. The American top minnow (*Gambusia holbrooki*), known today as the mosquito fish, was already selling in pet shops, and its penchant for mosquito larvae (wrigglers) was well known in Hawaii. Although there were fears that predators would exterminate it, mosquito fish were released around Brisbane in 1925, in Sydney the following year and in rural New South Wales the year after that. Brisbane's hatchery became a broodery for this species (along with guppies). The little fish thrived, and Australia's watercourses were changed forever.

Gambusia has proved one of the most successful of all aliens unleashed upon Australia. A live-breeder (unlike our native fish), quick-growing and remarkably tolerant of pollution, salinity and temperature extremes, it now flourishes in every mainland state, and in some remarkably remote places, including outback mound springs. I will never forget my surprise upon scooping some up in a remote muddy wetland in south-west Queensland, two days' drive from the nearest bitumen. This fish dominates my local Brisbane creek. I have also seen it in the Yarra in inner-city Melbourne and in rivers north of Perth. I'm always pleased to net a stream and not find these aquatic 'cane toads'. I have kept mosquito fish in tanks alongside native fish and they show no special talent for catching wrigglers. All of our small native fish are keen consumers of mosquito larvae, as the experts realised long ago. The tag 'mosquito fish' is false advertising of the worst kind; some biologists prefer to call them plague minnows. This impostor is likely to be useful only in polluted, weedy waters – situations native fish avoid. Success for *Gambusia* is guaranteed by disturbance to banks. In Queensland and New South Wales, rainforest along streams has been replaced by African para grass (*Brachiaria mutica*), a pasture plant. The unshaded water is warmed by the sun, and the para grass trails into the water, slowing down the flow, creating ideal plague minnow habitat – warm, weedy, sluggish water. Native fish shun this environment.

Health officials battling malaria and yellow fever took this fish everywhere. From its native North America it was liberated in Europe, Africa, the Middle East, India, South-East Asia, China, Japan, New Zealand, New Guinea and parts of South America. In North America I was not surprised to find it the most abundant fish in its native habitat in the south-east, thriving in cypress swamps and roadside drains, and swimming among alligators in the Everglades.

Mosquito fish displace native fish by eating their eggs, stealing their food and nipping their fins. They dominate shallow edges and sun-lit pools, often proliferating in degraded streams, forming large aggressive shoals but avoiding deep shady stretches. Unlike trout, they don't usually eliminate other fish, but in some heathland lakes they seem to have replaced ornate sunfish (*Rhadinocentrus ornatus*), and in outback mound springs they are reckoned a threat to rare gobies (and perhaps invertebrates). They also eat small tadpoles, and biologists blame them for an alarming decline in golden bell frogs (*Litoria aurea*), now an endangered species in New South Wales. These frogs rarely persist around *Gambusia*-infested waters.

Trout and mosquito fish are only two of our aquatic scourges; there are many more. Carp from Asia now dominate our biggest river, the Murray–Darling. During a recent survey of the river they comprised a whopping 93 per cent of the 4000 fish sampled: carp, carp and more carp. In Victoria, carp and goldfish now outnumber native fish. Carp weren't a problem for Australia until a vigorous new strain, the 'Boolara', was developed in 1961. It was soon recognised as a threat and 1300 dams in Victoria were poisoned in a valiant effort to defeat it. But some of these superfish slipped into the Murray near Mildura and the river's ecology is now in ruins. Exotic fish are doing great injury to Australia, but because their crimes are committed behind veils of water, out of sight and mind of most people, they escape the scrutiny they deserve.[1] It behoves us all to take them more seriously. The next chapter will have more to say about underwater invasions – those brought about by plants, snails and fish introduced through the pet trade.

1. The worst extinction event in modern times was triggered by a fish. After the Nile perch (*Lates niloticus*) was stocked into Lake Victoria in the 1950s, more than 200 species of smaller cichlid fish disappeared into extinction. The Queensland government very nearly introduced Nile perch (a close relative of the barramundi) into Queensland.

9

Wet Pets

Aquarium Escapes

> 'Before long it will be difficult to know what exotic fish are in Australia.'
>
> **Patricia Kailola, Bureau of Rural Resources, 1989**

In principle the Commonwealth government does not allow live fish to be brought into Australia; the disease risk is simply too great. A scientist wanting to bring in a few gobies for research would need to convince the minister for Primary Industries to issue a special permit, then keep the fish under permanent quarantine in approved premises. But one industry is exempt from this judicious rule. The aquarium trade is free to bring in some *6 to 10 million* live fish every year. Australia's quarantine policy is thus glaringly inconsistent, a point made by Canada during its successful appeal to the World Trade Organisation over Australia's ban on foreign salmon meat.

Most of those millions of aquarium fish come from South-East Asian fish farms where diseases run rife, epidemics are sometimes uncontrollable and the death rate runs to 40 per cent or higher. In

1990 alone diseases cost the farms more than US$1500 million. Australia's only protection from these dangers is a two-week quarantine period; if the fish look fine after a fortnight they are free to be sold. The policy – a compromise designed to protect imports worth a mere $2 million a year – is a nonsense, for there are plenty of diseases that remain dormant longer than a fortnight, and others, like the bacterium *Edwardsiella ictaluri*, that live harmlessly and invisibly on some hosts while killing others. Some of the high-risk bacteria are even a threat to humans. In recent years more than fifty fish diseases have turned up during quarantine checks; over the years, others have entered through airports and reached waterways where they now kill native fish. Experts suspect that many diseases found among native fish today are exotic imports that came in like this.

Diseases are just one concern (which we'll return to later). Aquarium fish, plants and snails also escape into local waters and become pests. Hundreds of fish species are legally imported; many more are smuggled in, and scores of waterweeds circulate in the trade, posing such serious risks to the environment that if keeping aquarium fish were a newly emerging hobby the government would probably enact strict laws to control it.

Fourteen species of snails alone have been intercepted at airports, on waterweeds or in bags of fish. Others have sneaked through unnoticed or been smuggled in. Now there are masses of exotic snails in our waterways and others circulating among fanciers. Some of them pose a health risk, serving as hosts for liver flukes and bilharzia. Western Australia until recently was free of liver flukes (*Fasciola hepatica*) – the native snails don't act as intermediate hosts – but the American ribbed fluke snail (*Pseudosuccinea columella*) now thrives in irrigation channels, and cattle near Perth have suffered fluke infection. Another snail, *Lymnaea rubiginosa*, often intercepted by the Quarantine Service but not yet established in the wild, is a host to the Asian liver fluke (*Fasciola gigantea*), and other parasite hosts have been intercepted (for example *Biomphalaria straminea*, host of *Schistosoma mansoni*, which attacks people, causing debilitating illness). If these snails ever escape into our streams while carrying flukes public health will be at risk.

The snail the Australian Quarantine and Inspection Service (AQIS) fears most is not a worm-carrier but a likely crop-cruncher; it is the mystery snail (*Pomacea bridgeii*), a pretty South American species smuggled in illegally more than twenty years ago and now sold by pet shops everywhere. This snail is almost identical to the golden apple snail (*P. canaliculata*), a notorious Asian rice pest (see chapter 32), and may in fact be the same creature. It is banned in rice-growing regions of New South Wales, and pet shops are inspected to ensure they are snail free. Its best chance of going feral is in north Queensland; a scare was raised a few years ago when empty shells were found beside the Ross River. Mystery snails come in a variety of colours like boiled lollies – yellow, banded, translucent – and several incipient pests may be involved. Genetic work is under way to study the problem.

If snails are a worry then aquarium plants may be a worse one. Water plants rank among the foulest of weeds for their penchant for proliferating in pollution-enriched dams and canals, which they often choke completely. Aquarists boast that hundreds of species circulate among fanciers, many of them illegal imports, some worth hundreds of dollars apiece. Most species suitable for aquaria are already in the country, often the progeny of illicit imports from Asia or Latin America. At a meeting in Brisbane in 1997, the pet industry warned government weed experts that the level of underground trade was enormous. That meeting was called because someone slipped cabomba (*Cabomba caroliniana*), a wispy waterweed from America, into one of Caloundra's water supply dams. A backyard grower probably planted the weed with the idea of selling the harvest to pet shops. The city spent $300 000 draining the dam in a futile attempt to remove it. Cabomba is turning up in wetlands all over northern and eastern Australia, presumably wild-planted by growers. First reported in the wild in 1981, it now ranks as one of Australia's twenty worst weeds (see appendix II).

Lagarosiphon (*Lagarosiphon major*) and water pennywort (*Hydrocotyle ranunculoides*) are problems too. The wort washed out of a garden pond in the 1970s into the Canning River in Western Australia and the government is now battling to suppress it. Lagarosiphon has not escaped into streams yet, but in New Zealand it chokes hydroelectric systems and reservoirs; the one Australian outbreak, in a farm dam near Melbourne, was eradicated in 1977. Declared noxious throughout Australia, this weed does not legally exist here, yet it turns up from time to time nonetheless. Early in 1999 a nursery chain in Victoria and Tasmania was found to be selling it, along with cabomba and other noxious waterweeds (*Elodea* species and parrot's feather *Myriophyllum aquaticum*, a common weed in my local creek), as oxygenators for water gardens.

Australia all but banned new waterweeds in 1997 when AQIS brought in the Weed Risk Assessment system (see chapter 40). But AQIS can't stop weeds entering illegally in plastic bags through the mail or control the hundreds of varieties already here. After the cabomba fiasco, the Queensland government in 1997 proposed an approved list of fifty safe species, a ban on the rest and an accreditation scheme to stop wild-farming 'backyarders'. The big growers resent wild-farming because it undercuts their prices, but they oppose any list. 'Restricting trade to only 50 species where at present over 300 are freely traded would severely limit the industry and feed a thriving black market', they argued in a submission by the Pet Industry Joint Advisory Council. Fortunately most aquarium weeds appear to be too delicate to take off in the wild. Some of them require very warm water and aquarists have trouble keeping them alive – hence the high prices.

The waterweed AQIS fears most is actually a seaweed, *Caulerpa taxifolia*. A German aquarium created a mutant strain for fish tanks and sent pieces to aquaria around the world (though not to Australia). This mutant escaped into the Mediterranean from the Monaco Aquarium and has invaded 3000 hectares of sea floor affecting five countries. Growing seven times bigger than the parent species, it blankets the sea bottom, eliminating animals and plants and depleting fish stocks. We must hope this horror has not been smuggled into Australia. Imported waterweeds have also brought us a tiny flat-worm, *Girardia tigrina*. Only 1.5 centimetres long, it no doubt came in attached to waterweeds, and now it lives in our waterways.

Pet fish win their freedom when owners tire of them and dump them in streams or down drains, or when ponds overflow after heavy rain. It happens all the time. Tropical fish frequently turn up in Melbourne sewers; the Canberra sewage works used to keep an aquarium full of the fish scooped out of its sewage ponds. Some of the fish that turn up are illegal species here. (In Hawaii piranhas have appeared in dams, and in Florida pet crocodiles [caimans] released into canals now breed alongside exotic fish.) North Queensland has most to lose from these escapes because most aquarium fish are tropical and rarely survive further south. Alan Webb, a fish biologist at James Cook University in Townsville who monitors north Queensland wetlands, is troubled by what he sees. New fish appear every few years. In 1985 six feral species were identified; now there are thirteen. The Ross River in Townsville carries six exotic cichlids as well as guppies, platies, swordtails and mosquito fish. The latest arrivals – the firemouth, green terror and Burton's haplochromid – turned up in 1997, though it is too soon to say if they are firmly established. Alan believes most of these fish flush into waterways when outdoor fish ponds overflow during torrential monsoon rains, since newcomers appear after particularly wet years.

The new fish are all cichlids, members of a family (Cichlidae) that practises parental care of the eggs. The most feared are tilapia, stocky-looking fish that grow to a hefty thirty-six centimetres. Remarkably tough, tilapia tolerate saline water, and they are turning up as uninvited guests all over the place. Because of their history as pests overseas they were banned in Australia in 1963, long before they went feral, but pet shops were still selling them in 1977, the year they were found living wild in a large dam in Brisbane. They have since colonised wetlands around Townsville and Cairns and the Gascoyne River in Western Australia, and they are showing an awesome ability to multiply in warm, shallow dams. When a Port Douglas resort found tilapia in its lake, Queensland Fisheries poisoned the waters – and up bobbed twelve tonnes of dead fish. Tilapia now dominate streams entering Brisbane's Leslie Harrison Dam. Overseas, where they are farmed for food, they are eliminating entire fish communities. In New South Wales, Vietnamese immigrants are illegally stocking them in dams as food. We will be hearing more about this fish in future. Live-bearing fish from the Americas (family Poecilidae) have also taken off. Guppies, platies and swordtails are now common in Queensland streams; the mosquito fish belongs in this group.

Escaping fish pose less of a problem the further south you go. Brisbane, with its brisk winters, supports fewer escapes than Townsville, though numbers are growing. Swordtails are common, platies and tilapia took off during the 1980s, and the blue acara, a cichlid, appeared a couple of years ago. Carp have been seen as well. Guppies, sailfin mollies and rosy barbs established colonies in the past but they have apparently disappeared, snuffed out by the nippy winters. In southern Australia the main escapes are goldfish and carp, along with a newcomer, the oriental weatherloach, a 20-centimetre burrowing fish that likes cool water. First recorded in 1984, it is now found in waters around Melbourne, Canberra, Sydney, the Snowy Mountains and elsewhere, and is spreading so fast it may one day colonise the whole Murray River system. Its entire distribution results from aquarium releases or pond escapes.

Long before it woke up to the dangers of waterweeds, the Quarantine Service decided aquarium fish posed a risk. Fish imports began booming in the early 1960s, when concerns were voiced about the entry of undesirable immigrants such as piranhas and electric eels. In 1963 the Department of Primary Industries put out a list of 700 approved species and banned everything else – as environmentally harmful, aggressive, poisonous, disease carrying or for other reasons.

Some years later, when eighty species were found to make up 96 per cent of imports, AQIS proposed a much shorter 'List of 100' (actually about 200). But importers howled, and because AQIS wants to discourage smuggling, Australia has kept to a list of several hundred approved fish (Schedule 6 is available on the Internet and updated regularly). This list is not popular with fish fanciers, who say it is too restrictive, although it lists several fish that have gone feral, including the blue acara, sailfin molly, platy and other fish that have escaped overseas but not here (yet). In any event, fish keepers breed plenty of species not on the list, which were either smuggled in or imported before 1963. And can we expect our quarantine officers, who are not biologists and who work with fresh foods as well as fish, always to distinguish the hundreds of 'good' fish from the banned ones, especially those little silvery things darting around in their hundreds in plastic bags? AQIS tells me staff training is improving and computer identification is on the way.

Smugglers sneak banned fish into Australia regularly in courier parcels, hand luggage, by sleight-of-hand at quarantine offices, by paying bribes, or by deliberately mislabelling shipments. Alan Webb knows of fish eggs brought from Thailand in hollow biros. AQIS regularly intercepts piranhas – I am told you can order the piranha species you want and have it delivered from overseas within fourteen days. Fanatical collectors with contempt for the law pay huge sums for banned fish, and pet shops sometimes participate in the smuggling. In 1997 a prominent aquarium owner in Brisbane was fined $5000 after postal staff noticed a leaking package from Japan – it contained Amazonian zebra catfish worth up to $450 apiece. The previous year two Sydney importers sneaked in 600 illegal armour-plated catfish (*Ptergygoplichthys*) and arawana (*Osteoglossum*) with their legal shipment – they simply swapped them for legal fish during the drive from the airport to the quarantine office. An AQIS officer noticed the changed seals on the containers. The men were fined $2500, considerably less than the fish were worth.

To help deter smuggling Alan Webb believes the Schedule 6 list should be revised. Some banned fish are unlikely to go feral because their water requirements (temperature, pH) are very precise. A more credible list would win greater respect from the trade. It is probably significant that most escaped fish are either live-bearers or hardy cichlids that practise parental care of their eggs.

More dangerous to the environment than the fish themselves may be the diseases they carry. Diseases can escape when sick fish are released, used as live bait when fishing (as goldfish often are), or when aquarium water or dead fish are flushed down drains. The problem is so serious for Australian aquaculture that in 1992 AQIS commissioned a major review. The two subsequent reports, written by J.D. Humphrey for the Bureau of Resource Sciences, and the summary by M.J. Nunn, run to 770 pages and they are not a comforting read. Live fish pose a much greater disease risk than salmon meat (which we ban), and 'live ornamental finfish are a special case because they are known or potential vectors of diseases of high quarantine importance, are traded widely internationally, and are imported in large numbers into Australia each year'. The threat is only taken seriously today because aquaculture is of growing importance, and because most of the world's worst diseases have yet to reach Australia. But as Humphrey notes, 'In general, few resources are expended and little has been done to control the spread of disease within Australia by ornamental finfish.' Australia desperately needs more fish disease experts to support this work.

Fish imported for aquaria are kept under quarantine for two weeks on the *importer's* premises. 'Some current quarantine premises are unsatisfactory and present a high risk of transfer of exotic pathogens to fish stocks outside the quarantine facility', says Humphrey. 'Examples include: direct access between quarantine areas and retail/wholesale areas, with no provision for, or failure to use appropriate washing facilities; transfer of potentially contaminated water; absence of protective clothing; and mixing of water from tanks containing new arrivals and those of fish to be released from quarantine.' Zelco Begic of AQIS told me that any fish dying during the two-week wait are frozen and handed over to quarantine inspectors for incineration or deep burial. Sometimes airport delays result in many deaths. No attempt is made to post-mortem the dead unless something unusual is suspected, but test results are slow to obtain anyway and Australia is short of disease experts and laboratory facilities. 'For many diseases,' Humphrey states bluntly, 'techniques for detection of the carrier state do not exist.' Fish that survive the fortnight in apparent health are free to be sold, even if, Zelco admitted, half their tank-mates die. Sick fish are not automatically killed; if the ailment can be identified and treated they may then be sold.

This system, introduced in 1981 and relying so much on the integrity of the importers, is farcical. By contrast, when livestock and birds are imported, high-security government quarantine stations are used, trained vets are on hand and quarantine periods may run to years. When some turkeys were imported into Australia recently – the first since 1948 – they came in as eggs from a disease-free flock in Canada, were hatched and reared in the microbiologically secure quarantine station at Torrens Island, then quarantined for another twelve weeks as chicks, with daily inspections for disease.

The number of diseases that have entered Australia with fish is anyone's guess. Some experts believe most of the diseases in our streams were imported like this; others that whitespot, chilodonella and other pathogens are merely native strains of cosmopolitan diseases. Research is under way to find out. No one is arguing about the goldfish ulcer disease (caused by *Aeromonas salmonicida*) that struck a Gippsland goldfish farm in 1975, entering on imported Japanese goldfish. The farm treated the disease but sent infected fish all over Australia, and the bacteria escaped into the wild, probably on live goldfish used by fishermen as bait. Silver perch can catch this disease; probably other native fish can too. Because Atlantic salmon are susceptible the Tasmanian government now controls goldfish imports into that state. Another example is an exotic tapeworm (*Bothriocephalus acheilognathi*) introduced with carp, which now kills western carp gudgeon (*Hypseleotris klunzingen*). But the most frightening suggestion of all is that the chytrid fungus infecting Australia's frogs (see chapter 16) came in on aquarium fish. Some diseases infect both fish and frogs (or tadpoles), and the theory that the pet industry has brought about the extinction of five of our frogs has been written up in the international journal *Conservation Biology*, although at this stage it remains pure speculation.

Instead of importing aquarium fish, Australia could breed her own. To a large extent this is happening. Local farms supply 5 to 7 million fish each year, almost half the market. Imports have dropped 5 million from their 12-million peak in 1975–6. But the industry relies on backyard operators and is constrained by the small size of the market. The National Task Force on Imported Fish and Fish Products has suggested short-term funding to hasten development and improve disease management. We should certainly breed all our goldfish, guppies and gouramis – three fish Humphrey identified as major disease carriers. Yet local fish farms could ruin our waterways if their stock escapes. Fish should never be bred in outdoor ponds that can overflow after rain. Carp that escaped into Lake Hawthorn near Mildura in 1964 ruined the ecology of our largest river, the Murray.

AQIS is reviewing its policies on fish and we must hope it decides to tighten the rules. At the very least we could follow New Zealand's lead by increasing the quarantine period to six weeks. Quarantine should also extend to marine fish, which are brought in freely at present after a quick airport inspection. Humphrey noted 'major inconsistencies' between quarantine for freshwater and marine fish. We need government disease experts, diagnostic laboratories and quarantine aquaria. How bizarre it is that we have superb quarantine stations for imported cats, dogs and birds, and a system for fish that relies on luck and trust.

10

When Beauty is the Beast

Garden Plants Run Amok

'We live in a rainbow of chaos.'
Paul Cezanne

Some of our weeds are strikingly beautiful. Water hyacinth, with its delicate, pastel-blue flowers framed by lush foliage, is an aesthetic delight; morning glory and salvation jane both bloom into glorious celebrations of the colour purple. When such weeds invade our wastelands and wilderness they blanket the land in a quilt of colour. Beauty was their passport to Australia. From distant lands they were brought here to brighten our gardens, parks and ponds, to lend colour and shade to a new land.

Garden plants, in all their glory, now dominate our 'worst weeds' lists, making up a staggering 30 per cent of all our noxious weeds and accounting for seven of our eighteen worst environmental weeds. Of the weeds emerging in recent years, two-thirds are garden escapes. Victoria alone has more than six hundred of them, and in many national parks they pose the main management problem. Garden plants behave so badly partly because there are so many of them. In Queensland alone some four thousand foreign species are grown, more than all of Australia's food, pasture and timber plants combined. Millions of them grow in gardens close to bushland. They represent one of Australia's most pressing problems but one of the least acknowledged.

Those seven garden plants on our worst weeds list behave like invaders from Mars. Rubber vine (*Cryptostegia grandiflora*) loops up trees on long, latex-oozing stems, smothering vast areas of riverine rainforest in the monsoon tropics. Brought in from southern Madagascar in the 1870s, it is perhaps Australia's worst vegetable scourge, having claimed 350 000 square kilometres of Queensland. Blue thunbergia (*Thunbergia grandiflora*) from India is invading Wet Tropics rainforests around Cairns, overpowering trees up to forty metres tall, sprouting from bloated, 70-kilogram tubers. In Central Australia, athel pine (*Tamarix aphylla*) has claimed hundreds of kilometres of outback riverbanks, deposing the native river red gum and changing the hydrology forever. According to *Plant Invasions*, 'It is capable of displacing native flora, destroying resources for fauna, lowering the watertable, salinising the soil and changing river flow and sedimentation regimes.' In the Northern Territory, Jerusalem thorn (*Parkinsonia aculeata*) forms dense colonies of spiny and impenetrable bush several kilometres wide; and in northern and eastern Australia, water hyacinth (*Eichhornia crassipes*) smothers lakes, rivers and dams, wiping out aquatic life and impeding boats. In southern Australia, bridal creeper (*Asparagus asparagoides*) and boneseed (*Chrysanthemoides monilifera*) now dominate the understorey in many eucalypt woodlands, replacing a wealth of native shrubs, grasses, orchids and lilies.

Apart from these seven there is lantana, the prickly claimant to 4 million hectares of eastern Australia, the camphor laurel and privet that have conquered northern New South Wales, the blue periwinkle that lines hundreds of kilometres of waterways in Victoria, the prickly pears, and all the other colourful garden escapes in our forests, on our beaches and in wetlands, proliferating, flowering, seeding, transforming the land forever.

Australia's backyard gardeners unfortunately have no sense of how bad it all is. Weeds? They are those little plants that sprout among the garden shrubs, they think. The truth seems inconceivable – that their English ivy, arum lily and Spanish heath are themselves the weeds. As an environmental consultant working around Brisbane, I have come to loathe these colourful invaders. Too often I have stumbled upon gullies and glades choked by Easter cassia (*Senna pendula*), Singapore daisy, Brazilian nightshade (*Solanum seaforthianum*) and all the other subtropical scourges I know so well. Many are the reports I have written for local councils warning about umbrella tree, asparagus fern, creeping lantana and the rest. I have served as an expert witness in the Planning and Environment Court, warning that a new housing estate would send waves of new weeds into nearby forests. More than anything else, it is painful images of so much disfigured bushland that have inspired me to write this book. Just last week I visited a beautiful woodland where young asparagus ferns – the advance troops – are staking out their positions among the kangaroo grass clumps on the hills. I know only too well the unfolding scenario – the proliferation of asparagus, the arrival of umbrella trees, then jacarandas and mother-of-millions. I pull out an asparagus fern in disgust and walk on, trying to forget.

Our cities are fertile breeding grounds for such weeds. A plethora of plants are grown in the bushier outer suburbs where gardening is so popular. When unruly ones are pruned back, the clippings that won't fit into the bin are often heaved into nearby bushland, to set down roots or shed vast quantities of seeds. It is ironic that the plants most likely to go feral, those that misbehave in gardens, are the very ones most often dumped in this way. If the forests fringing our cities were healthy this might not matter so much, but they are already degraded by pollution, dumping, run-off, fires, erosion, clearing and logging. Most native plants are adapted to infertile soils and dislike all the soapy seepage and dog do. Nutrient-loving invaders quickly take their place, along gullies especially, where rushing stormwaters strip back foliage and pollute the earth. Birds and bats make matters worse by devouring brightly coloured berries and spreading the seeds. They are especially troublesome during droughts, when the only fruits around are likely to be the privet and asparagus fern growing in overwatered yards. Shrubs with bright berries make popular garden features, and not surprisingly, many of our worst weeds - lantana, bitou bush, bridal creeper, camphor laurel, asparagus fern, privet, umbrella tree, sweet pittosporum and so on and so on – are fruit-bearers.

In most of the national parks surrounding our cities, ornamental weeds are a major management problem, usually *the* major management problem. This holds true for Ferntree Gully in Melbourne, the Blue Mountains outside Sydney, Mount Glorious near Brisbane and Belair in Adelaide. In Ferntree Gully, where 173 different weeds are invading the park, 32 are 'significantly threatening or endangering the viability of the native vegetation', and 30 of these are garden or park plants. The villains include trees such as sycamore and English holly; the shrubs Spanish heath and English broom; creepers like English ivy and wandering jew; and groundcovers like three-cornered garlic. Ivy and wandering jew weave thick mats that displace wombats and lyrebirds, threatening their very survival. In the Blue Mountains, naturalist Alex Hamilton first witnessed the problem back in the 1880s. 'All sorts of plants, commonly looked on as garden flowers,' he wrote, 'escape into the bush here, become weeds, and beat the natives out of the field.' Major weeds there today include Scotch broom, holly, buddleia and small-leaved privet.

Around Sydney, where a unique heathland flora, a national treasure, evolved on infertile sandstone, the soil enrichment has been measured. Annemarie Clements, in her study 'Suburban development and resultant changes in the vegetation of the bushland of the northern Sydney region', found that the soil phosphorus level in bushland near housing, now weed infested, was fifty parts per million higher than it should be. And the ground was wetter. Extra water was running off roofs, roads, courtyards and paths, and overflowing from swimming pools and septic systems. Nearly every weed in Robin Buchanan's book *Common Weeds of Sydney Bushland* is a garden escape.

But Melbourne is our weediest city, a recent study finding that 90 per cent of its bushland is badly invaded, with 50 per cent totally overgrown, mostly by garden escapes. Local councils are losing the weed war, control proving too labour-intensive and costly. At Sandringham Foreshore Reserve near Melbourne, a venue for naturalists' rambles last century, botanist Jamie Kirkpatrick compared changes over sixty years, from 1911 to 1971, and found that many natives had disappeared, with garden plants taking their place. In his study 'Plant invasion and extinction in a suburban coastal reserve', he noted, 'Some of the garden escapes, such as arum lilies (*Zantedeschia aethiopica*) and garden geraniums (*Pelargonium 'domesticum'*), strike a rather bizarre note among the sombre native vegetation.' Indeed.

Such problems, although worse around cities, fester everywhere. Country towns and even remote homesteads can serve as beachheads for ecological invasions. Naturalist Samuel Dixon noticed this back in 1892. 'Many garden plants are spreading rapidly, and becoming weeds,' he wrote, 'so that it is now quite common to find some of them in secluded localities to which they have spread from a selector's garden'. One example is athel pine, not considered a weed until the remarkable wet year of 1974 when floods carried its seeds from homesteads around Alice Springs hundreds of kilometres along the Todd, Ross and Palmer rivers. After one bout of rain it emerged from obscurity as one of Australia's most baneful weeds.

Seashores suffer more than most habitats because we love to live by the sea, and because dunes, pummelled by wind, waves and wandering feet, are easily invaded. When George Batianoff and Andrew Franks of the Queensland Herbarium surveyed Queensland beaches (a pleasant-sounding task), they found 105 species of invading ornamentals (not a pleasant outcome). Weeds, including non-ornamentals, now account for about half of all plant species sprouting along the Queensland coast. Serious transgressors include mother-in-laws tongue, agaves, pink periwinkle and gloriosa lily. The authors also saw plenty of signs of gardeners dumping plants. 'In some instances these illegal garden dumping areas resembled compost pits', they noted. 'At these regular dumping sites soil fertility and organic matter increases, further favouring the establishment of invasive ornamental species.' Their subsequent report, 'Invasion of sandy beachfronts by ornamental plant species in Queensland', makes fascinating reading.

Small rainforest remnants also fare badly. Wingham Scrub, a precious nine-hectare remnant of floodplain rainforest near Taree, was subsiding under garden vines until the National Trust raced to the rescue in 1980. Cats-claw (*Macfadyena unguis-cati*), with its brilliant yellow flowers, was trailing nets of stems over the ground, and rising high into the canopy, draping rainforest trees under so many stems their crowns tore apart. One giant stinging tree was holding up 560 vines, an intolerable burden. Madeira vine (*Anredera cordifolia*) groped through the mid-storey at a metre a week, sprouting hundreds of aerial tubers, and a third creeper, wandering jew (*Tradescantia albiflora*), carpeted the forest floor, denying seedlings a chance to sprout. The National Trust sprayed up to 160 litres of herbicide a year to break the exotic hold.

In other rainforests, Dutchman's pipe (*Aristolochia elegans*) is killing off one of Australia's largest, rarest and most celebrated butterflies – the Richmond birdwing (*Ornithoptera richmondia*). This butterfly lays its eggs on the alien vines, mistaking them for its proper food plants (which are closely related) and dooming its caterpillars to an early death. The Cairns birdwing and big greasy also fall victim to this vine. At Burleigh Heads National Park on the Gold Coast curtains of choking pipe vine were hacked away for the butterfly's sake. Even in Australia's alpine national parks, garden plants are multiplying. The invaders of most concern include yarrow, cotoneaster, sycamore, black willow (*Salix nigra*) and alstroemeria (*Alstroemeria aurea*). Some of these weeds have spread from ski resort gardens, more than thirty species sprouting around Thredbo.

There are many more of these weeds than we think, partly because our tastes in gardening have changed so much. When the Reverend Tenison-Woods was warning about weeds back in 1881 the garden subjects that concerned him were lantana, gorse (*Ulex europaeus*), billygoat weed (*Ageratum houstonianum*), purple-top (*Verbena bonariensis*), sweetbriar, red cotton bush (*Asclepias curassavica*), mullein, pink periwinkle and evening primrose (*Oenothera*). Few of these plants earn a place in gardens today and to those who know them as weeds, their origins are largely forgotten. The same is true of hemlock, Scotch thistle and the many other species mentioned in chapter 6. The contribution of garden plants to our weed problems is always underestimated. We are locked into an absurd cycle of introducing new garden plants (new weeds-to-be) whenever older ones become weedy and fall out of favour. And the problems are getting worse. More and more Australians are opting to live near bushland on acreage blocks beside rivers, beaches, rainforests and national parks. New garden plants keep entering the country and more and more of our established ones are escaping.

Australians urgently need to adopt a new gardening ethos. We must accept that gardening within a kilometre or so of bushland entails an ecological responsibility. Weedy species should not be grown. New garden plants should be treated less like exciting new products invented to brighten our lives and more like wild organisms harbouring the drive to escape. Local councils should add more garden plants to their noxious weed lists. Nurseries should learn their local weeds and stop stocking them; eco-friendly garden centres could earn industry accreditation. The nursery industry, or at least its peak body, the Nursery Industry Association of Australia (NIAA), is well aware of the problems, for the simple reason that nurseries suffer whenever garden plants are banned. In February

1999 the NIAA, along with the federal Weeds Cooperative Research Centre, put out a draft strategy for tackling garden escapes. 'Garden Plants Under the Spotlight' contains many excellent recommendations, including a proposed ban on the sale of fifty to one hundred of the worst garden plants, with many more to follow. Plants they suggest banning include ivy, arum lily, camphor laurel and several willows.

The problem here is that Australia is so vast and the numbers of problem plants enormous – exceeding 860 species according to the Strategy. David Cooke, a government botanist in South Australia, has expressed a growing concern: 'I believe the gardening public will accept that a few popular plants are too dangerous to grow. But as the list grows longer, it will be taken less seriously. A point will be reached where a gardener, told that yet another familiar plant is an environmental weed, will respond with an exasperated laugh.'

The banning of plants certainly incites reaction. When Melbourne's Eltham Shire proposed in 1994 to ban sixty weedy garden plants from nurseries and home gardens, ivy, holly and cherry-plums included, there were howls of protest. Columnist Mary Lord, in the *Melbourne Weekly*, conjured up spectres of snooping council inspectors directed by a lunatic fringe. 'In Eltham, at least, this would be the end of the vegie patch and the herb garden, just for starters,' she claimed with considerable exaggeration. 'I think this sort of nonsense should be stopped before it takes hold and we are forced to organise mass demonstrations in support of gardeners' and house-holders' rights while we can.' The proposal was dropped. A couple of years later a parliamentary committee on weeds happened to visit a nursery in Eltham and found them promoting a major noxious weed – English broom – as their 'Plant of the Month'. The committee found that 'many serious agricultural and environmental weeds are sold throughout Victoria under horticultural and trade names'. Even rubber vine can still be bought from nurseries in New South Wales.

We can probably learn from a recent New Zealand success. The Northland Regional Council has produced *The Good Plant Guide*, endorsed by the nursery industry, which lists hundreds of ornamental plants, both native and exotic, that are safe to grow. By emphasising good plants the approach is friendly and positive. And the good news is, hundreds and hundreds of garden plants *are* safe to grow, especially the many hybrids and double-flowered forms with reduced seed-set and vigour. As Alex Hamilton noted with relief back in 1892: 'Of those introduced for ornamental purposes, a large number do not spread to any extent: they are children of civilisation and show no tendency to become feral.'

11

Seeking Greener Pastures

CSIRO Welcomes Weeds

> 'The early promise shown in Departmental trials has been fully substantiated and it seems obvious that this legume will add greatly to the development of pastures and the livestock industry in northern and coastal areas.'
>
> **G.H. Allen, senior agronomist, praising glycine, one of the many pasture disappointments that are now major weeds, 1961**

The northern hairy-nosed wombat is one of the rarest animals on earth. About seventy-five survive, all in Epping Forest National Park in central Queensland. This reserve is in the throes of invasion – it is drowning under a sea of African buffel grass (*Cenchrus ciliaris*). Buffel has overrun more than half the park, engulfing many wombat burrows. Park rangers were alarmed until they found that hairy-nosed wombats will eat buffel grass. It now makes up almost half their diet.

Buffel grass is one of Australia's eighteen worst environmental weeds. Probably introduced with Afghan camels into central

Australia, it is now naturalised over 200 000 hectares of inland Queensland and covers large parts of four other states. It is still spreading, displacing inland rainforests, outback oases and spinifex plains, and looming as a threat to rare plants and animals. It is one of many African pasture grasses that are 'Africanising' the world, invading grasslands wherever cattle are grazed.

In 1995 the Queensland Conservation Council asked me to represent them on a committee with an interest in buffel grass – the 'Task Force to Promote the Responsible Use of Introduced Tropical Pasture Plants'. The CSIRO had formed this group to counter complaints that pasture plants they had introduced were becoming weeds. The task force had met three times before, and minutes of earlier meetings mentioned a new way to predict a plant's weediness, developed by the Australian Weeds Committee. This 'risk assessment package' was supposed to predict whether a new pasture plant up for release would become a pest. The task force was upset because, when tested on some existing pasture plants, the package had rejected ten of them. According to the minutes 'the package was seriously at fault in rejecting some economically significant pasture plants of relatively low weed potential'. One of the rejected plants was buffel grass.

I could understand their ire. Buffel is a mainstay of the northern grazing industry and one of the most important grasses ever brought to Australia. The Queensland government sells *The Buffel Book* to help you grow it. But however useful it may be, buffel can hardly be passed off as having 'relatively low weed potential'. By the federal government's own reckoning it is a worse environmental weed than lantana or prickly pear. *Plant Invasions* is blunt: 'It is an aggressive coloniser of moist habitats such as run-ons, river levees and alluvial pans where it forms dense monocultures displacing native grasses and sedges and altering the fire regime.' Even the Queensland government's book *The Weeds of Queensland* describes it as a crop weed. It is also a major weed in Arizona and Mexico.

Before attending my first meeting I phoned the task force

convener, Dr Bob Clements, head of the CSIRO Division of Crops and Pastures, and suggested his task force had erred by denying that buffel grass was a weed. I did not want to see the assessment package rejected for actually doing its job. Bob flatly contradicted me. Buffel grass, he said, was a valuable pasture grass. I readily agreed with this, but said that buffel, whatever its merits, *also* behaved as a weed. Bob would not accept this idea but said I was welcome to express it at the meeting, which I did.

That meeting was lively. None of the CSIRO scientists seemed to appreciate my concerns. The Landcare representative even threw me a challenge – to name *one* pasture plant that had ever become a weed. In the years since, I have realised that almost every pasture plant ever used has become a weed. I wish I had thrown the challenge back at him: to name one pasture plant that had *not* become a weed. But I never got the chance. Bob left the division and the task force never met again.

Pasture plants create a quandary for Australia because the conflict of interest they raise is so stark. Unlike most plants imported into Australia, they are expected to become naturalised, to form self-sustaining populations in the field and under grazing pressure. To do this well they must behave like weeds – be quick growing, aggressive, prolific seeders, tolerant of drought and grazing, and not too tasty to stock. Pasture scientists know full well that ecologically, pasture plants resemble weeds. One of them, J.J. Mott, even wrote a chapter glorifying the fact under the outrageous title 'Planned Invasions of Australian Tropical Savannas'. Only the narrowest of lines separates a plant naturalised on a farm from a plant that crosses the fence, and virtually all pasture plants do cross that line, becoming weeds of roadsides, crops, gardens, wetlands and forests.

The problems are perhaps worse in Australia, the only continent without native hoofed animals, than anywhere else, for our plants are not designed to withstand intense grazing and trampling. Pasture scientists realised this long ago and began importing hordes of grasses from Africa (where grazing pressure from antelope, elephants and other large mammals is intense) and legumes (to improve soil nitrogen) from Latin America. Between 1947 and 1985 more than 460 exotic pasture grasses and legumes were trialled in northern Australia alone. Only twenty-one of these proved useful but sixty ended up becoming weeds, including twenty of those twenty-one useful plants. Most of these figures come from a landmark 1994 study by the CSIRO's Mark Lonsdale, who points out that thirteen of those sixty weeds are now major crop pests. Sugarcane growers do constant battle with para grass, green panic, centro, setaria, siratro, Rhodes grass and the latest invader, hymenachne.

The *Plant Invasions* report was the first to raise alarms. Its list of our eighteen worst weeds includes six pasture grasses and two fodder and shade trees. Many scientists were shocked by this list, and the pasture industry was thrown on the defensive, the CSIRO forming the committee I sat on. Although it never met again, the following year Dr Bryan Hacker, project leader of plant introductions, asked me to address a CSIRO workshop on the future of pasture introductions. This was a generous gesture coming from one who has done so much to bring 'weeds' into the country. Bryan even paid me to attend the two-day event. My paper, 'Tropical Pasture Plants as Weeds', was blunt. I pointed out that each of the five main groups of pasture plants – grasses, aquatic grasses, woody legumes, twining legumes and herbaceous legumes – has spawned major problems for Australia.

African grasses are probably the worst offenders because they are altering the very way our forests burn. They are valued by farmers because they grow so tall (gamba grass [*Andropogon gayanus*] up to four metres), but if left uneaten they dry into thick stands of straw, and what should have been fodder for a cow ends up as fuel for a fire – a very hot fire. All over Australia infernos are raging. In the monsoon tropics, the boundaries between rainforest and woodland were once defined by Aboriginal burns; now the rainforest is on the retreat. Around Darwin, mission grass (*Pennisetum polystachion*), a useless introduction that generates flames up to five metres high, is blamed for a 60 per cent loss of monsoon rainforest. Rated one of the world's worst weeds, it looms, along with gamba grass, as a major threat to Kakadu. In Queensland buffel is scorching back dry rainforest in many national parks, including Carnarvon, Palm Grove and Expedition Range. In coastal Queensland its place is taken by molasses grass (*Melinus minutiflora*) and guinea grass (*Panicum maximum*), two hot-burning coastal grasses.

The combustibility of these grasses is their secret weapon. They torch back nearby plants then resprout quickly from roots and shed vast loads of seed. In this way they advance across a wide front, taking out native grasses, shrubs and seedling trees. In King's Park, in Perth, veldt grass (*Ehrharta calycina*) fires are even killing off tuart (*Eucalyptus gomphocephala*), the largest local tree. In central Australia, countless herbfields growing on river flats – the oases of the outback – have vanished under the tide of buffel. At Simpson's Gap National Park near Alice Springs, a new ranger from the north decided the park needed a burn, and in one fell swoop the sedges along the river flat were ousted. Buffel grass seeds prolifically but the seed is not eaten by birds, which helps explain why some of our finches and parrots are declining. Our native grasses are fire adapted too, but they cannot match the ferocity of these invaders.

Grasses that grow or trail in water are another curse. Para grass (*Brachiaria mutica*) now blankets 100 000 hectares of wetlands in Queensland. In Kakadu it is displacing water chestnut, a staple food of magpie geese, and if left unchecked Kakadu could lose its star attraction, the large flocks of waterbirds. Para grass also thrives in Queensland along streams where rainforests grew long ago, and by trailing into the water it takes over the habitat of native fish while creating ideal habitat for American mosquito fish and swordtails. It dominates my local creek, as it does most streams around Brisbane.

Cattlemen like para grass because it provides feed in swampy ground and in shallow pools. But they also want grasses that will grow in deeper water so they can pond their swamps to utilise 'wasted' land. In tropical Queensland, the Department of Primary Industries promoted hymenachne (*Hymenachne amplexicaulis*) and aleman grass (*Echinochloa polystachya*) as ponded pasture grasses, and hymenachne has since won a place on Australia's list of twenty worst weeds (see appendix II). It is fast invading cane farms and wetlands.

Legumes are nearly as knavish as grasses. Among the woody legumes, prickly acacia (*Acacia nilotica*) has exploded into a disaster of international proportions. It is turning the Mitchell grass downs of inland Queensland into an African vista of prickly thickets; half a million hectares are infested so far. Mesquite (*Prosopis* species) could colonise 4 million hectares in western Queensland, although the state government is determined to eradicate it, and leucaena (*Leucaena leucocephala*) is marching along riverbanks in Queensland and the Northern Territory. The CSIRO, to their credit, did not introduce these plants, except for some leucaena varieties; state departments of agriculture brought in many pasture 'weeds'.

Twining legumes from Latin America tell another sorry story. Glycine (*Neonotonia wightii*), for example, chokes riverine rainforests, and siratro (*Macroptilium atropurpureum*) invades beaches, mangroves, forests, roadsides, even vacant allotments in the very heart of Brisbane. These vines, now major crop weeds as well, are the legacy of an experimental approach by the CSIRO in the 1960s. Although vines had never shown their worth as pasture plants elsewhere, the CSIRO thought it would give them a try. In trials conducted during good years they seemed ideal, but when sown by farmers and subjected to drought and typical overstocking rates, they vanished from pastures but escaped into the wilds. Now grown only on a minor scale, they are acknowledged as a big disappointment. Australia would have been much better off without glycine, siratro, calopo (*Calopogonium mucunoides*), kudzu (*Pueraria lobata*), and centro (*Centrosema pubescens*), most of which are spreading fast. Calopo is described in a Northern Territory weed book as an 'extremely vigorous climber that smothers out supporting vegetation' and that is 'not often grazed'.

Herbaceous legumes may be the least troublesome group. Townsville stylo (*Stylosanthes humilis*), an accidental introduction of great value, invaded vast areas of Queensland before it was pushed back by a disease, anthracnose. One government employee used to broadcast its seed from trains during long journeys. Like phasey bean (*Macroptilium lathyroides*), sensitive mimosa (*Mimosa pudica*) and other herbaceous legumes, it still behaves as a weed, often sprouting along roads, but only invades heavily disturbed ground.

All legumes are diabolical invaders because they capture atmospheric nitrogen, thus fertilising the soil and preparing it for other weeds. In Australia it is often the soil's infertility that keeps weeds out. After legumes are sown, nitrogen-loving grasses invade, choking them out and ruining the pasture. Weed lists in Australia, and around the world, are dominated by the very plants – legumes and grasses – that scientists value for pastures.

The goal of pasture science is to create stable and productive pastures of grasses and legumes. But I have come to doubt whether these ever exist outside the confines of agricultural research stations. At a recent CSIRO meeting where I was again invited to speak about weeds, I put the question: Do stable grass–legume pastures really exist? Are you chasing an achievable goal? There was only one reply, from Bryan Hacker, who said he could show me a stable pasture system hundreds of years old – in England! Even in articles written by pasture boffins the doubts sometimes show through. The boosts in productivity achieved by legumes are only transient, it seems, the pastures soon deteriorating ('running down'). Myers and Robbins, writing in 1991, explain that 'the benefit that N-fixing legumes pass on to associated grasses is usually less than commonly believed. Seldom can they completely prevent productivity decline.' To manage for run-down Myers and Robbins recommended lower stocking rates. In another article, Cavaye and colleagues warn that 'there is a run up in pasture growth after development and this can give an unrealistic expectation of production'. Pasture decline is normal and predictable, they say. These and other articles imply that farmers are often unrealistic, demanding too much from their land. The real problem is not run-down but overstocking. New pasture plants soon become scapegoats when pastures fail, leading to renewed calls for newer and better plants.

But farmers show no signs of tempering their expectations. Peter Emmery, a grazier, spoke for the producers at the CSIRO workshop. Cattlemen, he said, 'seek both new legumes and grasses, with generally low costs of establishment, giving both increased production per head and per hectare. Producers gave a long list of desirable pasture characteristics they felt were not adequately provided for in existing pasture cultivars.' Graziers, he commented, are 'very worried about conservationists' activities to stop or restrict introductions and releases'.

I have come to wonder whether pasture plants have done Australia much good at all. Do they just appear to be useful because they extract (then exhaust) the nutrients of the soil more quickly than better-adapted native plants? Where are the Australian examples of stable grass–legume pastures? Do the gains to the cattle industry outweigh the costs to crop farmers and conservation? Should the beef industry pay for the problems it is causing? How much is our meat costing the environment? I can't answer these questions, but I believe they are important. Mark Lonsdale has also raised them. 'It is difficult then to assess whether there has been a net gain to the nation from these introductions,' he wrote in his 1994 paper, 'and whether the benefits of the 21 useful plants have outweighed the costs of the 60 weeds.' In the Northern Territory the cattle industry creates only $1.48 profit per hectare, compared to the $120 a year it costs to control woody weeds. Were the cattle industry to close down tomorrow we would still be paying for these weeds in a thousand years.

At the workshop I recommended that no more grasses, twining

or woody legumes be introduced, and argued for more research into native alternatives, some of which show promise. Albizia (*Albizia lebbeck*), a native tree in northern Australia, was shown by the CSIRO's Brian Lowry to provide excellent fodder, but he is frustrated that no one, until very recently, has shown interest. Pasture scientists Silcock and Johnston note, 'Special plants from overseas have received considerable attention without great success in difficult environments but native plants have received only cursory attention.'

The workshop proved an interesting though trying experience. The CSIRO's funding is down because its work has not met expectations. Some speakers argued we have enough pasture plants already; we just need to use them better. One scientist argued the opposite – that hundreds of wonderful plants, growing on remote mountaintops and plateaus in distant and dangerous lands, await discovery. He showed slides of expeditions on camel-back. The spirit of acclimatisation lives on! I almost fell off my chair when Bryan Hacker mentioned a promising new mimosa (*Mimosa* species) that the CSIRO has dabbled with. Australia has four exotic mimosas, three of them declared noxious weeds, and one of them the worst weed in the Northern Territory (*M. pigra*). Bryan even flashed up a slide of this plant flowering in a pot at the CSIRO's farm near Brisbane, against a backdrop that looked like a forested stream. An interstate scientist queried the complete lack of security, commenting that this would not be permitted in his home state.

Two speakers were hostile to my point of view, though most were friendly, and two CSIRO scientists, John McIvor and Sue McIntyre, also delivered a paper warning about weeds. Although some participants tried sincerely to incorporate these inconvenient views, there seemed to be little appreciation of the scale of the problem. Opportunities for reconciliation were so limited that by day two I was keen to escape and forget the whole exercise. A year and a half later the papers from the workshop appeared in the journal *Tropical Grasslands*, and a pleasant surprise awaited me. Bryan and two other scientists had written a summary that proved more conciliatory than anything said at the workshop. After acknowledging my concerns, they commented: 'As introducers of exotic forages, the ATFGRC[1] clearly has a responsibility to the community as a whole, and not just to the graziers. Experience with those species so far introduced and their tendency to develop weed potential, indicates that twining species, and trees or shrubs which are capable of growing above browse height, should not be introduced. Caution should also be exercised in introducing grasses, particularly those species which develop a large bulk of standing biomass.'

But despite these judicious comments, they closed the paper by setting out the CSIRO's future priorities, which included an extraordinary shopping list: new legumes for clay soils, salt-tolerant grasses, a leucaena-like shrub and so on. Some of these plants, the salt-tolerant grasses especially, will certainly become treacherous weeds. Signs of compromise *were* evident, just. They proposed that native pasture plants be studied as well, and suggested that any leucaena-like shrub be sterile, or nearly so, 'to prevent it developing into an environmental weed'. Bravo, I suppose.

The last pages of that issue of *Tropical Grasslands* announced the release of four new pasture cultivars. Conservation sentiments peep through here too. The descriptions of a new digitaria (*Digitaria milanjiana*) and centro (though not the others) mention that they did not become weedy during pasture trials. Would their release have been stopped if they had? They noted that digitaria behaves as a crop weed in Africa, and it *has* escaped into the wild in a minor way in Queensland and New South Wales. I spoke to one leading weed expert who expects it to become another eco-disaster.

The CSIRO must also stop trialling plants in unsafe places – alongside streams and roads especially, where seeds are easily borne away. In the past, old trial plots have sometimes been abandoned, allowing seeds to spread as they might. One pesky little legume, *Indigofera circinella*, has been traced back to pasture trials conducted in the 1970s. It was not released as a pasture plant but is now spreading as a lawn weed in Brisbane. I have argued that some trials, for palatability for instance, be conducted overseas, but though this may seem like common sense the proposal was not well received.

This chapter has focused on tropical pasture plants, but I could have taken the same script and instead featured temperate pasture plants – lupins, medics, clovers, paspalum, reed sweetgrass and tagasaste – or plants sown for land reclamation such as kochia, bitou bush, lovegrass and marram grass. Bitou bush, for example, planted to revegetate New South Wales beaches after sand-mining long ago, is now that state's worst weed; whole banksia woodlands have subsided under its succulent stems.

Ironically, agronomists did not introduce our most important pasture plants. Buffel grass apparently came in with the Afghan camel trains in central Australia; Townsville stylo arrived with hay brought to Point Stuart in about 1916; leucaena was imported by the Darwin Botanic Gardens last century as a coffee shade-bush, and subterranean clover appeared in southern pastures long ago. The CSIRO has contributed to the process by introducing new varieties of these and related plants.

I have spent enough time with CSIRO pasture scientists to know that they are decent, caring people. Some of them regard themselves as conservationists. But their research has been so compartmentalised, their brief so narrow, that they have failed to consider all the consequences of their work. I hope that they learn to broaden their outlook, to accept the point made by Bryan Hacker and his colleagues after the workshop – that pasture scientists owe a responsibility to the community as a whole and not just to graziers. If not, history may judge them as harshly as the nineteenth-century acclimatisation movement is judged today.

1. The Australian Tropical Forages Genetic Resource Centre of the CSIRO.

INVASION BY STEALTH

As world trade and tourism grow, more pests travel the globe, hitching rides on ships' hulls, in ballast water, in cargo and in transported soil. Some of the most alarming invaders are the exotic diseases infecting our animals and forests.

12

'Every Creeping Thing'

Species on the Move

> 'Every beast, every creeping thing, and every fowl, and whatsoever creepeth upon the earth, after their kinds, went forth out of the ark.'
>
> **Genesis IX**

We are always on the move these days, it seems, and everywhere we go, plants and animals go too. Seeds fasten to our socks and car wheels, spiders and snails snuggle in cargo, beetles bore into foodstuffs, barnacles attach to hulls and parasites travel in our tissues. We will never understand the feral future without appreciating how easily, and in how many ways, exotic species spread.

When the First Fleet dropped anchor in Sydney Harbour, a veritable Noah's ark of live cargo clung to the hulls of the seven ships, hidden below the waterline. Barnacles were probably present, and algae, sponges, mussels, crustacea and worms. In that age of wood, shipworms bored into hulls, the holes affording havens for other creatures. Heavily fouled ships carried entire communities of sea life beneath them. Nowadays hulls are cast from steel and toxic anti-fouling paints are applied; they slice through water at high speed, but determined hitchhikers still accompany them. On modern supertankers specialised seaweeds and slime films (made up of diatoms, bacteria and algae) are regular unpaying passengers. Hull foulers need to be very tough to survive the toxic paint, enormous water drag and dramatic shifts in temperature as ships race round the globe, but many persevere, congregating wherever paint has been poorly applied or stripped away. They are a blight on shipping because they increase friction, slowing ships and increasing fuel costs. A mere millimetre of slime can reduce a ship's speed by 15 per cent.

This hull travel has left a pernicious legacy. In the shallow waters of New South Wales the dominant lace coral today is an American import, *Watersipora arcuata*, which is also found in southern, western and northern Australian waters, right up to Torres Strait. Dozens of other hull travellers live in our seas, including sponges, more than twenty kinds of mollusc, crustaceans, seasquirts, worms, anemones and seaweeds. More than a hundred exotic species lurk in Port Phillip Bay alone, most of them hull hikers, including codium (*Codium fragile tomentosoides*), a seaweed that explodes into large algal blooms. Australian authorities are realising that hull invaders may be an underrated and growing menace; in 1998 the Quarantine Service began a major study on the issue. AQIS now forbids foreign ships from scraping down their hulls in our ports unless the work is done on land and the wastes disposed of safely. But hull invaders will continue to pose problems. Today's anti-fouling paints contain tributyl tin (TBT), a very successful deterrent and one of the most toxic substances known. But TBT is blamed for deformities in oysters and deaths of other harbour species, and Australia has banned its use on boats under twenty-five metres. The International Maritime Organisation is under pressure to ban it worldwide. There is no satisfactory substitute.

Sea creatures travel with ships in other ways as well. AQIS warns that Japanese kelp can spread via 'anchors, ship's refuse, nets and fishing lines, water on boats or in the case of the Japanese longliners, within the frozen bait brought to Australia from Japan'. When anchors are weighed, the chains often drag up live detritus, including species that may survive in wet anchor wells. AQIS asks of visiting ships that 'Subject to accessibility, all sources of sediment retention such as anchors, cables, chain locks and suction wells should be cleaned routinely to reduce possibility of spreading contamination'. Fish and other creatures can travel in outlet pipes and in sea-chests, the entry chambers to ballast outlets. Japanese kelp, one of our worst weeds, probably entered Australia in ballast water. Ballast, the water taken onto ships to maintain stability when cargo holds are empty, is now Australia's main source of new sea creatures and the subject of a later chapter.

Animals on ships draw more attention if they travel within the hull. Rats, cockroaches and weevils were familiar if unwelcome guests aboard ships in earlier times. Midshipmen of Nelson's Navy on long voyages were said to have supplemented their meagre rations of spoilt salt beef with fresh rat, and some mischievous old salts swore by the protein boost offered by the weevils that infested ship's biscuit. German cockroaches were a problem for the explorer D'Entrecasteaux, visiting Australia centuries before Mortein was invented; his botanist, La Billardière, complained: 'These insects, not contented with our biscuits, gnawed our linen, paper, etc. Nothing came amiss to them ... They continually disturbed our repose, by passing over any part of our bodies which happened to be uncovered.'

All ships of size carried rats. A Tasmanian newspaper last century reported: 'The number of rats leaving the convict ship now tied up in the Bay has to be seen to be believed.' When a supply ship ran aground on Lord Howe Island in 1921 the escaping rats proliferated, and drove five native bird species to quick extinction. Early this century rats and their fleas brought plague to Australia; it struck Sydney, Brisbane, Melbourne, Adelaide, Perth and many country towns, reaching as far inland as Kalgoorlie. A worried Heber

Longman, director of the Queensland Museum, declared in 1919 that 'Only by persistent propaganda may we keep our White Australia free from the Black Plague'. He was applauding the Queensland Health Department's slogan: 'No food, no rats; no rats, no plague'. By the 1920s ships were fumigated and the plague, though not the rat, problem was solved. During 1926 fourteen vessels from Calcutta were found to be harbouring between them at least 454 rats.

Even birds travel on boats. The Asian house crow, a handsome grey and black creature, often reaches Australia on ships from India and Sri Lanka. C.G. Hylton witnessed one such excursion in 1926:

> On leaving Colombo it was noticed that three Colombo crows had settled on the main yard-arm of the ship, and remained there, with an occasional fly around the vessel, until nearing Fremantle, when one of the birds died. The other two, however, as soon as they were sufficiently near to land, left the vessel and flew ashore.
>
> What they lived on during the voyage from Colombo to Fremantle it is very difficult to say; possibly they managed to steal some of the meat which was thrown out for the cats, of which there were a good many on board. Otherwise, it is hard to realise that they lived for nine or ten days without food or water. The incident was of interest to most of the passengers, and a number of us daily looked at the yard-arm to see if the intruders still remained and enjoyed their trip to Australia.

A Colombo crow was spotted on shore a year and a half later.

Over the years, many Asian house crows have been shot in Western Australia. A vigilant state government has stopped them from establishing. I have watched these birds in India and Sri Lanka scratching about in urban filth and I'm glad we don't have them here. But they are not the only birds that travel well on boats. In 1996 an AQIS inspector was kept busy stalking a clutch of sparrows that reached Port Hedland on a Taiwanese boat. Western Australia, so far, remains sparrow free – another tribute to that state's vigilance.

Most stowaways are less obvious because they are smaller and because they travel concealed within cargo. Spiders and insects travel with impunity inside machinery parts or within containers. The brown widow (*Latrodectus geometricus*), a dangerous spider (closely related to the redback) that turned up in Brisbane a few years ago, was probably a cargo stowaway, as was the crazy ant (*Anopholepis gracilipes*) that invaded Christmas Island, where it is killing millions of native red crabs (*Gecarcoidia natalis*) in a major ecological tragedy. Beetles and moths bore into biscuits, cakes and other foodstuffs; a friend of mine found live grubs in a fruitcake she imported recently from America. Mosquitoes travel in rainwater trapped in tyres or wherever pools of water collect. Most of the smaller exotic animals in Australia – the European wasp, daddy-long-legs spider, slugs, the Asian house gecko and so on – probably entered with cargo. Each year a million freight containers enter Australia, and those that contain no organic produce or wooden palettes are not sprayed.

Hundreds of different weeds have entered on ships – in mud stuck to vehicles and machines, on the shoes of passengers and crew, as impurities in crop seed and stock feed, and in the past, in pot-plant soil, fodder and packing straw. As one Tasmanian observer, the Rev. W. Spicer, noted in 1877, 'Every bushel of corn that has been at any time imported from England was almost sure to contain the seed of the common spurrey provided it had been grown in a sandy soil. Every pet canary bird, that arrives in the island, conveys in its food-tin the germs of the pretty canary grass, which now abounds on our rubbish heaps and road sides.' Serious weeds that have entered as seed contaminants include Siam weed (*Chromolaeana odorata*), which arrived with pasture seed in the 1970s, and golden dodder (*Cuscuta campestris*), which came in with basil seed in 1981, 1988 and again in 1990.

Tenison-Woods recorded in 1881 how fumitory (*Fumaria officinalis*), a delicate weed with tiny pink flowers, reached western Victoria in fodder:

> The late Mr. George Crouch, of Portland, assured me that he, had never seen the plant in Australia until 1853. In that year he was surprised to find it growing abundantly in his own garden, with some other new weeds. He had bought some hay from a merchant vessel discharging at Portland, Victoria. It was English hay, the surplus of a supply for some valuable stock brought out in the vessel. Mr. Crouch assured me that wherever the hay had been stored, the Fumitory began to grow, and it is now widely spread over the colony.

Botanist P.M. Kloot estimates that ninety-two weed species have entered South Australia attached to stock, fifty-seven as seed contaminants, forty-three in dry ballast, eleven on footwear and six in fodder. During droughts, weeds are often redistributed around Australia in emergency feed. During the 1980–81 drought in New South Wales, imported hay bales were found to contain more than a hundred kinds of weed seed, many of them noxious. A typical bale harboured almost 70 000 seeds of twenty-one different types. Mintweed (*Salvia reflexa*), a sweet-smelling noxious pest, probably entered Australia in drought fodder brought from the United States in 1902.

Seeds also travel in the mud adhering to the tyres and undersides of cars. N. Wace sifted seeds from sludge washed off cars at a Shell Canberra Superwash and was able to germinate more than 18 000 seedlings of 259 species – a surprise mix of weeds, native plants, garden plants and crops, with weeds predominating. A vehicle check at Kakadu National Park found that 70 per cent of cars were bringing in weed seeds, including three species – fierce thornapple (*Datura ferox*), wild turnip (*Brassica* species) and Mediterranean barley grass (*Hordeum hystrix*) – not yet found in the park. Other weeds, notably bladder dock (*Acetosa vesicaria*) and Mossman River grass (*Cenchrus echinatus*), invaded Uluru via tourist traffic. On islands in the southern Great Barrier Reef often visited by tourists, more than half the plant species are weeds. On Lady Elliot Island they outnumber native species by fifty-eight to nineteen. Weeds also travel into reserves as seeds attached to slashers used to mow picnic grounds. The disease phytophthora (discussed in chapter 15) has invaded many reserves on road-building machines, and the Amazonian earthworm is spread into forests in mud stuck to logging trucks.

The quarantine significance of air travel was first recognised back in 1928 when an American inspector, Max Kisliuk, boarded a European zeppelin arriving in New Jersey, and upon seizing three floral bouquets, found a spider, moth's eggs, cocoons, thrips, an aphid, bugs and three plant diseases. With the explosion in air travel over recent years, planes have emerged as a growing source of pests. In several European countries, including France, Holland, Belgium and Switzerland, people living beside airports have been struck down by malaria delivered by tropical mosquitoes escaping from inbound aircraft.

The prospect of pests entering the country on aircraft has long concerned the Australian government. Planes that load at tropical airports close to swampy grasslands on hot steamy nights pose the direst threat. When Indonesian Airlines began flying into Darwin in 1973, some passengers were found to be ailing from malaria, and blood-filled anopheline mosquitoes were discovered on board. Since then all aircraft entering Australia have been sprayed. During the 1970s, fauna surveys were conducted on 300 incoming planes; more than 5500 dead insects were retrieved – an average of 18 per plane. Unsurprisingly, flies were the most frequent fliers (2877), followed by bugs (781), mosquitoes (686), wasps, bees and ants (381), moths (288) and beetles (285). A few grasshoppers, cockroaches, dragonflies, earwigs, lacewings and mayflies turned up as well.

In 1970, 882 000 passengers flew into Australia from overseas. By 1990 the figure had risen to 4.7 million, and by 1998 to 7.5 million. Tourism is emerging as the world's major industry, and more and more visitors are arriving from tropical lands where most of the world's pests lurk. About one in ten passengers declare quarantine items, but most illicit imports go undetected.

I know from my own experience how easy it is to carry things in unwittingly. I returned from one trip to Africa to find dried mud caked to my sandals. Examining it closely I found a trove of organic riches: bits of straw, grass seed husks, flakes of snail shell, four seeds and some fungal threads bearing spore heads – a forensic record of my trip lay scattered before my eyes. One of the seeds was nearly as big as a dried pea, and I thought of sprouting it to see what it was, but a tiny insect later drilled an exit hole in one side. The question I ask myself is, was it an Australian or an African insect? On my second African trip I took care not to bring back mud, but embedded in my thigh upon return I found a live African tick. I have returned from Asian trips with even more impressive souvenirs. Once I brought back malaria (quinine-resistant vivax); another time I found in my luggage a seed-pod of mesquite, one of Australia's worst weeds. I care enormously about pests, yet even I, despite my best efforts, find myself bringing them in. The world has grown addicted to travel and trade, and I doubt we will ever staunch the hidden flow of exotic pests. We have embraced a global economy that will inevitably globalise our ecology.

13

Soil Travellers

Worms, Seeds, Spores

'To see a World in a grain of sand . . .'
William Blake, 1863

Charles Darwin was so inspired by earthworms he wrote a book about them praising their role in ploughing the earth. 'It may be doubted', he wrote, 'whether there are many other animals which have played so important a part in the history of the world as have these lowly organized creatures.' But in much of Britain today, worms no longer plough the fields; they have vanished down the stomachs of a voracious predator, a flatworm from New Zealand (*Artioposthia triangulata*). First discovered in a Belfast garden in 1963, this worm is now found all over Northern Ireland and Scotland and in many parts of England. It pursues earthworms within their burrows, embracing them and secreting an enzyme so powerful they dissolve into a pleasantly digestible 'soup'.

Wherever these aliens appear, earthworms disappear. The rapid progress of this insidious little creature has been monitored on British TV and discussed at length in garden magazines. In the

Journal of Hydrology (1995), A.H. Haria warns: 'The impact on the environment may be devastating as soil erosion, waterlogging and pollution become more significant.' Where worms no longer aerate soil, flooding will worsen, and costs of flood control, says Haria, 'may reach astronomical proportions'. Other experts predict falling crop yields and the starvation of worm-eating shrews, moles, badgers and birds. Darwin must be turning in his grave.

These flatworms reached Britain in pot plants. In Scotland they first appeared in Edinburgh's Royal Botanic Gardens in 1965 after plants were imported from New Zealand. By 1992 they had invaded every major city botanic garden in Scotland, and most of her National Trust properties too, the Royal Gardens serving as an effective relay station. Helped along by the nursery trade, this worm will conquer continental Europe soon (it has already reached Iceland). An ecoclimatic study shows it could establish itself in most of north-western Europe, North America, Australia, southern South America and South Africa. In Australia it could claim the whole south-east from Adelaide to New South Wales. In New Zealand, ironically, it is neither common nor destructive, surviving only in remnant beech forests and in gardens in parts of the South Island.

The transport of soil around the world is a risky business – you never know what you are carrying. Seeds, eggs, spores, bacteria, mosses, liverworts, worms, mites, snails and slugs are common passengers in pots. Over the years, Australia has imported hundreds of weeds, fungi and animals in this way – even the flowerpot snake, an innocuous soil-dwelling reptile.

There's little doubt the dreaded cinnamon fungus (*Phytophthora cinnamomi*) that is killing our forests arrived here in soil. This fungus (the subject of chapter 15) latches on to plant roots and kills them for food. Other *Phytophthora* species (potato blight, for instance) destroy fruit and vegetables. Several foreign species lurk in our forests, preying on native plants. Their spores travel around on bulldozer blades and in pot soil. In Western Australia, many nursery plants are *Phytophthora* infested, a university study detecting eight species of the fungus in one nursery alone.

Other nasty soil travellers include the deathcap mushroom (*Amanita phalloides*), found sprouting around exotic oaks where its deadly flesh tempts innocent mushroom-pickers (two Australians have died so far), and the red-and-white fly agaric (*A. muscaria*), the hallucinogenic fungus often featured in European story books. The world's worst weed, nutgrass (*Cyperus rotundus*), depends entirely upon soil travel, since it almost never sets seed; its little tubers journey about in pot plants and loads of gravel. European heath rush (*Juncus squarrosus*) is probably a soil traveller too – in Australia it grows only on mountains in western Tasmania, probably from seeds brought in on bushwalkers' boots.

Australia's soils contain hordes of exotic earthworms, flatworms, slugs and snails. Indeed, alien earthworms now dominate our gardens and farms. When 1450 students dug up worms for the CSIRO Earthworms Downunder survey in 1992, they catalogued twenty-seven feral species and just five natives; six of the seven most common species were exotic. Australia has some 350 native earthworms, but they keep to the native forests, leaving disturbed ground, farms and gardens free for the forty-five exotic species.

Most earthworms help farmers by aerating soil, but the Amazonian earthworm (*Pontoscolex corethrurus*), is proving a threat to conservation. In Queensland rainforests it proliferates wherever forests are penetrated by logging operations or roads. Its eggs, laid in cocoons, are carried to remote places on tyres, bulldozers and probably by grubbing feral pigs. Although small – only about eight centimetres long – this worm multiplies quickly and appears to displace native worms. Unlike some native species it does not help wood decay or burrow deep enough to aerate soil.

Most snails, slugs, slaters and flatworms in our gardens are also imports. Remarkably, Australia is home to ten foreign species of slater, or woodlouse – tiny crustacea that crawl through mulch, devouring decaying matter. The delicate slater (*Porcellionides pruinosus*) has colonised all the mainland states, along with Norfolk, Christmas and Lord Howe islands. Australia now has one flatworm species from Brazil, one from Asia, one from New Guinea, two New Zealanders and four from places unknown. Our most common exotic, *Bipalium kewense*, found in gardens under rocks and boards, was first discovered in England (in Kew Gardens, as the name implies) but originated in the mountains of Vietnam and Cambodia.

The flowerpot snake (*Ramphotyphlops braminus*) is perhaps our most impressive soil immigrant. From India it has journeyed to many foreign lands, colonising Mexico, South Africa, Hawaii and New Guinea. During the 1960s it appeared in backyards in Darwin, then turned up in Katherine and Torres Strait. It is likely to spread south as far as Brisbane or Sydney. A completely harmless, worm-like reptile, it ploughs through soil in search of ants.

How many animals have entered Australia in soil? No one knows, but the figure must be astronomical. A single square metre of soil can conceal hundreds of thousands of insects, mites and worms and literally billions of microbes. Ever since the First Fleet arrived bearing fruit trees in pots, foreign soil has been flowing in. Hundreds, perhaps thousands of species of exotic insects, mites, fungi, microbes and weeds must have entered the country in this way.

Today the government does not allow foreign soil into the country. 'Soil is an absolute no-no', Carson Creagh of AQIS told me, yet it still flows into the country in bountiful amounts. Quarantine inspectors in Adelaide checking oil rig equipment brought from Texas in 1996 uncovered a staggering fourteen tonnes of the stuff. One bulldozer was carrying one and a half tonnes, and several more tonnes were located in the bases of the drilling platforms. A grader had to be dismantled and sliced open to remove the muck. Gangs of cleaners used a 2000 psi steam cleaner to blast away hardened mud. A 12-metre-long platform was so encrusted it had to be kiln-heated to extinguish all soil life. During this epic exercise South Australia's Environment Protection Authority protested about Texas soil washing into Adelaide harbour and demanded the operation be completed at a safer site, which was done. Another bulldozer, from New Guinea, harboured soil and 'enough live ferns and other plant material to stock a small greenhouse', according to AQIS. This machine was also painstakingly cleaned, but the recent Nairn quarantine review[1] noted that dirty mining and harvesting equipment was sometimes unloaded without even a cursory inspection. Australia also exports filthy vehicles. One shipment of locomotives sent to New Zealand from Queensland in 1997 was found to contain three tonnes of soil, complete with weed seeds, wasps and a few redback spiders for good measure. The growing trade in second-hand machinery is very worrying.

AQIS must also cope with importers who resort to trickery. In 1997 a container-load of mud-encrusted tyres was brought into Australia dressed up with tyre black to look new. The ruse was exposed when an inspector prodded a very shiny tyre and a lump of painted dirt fell off. Smaller amounts of soil come in as weights inside wooden artefacts (each year AQIS burns large numbers of Balinese carvings) and in smuggled pot plants. Plants have been found concealed in bizarre hiding places, from speaker boxes to tampon packets. Many smuggled pot plants go undetected and countless weeds and animals must get into Australia in this way.

Confiscated soil is treated like radioactive waste and buried in drums deep underground. All over Australia, alien seeds, spores and other life forms lie imprisoned below ground in AQIS depots. With increasing travel and trade, the numbers of foreign species in our dumps can only grow. Some seeds stay viable for tens and even hundreds of years, and spores can survive for hundreds and sometimes millions of years. All this might be a greater focus of concern if we did not have more pressing issues to address, including ballast, the topic of the next chapter.

1. Nairn, M.E., Allen, P.G., Inlis, A.R., & Tanner, C. (1996) *Australian Quarantine: a shared responsibility*. (Hereafter referred to simply as 'the Nairn review'.)

14

Ballast Blues

Something in the Water

> 'Ah! Little do you think upon the dangers of the seas.'
>
> **Martin Parker**

In 1991 I was walking along the shores of D'Entrecasteaux Channel in Tasmania when my companion, Chris Reed, pointed out some Pacific oysters (*Crassostrea gigas*) growing on the rocks. I was keen to sample one so I cracked open a shell and swallowed the contents. Minutes later my throat was burning and I was bent over, vomiting. I looked over to the oyster farm across the bay, where the same oysters – an exotic species from Asia – were growing, and wondered what had happened.

A few days later I knew. The TV news announced that the oyster farms in the channel had been closed by an outbreak of toxic dinoflagellates; fortunately no one had been poisoned, the announcer assured. The dinoflagellates, microscopic algae absorbed by the oysters, had probably come into the channel with ballast water from Japan. This was my first encounter with a ballast invader, and one I will never forget – I've not been able to enjoy oysters since.

When an unloaded cargo ship prepares for sea, seawater is pumped into ballast tanks to maintain a stable weight and balance. Animals and plants are often sucked into the tanks and if they weather the journey at sea they may be disgorged alive at the next port. The ballast enters through intake hatch covers – steel plates with holes 1 to 5 centimetres wide – so creatures have to be small to get in. To reach Australia alive they must survive the journey (seven to ten days for a Japanese bulk carrier, eighteen days for a woodchip ship) in airtight tanks in the dark with changing temperatures. Most die, but the hardiest survive.

Ballast invaders have the Australian government worried. Some nightmarish pests have managed to sneak into Australia by this means. During the 1980s we were visited by Japanese kelp (*Undaria pinnatifida*), a smothering seaweed, by the Northern Pacific seastar (*Asterias amurensis*), a horrific predatory starfish, and by the giant sea worm (*Sabella spallazanii*), a threat to scallop beds. More ballast-borne pests are sure to arrive in the coming years.

Ballast is a serious concern for Australia because we export so many bulk goods, such as wheat, woodchips, coal and minerals. 'By virtue of our international role as an exporter of commodities, Australia is a net importer of ballast water and therefore at greater risk from biological introductions than other trading nations,' admits a recent Quarantine report. Almost 98 per cent of Australia's trade is conveyed by sea, and the enormous ships that carry these goods arrive with cargo holds empty but ballast tanks full. A large bulk carrier holds 20 to 100 (sometimes as much as 140) million litres of water. Each year 150 million tonnes of foreign water – a volume greater than that held by Sydney Harbour – is disgorged into more than sixty ports by 10 000 vessels coming from 600 overseas harbours. Ninety per cent of it comes from the Asia-Pacific region, more than half from Japan, the world's main relay station for ballast villains.

Ballast is also a worry because our waters are isolated and relatively clean. Our sealife is very different from that of our northern trading partners. We are free of most of the pests and diseases that blight aquaculture in the north – the deadly oyster drill, for instance. As suppliers of clean shellfish and farmed fish we have a market advantage, but one we are likely to lose.

Before 1870 – when most ships were made of wood – sand, shingle and rocks, not water, were used as ballast. Captain Cook was probably the first to dump ballast in Australia when he unloaded Tahitian rocks at Endeavour River, but some ships still used solid ballast in the 1940s. Many small animals and seeds travelled to Australia concealed in this material. In South Australia, forty-three weeds are counted as solid ballast invaders, including the sea rockets (*Cakile* species) and sea spurge (*Euphorbia paralias*) that blanket beaches; also shellfish and crustaceans such as the European green crab (*Carcinus maenas*). Bitou bush, one of our worst weeds, apparently arrived with dry ballast dumped beside the Hunter River in about 1908, and alligator weed (*Alternanthera philoxeroides*), which blankets lakes in vast mats, was first seen sprouting on ballast dumped by service ships during World War I.

Not until the 1970s did Australian biologists peer into ballast water tanks to see what was there. Nets lowered into the tanks of Japanese containers yielded up tiny crustaceans, worms, molluscs, fish and eggs. Sixty-three species, most of them copepod crustaceans unknown from Australia, were recorded in a 1976-8 survey of thirty-one ships. Later studies probed the mud lining the tanks. If seas are choppy when ballast is loaded, fine sediment enters as well, to be shovelled overboard after the water is discharged. Dinoflagellate cysts, crustaceans and worms travel in this mud. More than 60 per cent of visiting cargo ships carry sediment, and half of these carry dinoflagellates. One Japanese ship was evidently filled during a toxic bloom – it contained more than 300 million cysts of the dangerous dinoflagellate *Alexandrium tamarense*.

More than fifty animals and plants have invaded Australia via ballast water, including fish, crustaceans, shellfish, worms, the seastar, kelp and four dinoflagellates. The number is small compared to the 2700 weeds in Australia, but numbers can be misleading.

Very few weeds are as menacing as the latest crop of ballast invaders, and in any case the number is probably an underestimate. Port surveys keep finding new species, and with the shift to larger, faster ships on tighter deadlines, more are likely to arrive soon. It is worrying to think that three of our worst ballast pests arrived only within the last twenty or thirty years.

In 1988 a forest of Japanese kelp was found near Triabunna, a woodchip port in Tasmania. Now one of Australia's worst weeds, this plant redefines the word seaweed. It can shed millions of spores in a day and grow a metre in a couple of months. Tens of thousands of these weeds were crowding the rock shelf at Triabunna to a depth of eight metres, infesting ten kilometres of coastline. Four years later the kelp zone stretched fifty kilometres to Maria Island and carried 400 tonnes of weed. In 1996 Japanese kelp appeared in Victoria, in Port Phillip Bay. In time it will invade all of our rocky shores between Wollongong and Cape Leeuwin (south of Perth), and biologists expect it to become 'a conspicuous part of the sub-tidal marine flora of southern Australian waters', displacing native plants and animals. Conservationists campaigning against woodchipping of old growth forests would be disgusted to know we received this pernicious plant in exchange. One enterprising company is now sending it back to Japan, where it is eaten as the delicacy called 'wakame'. Japanese kelp is also invading New Zealand and France.

The Northern Pacific seastar is a starfish forty centimetres across. When it appeared in the Derwent estuary in 1986 biologists rejoiced, thinking native starfish had reclaimed the cleaner waters. Six years later they learned the truth – that Tasmania had scored a new, merciless predator, one that swarms over the sea floor devouring everything in its path, dead or alive, including mussels, snails, fish, crabs, barnacles, worms, sea urchins, sea cucumbers, brittle-stars, ascidians, other seastars, even drowned dogs. Bivalve shellfish have all but vanished from the Derwent and the spotted handfish (*Brachionichthys hirsutus*), the first marine fish in Australia to win a listing as 'endangered', is threatened with extinction. Every few years starfish numbers explode, reaching densities around Hobart of nine per square metre and a total local population of about 28 million. Females can lay 19 million eggs at a time, and juveniles travel in ballast during their long planktonic stage. Seastars began appearing in Port Phillip Bay in 1995, evidently carried in on fishing gear or hulls. We will be hearing more about this rogue soon.

The other headline-grabbers, the most dreaded pests of all, are the toxic dinoflagellates. Microscopic planktonic algae, they form inert cysts that travel well in ballast. Three species (*Alexandrium catanella, A. minutum* and *Gymnodinium catenatum*) were found here in 1986, though they had obviously been here for some time, and a fourth (*Alexandrium tamarense*) was identified more recently. In still, polluted waters they proliferate into algal blooms, painting the sea red, yellow, blue, green or brown. Sydney experienced its first red tide in 1993. If filter-feeding shellfish suck in enough of them they become toxic to people, causing paralytic shellfish poisoning. Each year about 2000 people are poisoned worldwide, and many die. (One cheery AQIS biologist has told me I am lucky to be alive.) Fish, birds, dolphins, even whales, are also affected.

Toxic blooms are proliferating around the world, abetted by pollution and ballast exchange, and some scientists talk of a global epidemic. Australia takes the threat very seriously; AQIS has funded major studies on the subject. Their conclusion: these tiny plants threaten public health, the culture of fish, oysters and mussels, wild fisheries, other wildlife and tourism. They could poison pipis (the shellfish also called 'ugaris') and those who eat them on our beaches. The peril to oyster farms is real enough. Farms near Hobart are monitored regularly; there were costly closures in 1986, 1987 and 1991 for stretches of up to six months. In North America, some shellfish farms have closed for good. AQIS warns that reef animals – worms, starfish, sea urchins, shellfish, crustaceans – may also die: 'The lack of evidence of any significant wildlife impacts to date in Australia is comforting but probably not grounds for complacency.' In North America, humpback whales have died after swallowing the toxic algae.

AQIS plays down the threat to tourism but admits that a worst-case scenario could see tourists die, beaches close and costs to the community reaching $50 million a year. Tasmania's problems are already so bad that the marketing of her seafoods as pure and clean seems questionable. Tasmania receives less ballast water than other states – only 1 per cent of the country's total – but suffers enormously, presumably because her sea temperatures match those of Japan.

Other ballast-borne pests should be considered as well. The giant sea worm sprouts leathery tubes half a metre long, forming colonies that smother the floor of Port Phillip Bay. Asian mussels (*Musculista senhousia*), found in Perth's Swan River estuary, can grow at densities of over 3000 per square metre, displacing scallops. Japanese sea bass are large carnivores. We are lucky not to have the zebra mussel (*Dreissena polymorpha*), the number one ballast rogue in the United States. This unpleasant character carpets lakebeds at densities of up to 2 million per square metre, overwhelming the ecology, endangering native species, tainting drinking water and fouling pipes, boats, nets and beaches. In the Ukraine it became a pest in the cooling ponds of Chernobyl nuclear power plant.

Even more disturbing, the next wave of pests may well be diseases. AQIS knows of 148 pathogens that could arrive in ballast, including cholera and some dreadful fish diseases. IPN virus and furunculosis would devastate our trout and salmon farms. These diseases often infest fish farms overseas, especially farms located near ports. AQIS regulations stipulate that ballast water not be loaded within five kilometres of fish farms, but furunculosis can travel further than this in sea currents.

Ballast will remain a problem. Although many methods of treating the water have been proposed – including electrocution, heat treatment, ultraviolet light, ultrasound, ozone, poisons, salinity changes, filtration, oxygen starvation, reballasting and treatment at plants – none is both reliable and cheap. The shipping industry is now working with AQIS to find solutions, though it was not always so helpful. When AQIS at one point asked seventy shipping surveyors for details about ballast, only five cooperated, most others either refusing to help or failing to reply. At a 1994 ballast symposium Alan Taylor of BHP Transport put a self-interested industry view: 'Australia must not introduce unilateral guidelines for the management of its Coastal Ballast Water which are different from or out of step with the IMO [International Maritime Organisation] International Guidelines, nor any state, territory or port authority as it would create a ludicrous situation which could introduce a self imposed trade embargo on our international and national seaborne trade and hence have an adverse effect on the overall economy of Australia.' These concerns seem well founded. AQIS has obtained analyses showing that the costs of treating ballast water, at eighteen to twenty-five cents a litre, would exceed the profit margin on many shipments of iron ore and coal.

With 10 billion tonnes of water travelling the world each year, carrying about 3000 species of plants and animals, ballast is now an acknowledged global threat. Fortunately, there are countries, especially the US, Canada, New Zealand, Israel and Chile, that share Australia's concern. The IMO is adopting guidelines for ballast management based upon Australian protocols. The cost of trade will rise, but no one will be exempt. Ships all over the world will have to reballast at sea. They will exchange water in the ocean, either by emptying and refilling their tanks, or by pumping through large volumes of water ('flow-through' exchange). Open ocean carries fewer organisms than nutrient-enriched harbours, and oceanic species don't fare well in ports.

Reballasting is far from foolproof. A residue of the original water always remains, and even tiny residues can carry pests. The CSIRO looked at thirty-two vessels that reballasted at sea and found sediment in fourteen, some of it containing dinoflagellate cysts. And not all ships can reballast. Many large container ships, the older ones especially, lose stability. But most new ships are built with reballasting in mind. Ensuring compliance is another matter. AQIS is developing a method of verification.

Australia has led the way in ballast awareness. In 1990 we introduced the world's first voluntary guidelines. Ships bound for Australia are asked not to take on ballast from polluted harbours and shallow bays, or at night when bottom-dwelling organisms rise in the water. Ships should reballast at least 200 miles out to sea and avoid discharging in Australian waters. Captains must fill out Ballast Water Reporting Forms and are forbidden from dumping sediment except on land. AQIS tells me that most ships comply, and that Japanese captains are especially helpful and honest. The Quarantine Service adopts a risk assessment approach, targeting ships coming from high-risk areas, especially those whose port of departure has a similar climate. If they believe a ship contains dangerous ballast, it can be ordered to reballast at sea. AQIS is also addressing the problem of ballast movement between Australian ports. Recent surveys show an enormous range in the number of pests in different harbours. Port Phillip Bay has more than a hundred exotic plants and animals (both hull foulers and ballast invaders), while some ports have only a few species. The worst harbours are beginning to serve as relay stations for the transfer of pests. To counter this threat the federal government needs to gain control over interstate movement of ballast water, as recommended by the Oceans Policy, but the states will need to agree to this.

Reballasting is no magic answer, and better treatments may appear in future. One option may be to heat ballast water with waste engine heat, although ships would have to be redesigned to accommodate this. Water filtration is another possibility. The best hope is that one of these methods proves both practical and cheap and that world shipping moves quickly to embrace the new technology. That's a lot to hope for. In the meantime we can expect to see more and more aquatic pests heading our way, with AQIS working valiantly to turn them away.

15

Where Have All the Flowers Gone?

Phytophthora's Curse

'And this same flower that smiles today
Tomorrow will be dying.'
Robert Herrick

Once upon a time signs in our national parks told us to 'Take nothing but photographs, leave nothing but footprints'. The signs today warn that even our footprints can kill the forest, for the soles of our boots are spreading the deadly forest-destroying disease phytophthora. This fungus[1], one of the world's worst exotic pests, has changed all the rules. In many national parks, certain tracks and walking trails have been closed for fear of accelerating ecosystem collapse. Rangers are busy barricading hillsides and blocking roads in a rearguard action to save rare plants.

Phytophthora is stripping the colour from south-western Australia, a region renowned for its wildflower displays. One of the richest plant habitats on earth, it rivals the tropical rainforests in variety of plants. Over millions of years, shrubs have proliferated endlessly, especially nectar-rich banksias, dryandras and hakeas, foods of the unique honey possum. But the isolation that allowed the flora to flourish so freely has left the region uniquely vulnerable to introduced disease. Today the flowers that feed honey possums, pygmy possums, honeyeaters and jewel beetles are dying.

A swag of *Phytophthora* species have entered Australia, but cinnamon fungus (*Phytophthora cinnamomi*) is the most deadly. During the eighteenth century it migrated in pot plants from eastern Asia, where it was first discovered on cinnamon trees in Sumatra in 1922, to Europe, Africa and the Americas. By the 1870s it had entered Australia (also in pot plants, no doubt), where it long remained undetected. In 1930 it was found to be destroying Queensland pineapples; then it showed up on dying native shrubs near Sydney. But not until Frank Podger addressed a plant pathology conference in Toowoomba in 1966 was the scale of Australia's problem fully grasped. His revelations sparked so much concern that phytophthora dominated the next conference five years later.

Podger's achievement was to explain 'jarrah dieback'. Around Perth, jarrah (*Eucalyptus marginata*), a magnificent eucalypt, is the dominant tree, a vital source of timber as well as a critical habitat, occupying vast tracts of land. But in the 1920s jarrahs began dying mysteriously and by 1948 many forests had become graveyards. Scientists were baffled – until Podger found the fungus fastened to dying roots. The pale threads of phytophthora cling to fine roots and kill them. Heavily attacked plants are stem-girdled and die of thirst. Most kinds of plants are attacked to some degree. Trees usually survive, although phytophthora kills north Queensland rainforest, eucalypts in southern New South Wales and Victoria (6000 hectares lost in East Gippsland alone) and jarrah in the west. Shrubs usually suffer most, especially banksias and their relatives, heaths, peas and grasstrees. Orchids, sedges and most other monocots survive by replacing their roots. Besieged woodlands are transformed, their colourful understorey of shrubs giving way to monotonous sedges. In south-western Australia, 2000 of 9000 plant species may be highly susceptible. Nowhere else on earth are so many wild plants so vulnerable to exotic disease.

Cinnamon fungus strikes crops as well, notably avocadoes, pineapples, stonefruits, macadamias and some ornamentals; also pine plantations. It travels about in nursery soil, obliging farmers to deploy fungicides and grow phytophthora-resistant breeds.

In temperate Australia, the infertile soils are critical to its success. To absorb enough nutrients, native plants develop vast rootstocks, which render them easy targets. After heavy rain the fungus sheds millions of spores, which travel above and below ground with the flowing water. In the Grampians and Brisbane Ranges national parks, phytophthora was probably imported in gravel used to lay roads; in Wilsons Promontory a firefighting bulldozer was implicated. Boots, tyres, hoofs and paws carry them about as well. The fungus also creeps from plant to plant along adjoining roots. It does best in sandy, poorly drained soil low in humus, worst in moist mountain chains with fertile soils, where the micro-organisms in humus destroy it; in Victoria it keeps mainly to low foothills and coastal plains, although in north Queensland it steals deep into the rainforest.

On the banksia-rich sand plains north of Perth, phytophthora's advance has been chronicled from above. Aerial photos taken from a bomber during World War II showed treeless patches up to thirty metres wide that straddled tracks radiating out from strawberry farms. Later shots showed the infection reaching nearby swamps. Today more than half of the woodland has been obliterated, not by bombs, but by a stealthy underground scourge that advances a metre a year.

In the Stirling Range the fungus crept along management tracks pushed through in the late 1960s. By the 1970s four large dead patches, and many smaller ones, had opened up like spreading sores. Most of the park is now infested, with 90 per cent mapped as being at very high or high dieback risk. This park, whose lofty peaks attract heavy summer rains, is especially vulnerable, with more than a hundred susceptible plant species. Banksias, dryandras and grasstrees are hardest hit and seven rare species found nowhere else require urgent management attention. When I visited this park in 1998 dieback warning signs confronted me on every track. I felt uneasy about leaving the roads to look at banksias (most roadside banksias had died long ago), though at least the ground was dry. In a place of immense beauty I felt unclean and unwelcome. One cannot feel connected to wilderness in forests infected with this disease.

Phytophthora promotes erosion too. In the Grampians, where half the plant species – even kangaroo grass – are susceptible, steep slopes stripped bare by the disease are sliding away. Botanists Jill Kennedy and Gretna Weste, in the *Australian Journal of Botany*, have concluded that 'Erosion may be a serious problem on gravel or sandy slopes exposed to alternating wet and dry seasons. The water supply from the infested hills will carry infective zoospores and hence threaten the agriculture of the western plains.'

Rare plants are in particular danger. In the Grampians, four endemic species are threatened and some plants in Western Australia face imminent extinction; their seed has been collected for storage and the plants sprayed with phosphonate to keep the evil at bay. Many other rare plants are probably at risk, but no one is available to study them. Other plants, while not currently at risk of extinction, are losing local races, each defeat narrowing their genetic base.

More species of bird-pollinated flowers grow in south-west Australia than anywhere else on earth, so honeyeaters, bees and probably honey possums are suffering too. Other animals drop off in variety and numbers as the curse sweeps through, although some species, especially kangaroos, actually like the more open understorey.

Phytophthora is listed as a 'key threatening process' under the Commonwealth Endangered Species Protection Act and it has drawn an unprecedented response from national parks managers. Visitors are banned from certain areas in some parks and advised to practise careful hygiene in others. Disease-free areas in Stirling Range may be visited only in dry months, when spores are inactive. In Tasmania's

South West National Park the log bridge at the main entry from Lake Pedder has been removed, obliging hikers to wade through Junction Creek, thereby washing their boots; scrubbing brushes on each bank reinforce the message. The big challenge in Western Australia is to convince maverick four-wheel-drivers to keep out of the bush.

Fortunately, not all the news is doom and gloom. Some forests are bouncing back. In Victoria's Brisbane Ranges, where almost half the trees died in 1970-71, many plants have resprouted from roots or seeds, although others are not returning at all. In some places phytophthora appears to have vanished after running out of hosts. Its decaying victims feed micro-organisms, which then attack the fungus. Also, the disease may be losing virulence, its hosts developing resistance.

But no comeback is evident in Western Australia, where the dead are most numerous. Here, scientists hope phosphonate spray will save the day. Plants remain oblivious to attack by phytophthora, and this lack of awareness appears to give it an edge. Phosphonate jolts the plants into fightback mode so that they wall off the infection. The spray saved endangered banksias near Albany and is now under trial on rare heaths, although these are harder to help because their small, tough leaves absorb less spray. Phosphonate is also sprayed along phytophthora fronts in broad, 100-metre swathes to block its march, like using firebreaks against fire. Unfortunately, too much spray kills the plants themselves; also, phosphonate boosts soil phosphorous, which can kill heathland plants while encouraging weeds. In small reserves in the west, exotic grasses are already a chronic problem, invading wherever native plants fall away. But immunity conferred by phosphonate appears to be long-lasting, and the spray may be needed only once every ten or twenty years. At the very least it provides a breathing-space while other solutions (perhaps biological control) are sought.

It is important to recognise that cinnamon fungus is only one of many foreign diseases lurking in our forests. In the south-west another five kinds of *Phytophthora* (*P. citricola, P. cryptogea, P. drescheri, P. megasperma* and *P. nicotiana*) infect the flora, although at least one of these (*P. citricola*) appears to be native. These species are implicated in the death of banksias and rare shrubs, although not on an epidemic scale. There are twenty-two *Phytophthora* species in Australia, including twelve in north Queensland rainforests, but no one knows how many are native and how many are introduced, or how much harm they are doing. The well-known potato blight (*Phytophthora infestans*), which caused the Great Famine in Ireland in 1846, now infects native kangaroo apples (*Solanum aviculare*). Grape downy mildew fungus (*Plasmopara viticola*) is an exotic disease that strikes native grapes (*Cissus hypoglauca*), and the biocontrol agent blackberry rust (*Phragmidium violaceum*) now attacks native raspberries. In Western Australia, native fungi called cankers are spreading alarmingly in logged forests and in heathlands hit by hot summers.

I suspect there are many other introduced diseases now at work, skewing the structure of our forests and rewiring the ecology. Phytophthora wins the lion's share of attention because it is easily the worst, not because it stands alone. Plant diseases are a much-neglected problem and we are far from understanding the hidden role they play in Australian ecology. It's probably true to say of them that the more you seek, the more you'll find.

1. *Phytophthora* species have been reclassified as members of the kingdom Chromista, and strictly speaking they should no longer be called 'fungi', but, there being no other common name, I have retained this traditional epithet. They are certainly not plants.

16

The Sick and Dying

Exotic Diseases Strike

'Death, like birth, is a secret of Nature.'
Marcus Aurelius Antoninus

It began in Anxious Bay in South Australia in March 1995. A mystery epidemic struck the schooling pilchards (*Sardinops sagax*), killing them in their millions. East and west it sped at lightning speed, reaching Western Australia in April, New South Wales in May, and Queensland and New Zealand in June. Fish were dying along half Australia's coastline, from Geraldton in the west to Noosa in the east. Affected pilchards suddenly became listless, swam sluggishly and expired within minutes. The southern seas became vast graveyards bearing rafts of dead fish up to three kilometres long. Tonnes of stinking pilchards spewed up onto beaches. Travelling faster than ocean currents and faster than fish can swim, the disease may have been borne along by the hosts of gannets that came each day to gorge on the sea's unexpected bounty. By depleting the seas the virus must also have visited much suffering upon penguins, seals, squid and pilchard-eating fish.

After analysis, flesh of the dead fish was found to harbour a herpes virus of a kind never seen before. All the evidence suggests it entered Australia in frozen pilchards or fish meal fed to sea-caged bluefin tuna in South Australia. Each year Australian tuna farms import more than 10 000 tonnes of seafood from Japan, Russia, Peru, Chile and California – all without quarantine inspection.

But the consequences of this epidemic were trivial compared to those of another mysterious disease that struck the Queensland rainforests some years earlier. On this occasion several frog species disappeared completely, and others barely survived. Glen Ingram, a friend of mine from the Queensland Museum, was studying platypus frogs in the Conondale Ranges north of Brisbane when his subjects suddenly vanished. Glen was one of the last people to see a platypus frog alive. 'The gravid females used to sit in little pools among the rocks and you could come back again and again and find the same ones', he said. 'I'd always worry that an eel would come up and get them.' It impressed him that the frogs could elude snakes and white-faced herons as well as the eels. But something else got them instead. Glen still can't quite believe it happened. He has never written up his field notes, because, he says, 'it hurts too much'. I cherish my own dim memory of seeing this species, long ago, in a Conondale stream. The platypus frog (*Rheobatrachus silus*) was a national treasure – a remarkable aquatic frog with big webbed feet that stored its tadpoles in its tummy. Scientists were gearing up to study its stomach chemistry, hoping to find a cure for ulcers. The frog vanished in 1981, two years after the southern day-frog (*Taudactylus diurnis*), another species that Glen may have been the last to see. Day-frogs used to live in rainforest an hour's drive from my house.

In the years after these frogs disappeared Brisbane's biological community bubbled with theories. Had drought banished them to remote mountain retreats? Would drenching rains bring them back? We waited and wondered, but no frogs came back. Instead, other native frogs disappeared too, as the tragedy travelled north. The northern platypus frog (*R. vitellinus*), discovered only in 1984 in mountains near Mackay, vanished the following year. Then in 1992 two frogs in north Queensland, the armoured mistfrog (*Litoria lorica*) and mountain mistfrog (*L. nyakalensis*), went missing, and they now appear to be extinct too. On the other side of the world, in Central America, other frogs were vanishing, including the remarkable golden toad (*Bufo periglenes*) of Costa Rica, made famous by naturalist David Attenborough; today it survives only on film and in museum pickle jars.

Over the years the idea of a killer disease gained ground, especially after many sick and dying frogs were found in 1994 on Big Tableland in north Queensland. Their skins carried a new kind of chytrid fungus (genus *Batrachochytrium*), a single-celled organism that showed up in Panama and California as well. This fungus is turning up on dying frogs all over Australia, including green treefrogs in garden ponds. Fortunately, the only frogs going extinct appear to be stream-breeding species in montane rainforest; the fungus presumably does best in cool wet places. We can't yet say for sure that the chytrid is the killer, but the evidence is accumulating fast.

Exotic diseases are the most insidious and probably the most underrated of all our invaders. Largely invisible and easily overlooked, they play an ill-defined but probably major role in Australian ecology. If the case against the frog-killing fungus is ever proven and the missing frogs fail to return, this disease will emerge as a greater agent of extinction in Australia than any feral animal, except perhaps the fox.

Australia has lost more mammals to extinction – sixteen in all – than any other continent. When today's biologists try to explain this tragedy they hypothesise about foxes, rabbits, cats, cattle, sheep and fire but rarely invoke disease. Biologist Bill Freeland has been a notable exception in recent times. 'Was the disappearance of the thylacine from mainland Australia caused by a parasite brought to Australia by the dingo?' he speculated in 1993. 'Are the unexpected declines of three of four species of *Dasyurus* [quolls or native cats] due to susceptibility to toxoplasmosis or some other introduced parasite of cats, dogs or foxes (e.g., Rounsevell 1983)? The old reports of disease and epidemics among our native fauna . . . cannot and should not be dismissed as inconsequential anecdotes.'

I agree. Why is so little credence given to the many early anecdotes about marsupial disease? Is it because the subject does not lend itself to study and thus attracts no research grants? I have heard enough stories from old-timers and read enough old journals to take these accounts very seriously. Here are three examples.

Charles Blomfield, a pioneer in the New England Tablelands during the 1870s and 1880s, had this to say: 'I think the kangaroo rats died from a disease like the native bears did in 1895. Tiger cats and native cats also died from the same thing; this was before the foxes came.' Another pioneer, Amy Crocker, told of similar events on the Nullarbor Plain. She remembered the disappearance of 'lots of little Australian animals, even possums, which had gone before my day. My uncle told me of a strange virus which attacked all small marsupials, killing many species right out. Possums were among these. I think this occurred in the 1880s or 1890s. I was once told by a man from the eastern wheatbelt that a similar virus occurred there . . .' In 1923 prominent biologist Albert Le Souef claimed that diseases had carried off marsupials in their thousands: 'Well known instances of this occurred in 1898–9 and in 1901–2–3. Koalas, Native Cats (*Dasyurus viverrinus*), Rabbit Bandicoots (*Peragale*) and certain Phascogales were almost exterminated from Central Queensland to Victoria.' These were dire drought years and maybe Le Souef got it wrong. Maybe.

Evidence from Africa certainly supports a role for disease. In the 1890s rinderpest, an Asian cattle disease, galloped through the continent slaughtering millions of antelope, buffalo and giraffes. Africa never recovered. This disease, along with bovine tuberculosis, still kills big game today. Rabies and canine distemper have helped bring Africa's hunting dogs close to extinction. In 1994 a thousand lions on the Serengeti succumbed to canine distemper. Exotic epidemics – transmitted by village cattle or dogs owned by encroaching communities – pose a continual threat to the viability of some of Africa's game parks.

In Australia, the protozoan *Toxoplasma gondii* may be a serious cause of death. Brought in by cats, it now infects marsupials, even in remote outback deserts. Bandicoots, small wallabies and native rats often die after nibbling plants or insects tainted by cat faeces. Is it coincidental that most of Australia's extinct mammals were bandicoots, small wallabies and rats? Other scientists have suggested that viruses introduced in house mice may have killed off native rodents. A tapeworm (*Echinococcus granulosus*) introduced with dingoes long ago is also implicated in the deaths of an endangered bridled nailtail wallaby and a vulnerable Proserpine rock wallaby in Queensland, although it does not threaten the fate of either species.

Diseases have certainly taken a toll on our fish. Long before the pilchard disaster, in 1888 naturalist William Saville-Kent recorded a die-off of Australian grayling (*Protroctes maraena*) in Tasmania. The fish were said to 'have been seen floating down the rivers in thousands, covered more or less extensively with a cottony fungoid growth. So virulent and exhaustive was this epidemic that many, more especially of the southern rivers, were more or less completely denuded of their stock of this species and have so remained up to the present date.' Saville-Kent suggested that acclimatisers introduced the fungal disease Saprolegnia on trout or salmon eggs brought from England. The grayling recovered, but today another fish, the Macquarie perch (*Macquaria australasica*), is declining seriously; a disease carried by introduced redfin perch (epizootic haematopoietic necrosis, or EHN) is implicated in its demise. (Other fish diseases were discussed in chapter 9.)

Disease is also attacking one of the world's most exquisite birds – the Gouldian finch of northern Australia. This rainbow on wings, painted green, blue, purple and gold, was dedicated by the great naturalist John Gould to his artist wife Elizabeth. In 1865 he wrote of its 'great abundance' on the Victoria River, but today it is endangered, having disappeared under mysterious circumstances from most of its range. Researchers have discovered that many of the remaining wild birds harbour parasitic air-sac mites (*Sternostoma tracheacolum*) in their lungs. These mites infect aviary birds and presumably came into Australia with cagebirds; Gouldian finches appear to be unusually susceptible. Bird expert Stephen Garnett suggests an unknown disease struck the birds first and that the mites are now suppressing their recovery.

When diseases encounter new hosts the consequences can be catastrophic. Usually a disease and host evolve together; their relationship is stable both because the host develops resistance and because the disease gains no advantage by killing its host. Epidemics erupt when diseases encounter new hosts with no prior resistance. All over the world people are now bringing new pathogens and hosts together. Sometimes it is done deliberately, for biocontrol. Myxomatosis proved so devastating to European rabbits because it is a disease of *American* rabbits. But some rabbits in Australia inevitably survived and went on to found new, resistant strains.

Fortunately, our wildlife can also develop resistance to foreign disease. Australia's pilchards may one day develop some immunity to exotic viruses, although we know it hasn't happened yet because a second pilchard die-off erupted in late 1998. (Starting in South Australia, it soon reached Western Australia and Victoria, again devastating wild stocks and greatly depleting catches by fishermen.) Pilchards and rabbits are abundant and plenty of survivors remain to rebuild their numbers. But when an uncommon animal is struck, the whole population may succumb before it builds any resistance. Many of our rainforest frogs live in small colonies in the mountains, and this probably explains why these frogs, unlike fish, rabbits, and green treefrogs, disappeared completely. Some of the frogs that survived the epidemic are now increasing their numbers, and we must hope they are now disease resistant.

Environmental stress may be pivotal in such epidemics. Some

biologists argue that rising levels of UV radiation have made our frogs susceptible to disease (though tests show no rise in radiation in montane rainforest). Cassowaries are suffering bacterial infections that may well be triggered by environmental stress.

The disease picture is very complicated and often we don't really know what is happening. Mange is killing many wombats in southern Australia, but experts can't tell if the culprit, mange mite (*Sarcoptes scabiei*), is exotic (introduced by dingoes, dogs or foxes), a cosmopolitan mite that sometimes triggers epidemics, or a cosmopolitan mite that has struck hard because wombats are now stressed. The oldest wombat skins in Europe, from Nicolas Baudin's 1800–1803 expedition, carry apparent mange lesions, suggesting the disease is indigenous, but then again, these wombats may have caught mange from dogs on Baudin's ships. Mange epidemics date back at least to 1937, when thousands of wombats died in southern New South Wales, and the disease is considered a serious threat to small isolated wombat colonies, some of which are disappearing.

Of all the exotic invaders we can expect in the future, diseases pose the greatest threat. (After all, AIDS, anthrax, malaria, plague and influenza are all part of the exotic disease story.) If it is true that an exotic fungus wiped out five of our native frogs, then we should be very concerned about Newcastle disease, which could eliminate many of our bird species (see chapter 31). We could also lose crustaceans to the American disease *Aphanomyces astaci*, which has wrought disaster in Europe.

In 1993 the Australian Quarantine Service bowed to pressure from parrot-lovers and allowed in several South American macaws, the first since 1949. The macaws brought in two exotic diseases – internal papillamatous disease and proventricular dilation syndrome – that pose a risk to wild parrots. One reason for allowing in these birds was to discourage smuggling of macaws and the diseases that travel with them. But the example highlights the fact that foreign diseases will remain a serious problem for Australia in the interesting years to come.

17

The Price of Free Trade

Opening the Door to Invasions

> 'There is ... no one more didactic than a true believer in free trade, unless it is a true believer in the environmental cause.'
>
> **US Ambassador Michael Smith, 1994**

It is abundantly clear that free trade, the high ideal we are all asked to embrace, is a sure-fire recipe for a pest-infested world. The more we move goods around the globe, the more pests we move with them. Trade barriers are ecological barriers, and by breaking them down we open the floodgates to more pests. The rewards of free trade must be weighed against the huge costs of some of the problems it creates.

You might expect free traders to have thought all this through, to have put together cogent arguments to counter these concerns. During 1992 a bunch of them got together with environmental experts in the United States for a series of meetings, sponsored by the Council on Foreign Relations. One participant, C. Ford Runge, a professor in Agricultural and Applied Economics, wrote a book about it afterwards, *Freer Trade, Protected Environment*, with the subtitle *Balancing Trade Liberalization and Environmental Interests*. Such a book, you might expect, would devote several chapters to the emerging pest risks. In fact, exotic species do not rate so much as a paragraph! The index has no entry at all for 'pest', 'weed', 'exotic species' or 'disease'. Here we have a whole book about free trade and the environment – among the most momentous themes of our time – with nothing to say, nothing at all, about the crisis of emerging pests. Runge talks instead about how tariffs subsidise habitat destruction then draws the comforting conclusion that free trade delivers a greener environment.

Runge makes no claim to being a greenie, but Douglas Yu, a Harvard biologist, does. In a 1994 article in the journal *Conservation Biology*, 'Free Trade is Green, Protectionism is Not', he discounts concerns that factories will shift to poor countries where pollution standards are lax, that Third World countries will have to cope with toxic wastes without the resources to dispose of them, that companies will exploit free-trade treaties to overturn environmental regulations, and so on. Incredible as it may seem, exotics again get no mention.

But not everyone is asleep. Peter Jenkins of the University of Mexico pens a critique of Yu, stating the obvious – that 'increased international trade has the potential to cause more harmful exotic species introductions'. He cites two recent examples – the zebra mussel, the horrific ballast-borne pest discussed in a previous chapter, and the Asian tiger mosquito (*Aedes albopictus*), a disease carrier that reached North America in puddles in used tyres.

Yu is generous in his reply: 'Jenkins makes an excellent point, that the increase in trade due to free trade treaties can increase the rate of introduction of exotic species, with all their attendant problems. As far as I know, this is a new observation in the burgeoning literature on trade and the environment, and for that Jenkins' article is a welcome contribution.' Really? Has no one thought of this before? Are all the free traders and conservationists so incredibly blind they have given no thought to all those crop-devouring snails and poisonous dinoflagellates and farmyard epidemics? Is *Homo sapiens* a thinking ape? In fact, biologists have been warning about ballast-assisted invasions for some years, but the warnings have not been linked to the free trade debate.

I have sought out Australian writings on free trade and pests and here too the connection has not been made. The Australian Conservation Foundation, our peak conservation body, lists 'Trade and the Environment' as one of its ten priority campaign issues. A promising start, you might think. But its campaign is based upon the statement by the Asia-Pacific Pre-Summit Forum on Environment and Ecology held in the Philippines in 1996. That statement makes the important point that wealthy nations should not be allowed to exploit the Third World through trade, but again, no mention at all is made of pests in the 2300-word document. (In fact, rich nations have even more to lose from a free trade in pests than poor ones, since most new pests will travel from poor tropical countries to affluent temperate ones.)

Examples of pests travelling under liberalised trade regimes are easy enough to find. The western flower thrips (*Frankliniella occidentalis*), a native of North America, owes its spread around the world, during the 1980s and 1990s, to the burgeoning trade in cut flowers. It reached Australia in 1993, and the Australian Academy of Science fears it will soon spread commercial diseases, especially tomato spotted wilt, to native plants. Another example is the pilchard virus. Tuna farming is one of Australia's newer industries, set up to supply Japanese restaurants with sashimi. Tuna suitable for sashimi must be fed an oil-rich diet, and imported pilchards prove ideal. But some pilchards imported in 1995, and again in 1998, were apparently infected with the virus.

The burgeoning ranks of ballast and hull pests are a consequence of growing international trade already discussed. Ships were using ballast water a hundred years ago, yet several of our worst ballast invaders, notably Japanese kelp, the giant sea worm and the Pacific seastar, arrived only recently.

Some people can only grasp the significance of such events by attaching price tags to them. No figures can be put on these disasters, but there are several examples of trade-related pests for which estimates have been made. The zebra mussel that travelled from eastern Europe to North America in ballast water around 1986 is one. The US government estimates the cost of just clearing blocked water pipes at US$2 *billion*; the costs to industry, water supply, boating and fishing, from 1989 to 2000, could reach $6 billion in the Great Lakes alone. A jellyfish-like comb jelly, *Mnemiopsis leidyi*, that travelled in ballast in the opposite direction has helped destroy the Black Sea fisheries, at a cost so far of $350 million in lost revenue. This comb jelly recently entered the Mediterranean Sea, so its total costs to Europe will be much higher. The Office of Technology Assessment of the US Congress estimates that seventy-nine major exotic invasions into the United States have cost a total of $97 billion – an average of $1.2 billion per pest.

These are foreign examples, but a couple of guesstimates do exist for Australia. AQIS commissioned the Australian company ACIL to look at the economics of ballast water controls. ACIL, which published its results in 1994, estimates the cost of *not* controlling dinoflagellates at $40 to $400 million, with $200 million as a best guess ($57 million lost by fisheries, $137 million by tourism). By another estimate, the cost of existing ballast pests is $200 million a year in Tasmania alone. In 1994 poinsettia whitefly (*Bemisia tabaci* biotype B) reached Australia, having probably entered on poinsettia cuttings imported by a Coffs Harbour nursery. This form of whitefly attacks many kinds of crops and in California has cost $300 million in lost production since the 1980s. The whitefly is regarded by some experts as the world's worst agricultural pest, and the Australian Academy of Science, in a submission to the Nairn review, suggests a similar cost here. New South Wales flower growers estimate an annual cost to their industry of $30 million in pesticides alone.

We cannot budget for all our new pests because we don't know who they are. As agriculture bites deeper into the world's tropical rainforests more and more plant-chomping insects will switch diet from jungle fruits and leaves to crops, then migrate around the globe. Like the HIV virus, new pests and diseases will appear as if from nowhere. No experts in their wildest dreams could have foreseen that a flatworm from New Zealand (which is not a pest there) would one day devastate Britain's earthworms. In fact, New Zealand may have more than a hundred flatworm species for us to worry about.

To guard against future ecological chaos all nations should be tightening their quarantine regulations. But it won't happen, not only because the explosion in trade will make it impossible, but because free trade laws will intervene.

The World Trade Organization was formed in 1995, with Australia a founding member, on the fundamental principle that member nations dismantle their tariffs. The concern has been raised that countries will now use quarantine laws as surrogate tariffs to protect uncompetitive industries. A country wanting to protect its flower farms could ban imported flowers by exaggerating the pest risks. The WTO's Agreement on the Application of Sanitary and Phytosanitary Measures (SPS Agreement) is designed to prevent this. Members are required to 'base their sanitary and phytosanitary measures on international standards, guidelines or recommendations' and, although members 'may introduce higher levels of protection if there is scientific justification', they must – wait for it – 'take into account the objective of minimizing negative trade effects'. We are, then, expected to err on the side of trade, not caution. A temporary ban can be slapped on an import if it seems to be risky, but scientific evidence to back this up must be obtained 'within a reasonable period of time'.

In 1975, long before the WTO was born, Australia placed a ban on imported uncooked salmon – ostensibly because of the disease risk. Salmon-exporting nations saw this as a disguised trade barrier and in 1997 Canada asked the WTO dispute settlement body to investigate. Australia's argument was that imported salmon meat might carry any of twenty exotic diseases that could infect wild or farmed fish, if the meat were discarded near streams or used as bait. One disease, furunculosis, is so dangerous, Australia argued, that 'the entire salmon industry would probably cease operation in the event of disease outbreak'. Australia's case was weakened by one of its own draft reports, produced in 1995, recommending that some imports be allowed. A later report, drawn up after intense lobbying by Tasmanian salmon farmers, recommended a total ban.

In May 1998 the WTO delivered its verdict in this its first-ever quarantine case. In what the *Financial Review* described as an embarrassing blow for Australia, Canada's complaint was upheld. The WTO found Australia's quarantine policy to be inconsistent, because we allow in aquarium fish and herring bait, which both appear to carry greater risks than salmon. Live fish carry many more diseases than clean fish fillets, and bait is dangerous because it goes straight into the sea. Australia explained that risk assessments were under way on these two items, and that tighter controls might result. (Ironically, Australia allows in consignments of Canadian aquarium fish.)

Australia is now obliged to address this inconsistency. We can do this in one of three ways: by allowing in salmon meat; by bringing our quarantine laws into line through restricting (or banning) imported aquarium fish, fish food and bait; or by producing risk assessments that show salmon meat to be riskier. Dr Peter Beers of AQIS told me that salmon meat is considered riskier to Australia because it could harm our salmon and trout stocks, while most diseases carried by ornamental fish and bait would only kill native fish. In other words, salmon farms and trout fishermen matter more than our native fish stocks. In the wake of Australia's horrible pilchard die-offs this is a particularly unsavoury argument.

In July 1998 Australia appealed the WTO decision. (Canada counter-appealed, insisting the WTO judgment was too soft.) The appeal did not go Australia's way, and we are now locked into binding arbitration as the saga continues. The WTO cannot overrule Australia's laws and make it take salmon. But if Australia bucks the WTO ruling it could one day suffer retaliatory WTO-sanctioned trade bans. Our trading partners are not sympathetic. The US Ambassador to Australia, Genta Hawkins Holmes, accused Australia earlier in 1998 of too often 'using technical restrictions in order to provide protection for inefficient sections of agriculture'. Our ban on cooked chicken meat, imposed because of a perceived risk of Newcastle disease, has not proved popular. Nor has our ban on New Zealand apples (because of fireblight), which provoked New Zealand's minister of agriculture, Lockwood Smith, to claim that 'Australia possibly has the worst reputation in the developed world for using bogus quarantine reasons for excluding trading partners' products'.

The implications of all this are disturbing. Australia's quarantine laws are the subject of international ridicule. If, in the national interest, Australia decides to tighten its quarantine laws further by banning imports of most aquarium fish, fish products, cut flowers and garden plants (none of which seems especially likely), we can expect further international condemnation. Though such moves would remove the inconsistency over the salmon ban, they might well provoke complaints to the WTO from nations that export these products. Will we be forced to take appalling quarantine risks for the sake of our standing as a WTO member? Should other countries have this kind of influence over our quarantine laws? Free trade may be good for trade, but it is not, on the evidence I can see, good for Australia.

Postscript: On 19 July 1999 AQIS acknowledged Australia's inconsistency over fish imports and ruled to allow in uncooked salmon fillets, subject to strict health measures.

IV

AUSTRALIANS AS PESTS

Australia is not only a victim but also a major supplier of pests. Australian animals, plants and fungi have invaded every corner of the globe. Britain and New Zealand have curious tales to tell.

18

A Source of Perverse Pride

Australianising the World

> 'Travelling Australians are to be found in their ugly droves.'
>
> **Robert Milliken, 1973**

Australians suffer from an ecological inferiority complex dating back to pioneer times, when our fauna and flora were pronounced the most primitive on earth. To early naturalists, Australia was the land of living fossils, an archaic island inhabited by lowly marsupials – backward and foolish creatures that ought to be extinct. This notion, built upon British assumptions of racial superiority, pervaded biological thought right up to the 1960s and still holds some sway today. It explains, we are to understand, why Australia suffers so much from foreign pests and why marsupials, in particular, fare so badly against foxes and cats. Albert Le Souef, curator of Taronga Park Zoo, put it well in 1923: 'When animals of this class [marsupials] suddenly find themselves placed in competition with such advanced forms as the Fox, the Cat, and the Rabbit – types that are far ahead of them in the evolutionary scale – it is ... inevitable that they should go down before the invader.' Similar claims were advanced about the Aborigines, often by the same men.

If Australians knew how many pests we have unleashed upon the rest of the world, any vestige of inferiority would vanish. Our plants, animals, even our fungi do remarkably well overseas. Australian pests can now be found in most corners of the world, attacking crops, invading forests, advancing inexorably under the waves. To foreign lands we have contributed mammals, birds, reptiles, frogs, insects, spiders, crustaceans, snails, worms, trees, shrubs, vines, herbs, seaweeds, mushrooms, even a virus. We have probably more than a hundred expatriate animals (though most of these are confined to New Zealand) and almost as many exotic plants. We may take perverse pride from knowing that Australia is a major centre of evolution, one that has contributed its fair share to the homogenisation of global ecology.

Some parts of the world are now thoroughly Australianised. I have walked in tea-tree forests in Florida, in wattle groves in California, and through hakea and wattle thickets in South Africa. California now has nine exotic eucalypts, eleven wattles, four saltbushes, a tea-tree, Australian insects of many kinds, a marine worm and three crustaceans – two in the sea and one on land. Hawaii has many Australian plants and insects, as well as rock wallabies and lizards. (Native Hawaiians call our silky oak 'he-oka', 'oka-kilika' and 'ha'iki'.) Florida has trees, budgerigars, insects and a mildew. Even Mother England has an Australian contingent. New Zealand, the most Australianised foreign land of all, is dominated by its larger neighbour ecologically as much as economically. Australian invaders have reached some of the remotest places on earth. Redback spiders can be found on Tristan da Cunha, peaceful doves on St Helena and sweet pittosporum (*Pittosporum undulatum*), a rainforest tree, in the Azores – three of the most isolated island groups in the mid-Atlantic Ocean.

Our plants do best, unsurprisingly, in soils and climates that match Australia's. The Cape Province of South Africa has a mild Mediterranean climate and very poor soils, just like southern

Australia, and southern Florida is warm and humid like south-east Queensland. Plants have migrated freely in both directions. Groundsel bush (*Baccharis halimifolia*), native to Florida, has invaded tea-tree swamps in Queensland, and many plants from South Africa, including bitou bush, boxthorn, watsonia, spiny emex and soursob, are now weedy in Australia. In both countries I have seen wattles and bitou bush competing side by side.

Some of our plants have done great harm overseas. Australian trees and shrubs threaten the very survival of the Everglades in Florida and the 'fynbos' heathlands of South Africa. Cape Town is losing its water supply to greedy soil-drying wattles (and pines) and in Florida great paperbark-fuelled fires now roar across the land. Our animals are sometimes very harmful too. The brushtail possum is New Zealand's worst pest, and one of our burrowing crustaceans is eroding San Francisco Bay. In Europe, Australian beetles chew their way through carpets when given half a chance. In 1979 Ross River fever, a virus associated with marsupials, swept through Fiji, New Caledonia, American Samoa and the Cook Islands, infecting hundreds of thousands of people, including almost half the Fijian population. Tobacco blue mould (*Peronospora tabacina*), a fungal infection of native Australian tobaccos, now blights tobacco crops all over the world. In 1979 it destroyed US$250 million of tobacco in the United States and Canada.

Some of our species travelled abroad so long ago they were first given scientific names by men who assumed they were native to Europe or America. They include two crustaceans (*Iais californica* and *Arcitalitrus dorrieni*), three fungi (*Setchelliogaster tenuipes, Desomyces albus* and *Cryptosporiopsis eucalypti*) and two plants (*Chenopodium auricomiforme* and *Lepidium peregrinum*), variously 'discovered' in Britain, Switzerland, India and the United States.

Our first exotic export, so far as we know, was a louse. Long ago, some Indonesian sailors visited Australia and took home a dingo. Australian lice travelled back with this dog, making new homes in Asia, sucking the blood of village curs. Over time they spread west into Africa and east into the Americas, broadening their tastes to include jackals and coyotes. This louse (*Heterodoxus spiniger*) is now a great curiosity – the first Australian animal spread abroad by people. We know it originated here because every other member of its family (Boopiidae) lives in Australia and New Guinea and feeds only on marsupial blood. It probably evolved from a sister louse (*H. longitarsus*) found on kangaroos.

Australia's representation abroad would be much greater if our life forms had more chances to invade. Marsupials simply haven't had the same travel opportunities as rabbits and cats. Otherwise brushtail possums would probably be pests in Hawaii and California just as they are in New Zealand. Our canefield rat (*Rattus sordidus*), a native rodent despite its name, would thrive in the cane fields of Fiji and the Caribbean. As it stands, Australian wallabies have done remarkably well in reaching England, Scotland, New Zealand and Hawaii.

Most of our expatriate pests come from those groups that travel the most. Australian shrubs and trees make popular ornamental and timber trees abroad; many have become weeds. Australian saltbushes, grown for fodder, have won footholds in southern Africa, California and Hawaii. Australian plants are valued because they grow fast in poor soils (an adaptation to the most infertile continent on earth), and this predisposes them to becoming weeds. Some are now remarkably widespread. Blackwood (*Acacia melanoxylon*), one of our wattles, has gone feral on every continent except Antarctica and even has a toehold in south-west England. Sweet pittosporum is weedy in New Zealand, South Africa, Jamaica, Bermuda, Hawaii and the Azores.

Export of ornamental plants has allowed associated fungi and insect pests to travel abroad too. Our sap-sucking bugs have done especially well, showing a great facility for switching from Australian natives to orange and lemon trees. As many as 500 different fungi grow with eucalypts overseas, though not all are native to Australia. In southern China, Australian earthballs (*Scleroderma* species), found sprouting in eucalypt plantations, are sold in food markets.

Our marine exports include a barnacle, two crustaceans and some seaweeds, all found today in Europe, Africa or North America. On a more positive note, Australian insects (and a fungus) have proved their mettle in biological control programmes by reining in pesky Australian insects or plants in many countries. The following chapters will look at the story of Australians abroad in more depth, beginning in Britain, turning then to New Zealand, visiting the South Pacific and Florida and winding up in the capital of Ethiopia.

19

Colonial Revenge

British Wallabies and Budgerigars

> 'And in a wonderful way to the shingle-beds of the Tweed have the ends of the earth sent their colonists, emphasising the federation of the nations and the pride of the motherland.'
>
> **George Claridge on the spread of Australian plants to Britain, 1919**

In the English Midlands, at the southern end of the Pennine Way between the sprawling metropolises of Manchester and Sheffield, are the wild moorlands of the Peak District, where, if you are lucky, you may catch a glimpse of a wild wallaby. The wallabies have colonised the woods and moors, where they browse the heather, bracken, bilberry and grass. They are red-necks, descended from five animals freed from a menagerie during World War II. By 1962 they had multiplied to fifty or more, but a hard winter reduced their numbers again to about ten. However, the Peak Park Planning Board is so wallaby-proud it has bought up 400 hectares of their prime habitat, so their future looks hopeful.

Wallabies also run wild around Loch Lomond in Scotland, and in years past, small mobs lived in Sussex, on the Channel Isles and in Germany. Scotland's wallabies date back to four red-necks freed on an island in Loch Lomond by the Countess of Aran in 1975. There are twenty-six or more animals now, nibbling on heather, blaeberry and grass beneath stands of oak, birch and Scots pine. Both groups came from Whipsnade Zoo, from their free-ranging colony of over 400 animals. An older wallaby colony on Herm in the Channel Isles was butchered during World War I to feed British troops (one of our contributions to the Great War) and a fourth group of red-necks lived in Sussex from 1908 to 1972 but subsequently died out.

Although there are only a few dozen completely wild wallabies in Britain, the English take them very seriously. Their droppings are probed and their habits written up in scientific journals. The weighty *Handbook of British Mammals* devotes five pages to these marsupials, including pictures of footprints, a skull and a chilly-looking wallaby bounding through snow. No red-necks in Australia are as well studied as these two colonies. Have Britain's biologists nothing better to do?

These marsupials are a curiosity, nothing more, but Britain has other immigrants from down-under that are, like some of those loud-mouthed Aussies in Earl's Court, very annoying pests. Australia has received so many pests from the Motherland, perhaps it is only fair that we should have donated a few in return. A rampant water plant, a prolific barnacle and a carpet-chomping beetle all cause special concern.

The plant really has them worried. It's swamp stonecrop (*Crassula helmsii*), insignificant in Australia, a slightly fleshy herb with pink flowers found sprouting beside ponds and sluggish streams. Marketed as an ornamental in Britain, it has turned into a determined intruder found growing in pools up to three metres deep or floating as mats of weed. By blanketing ponds it evicts native herbs, posing a threat to rare plants. Very hard to eradicate, it sprouts from pieces of stem carried by birds or on boots, boats or nets. Biologists suspect this weed is aggressively displacing native plants.

A second Aussie pest is thriving in the sea. During World War II a tiny acorn barnacle, *Elminius modestus*, hitchhiked to England on a ship. First seen in Chichester Harbour in 1945, it has since colonised estuaries all over the south-east from Norfolk to Dorset, including the Thames, where it is now a dominant animal, living in densities of up to 200 per square inch. Charles Elton wrote in 1958, 'This barnacle is certainly able to get about on the hulls of ships, for it fouls them quite intensely, and was taken early on from a vessel going between Holland and England ... This is a tough and dominant species, able to occupy the shore in face of competition from other kinds of barnacles, though it does not replace them except in certain zones.' A great pest on oyster and mussel beds in southern England, it has gone on to colonise estuaries in France, Holland, Germany, Denmark and South Africa.

The 'Australian carpet beetle' (*Anthrenocerus australis*), having traded a diet of decaying kangaroos for floor coverings, has found its true purpose in life is to chew its way through carpets in England and Holland (and New Zealand).

Most other invaders are plants. Whenever wool is imported, the fleeces are cleaned of burrs and other seeds, which often find their way into English soil. Over the years more than twenty Australian plants – mainly saltbushes (*Chenopodium, Atriplex*), daisies (*Calotis, Cotula*), storksbills (*Erodium crinitum, E. cygnorum*) and grasses – have become established by this means. They include swamp wallaby grass (*Amphibromus neesii*), listed in Stace's *New Flora of the British Isles* as a 'rather characteristic wool-alien on tips, waste ground and fields'.

Wool aliens exerted a great fascination over English botanists early this century. Downstream from the mill town of Galashiels in Scotland, where fleeces were washed in troughs, Ida Hayward, a fellow of the Linnean Society, scoured the banks of the Tweed River from 1908 to 1917 seeking strange plants. Her best finds came from a delta of shingle where the Tweed met the Gala, or further downstream wherever eddying water cast up its load. Hayward collected an extraordinary number of exotics – 348 in all, coming from all parts of the globe. A colleague, George Druce of Oxford, wrote up her findings in a book, *The Adventive Flora of Tweedside* (1919), which sounds excruciatingly dull but has fascinating implications. From this source we learn that Australian daisies (*Helipterum floribundum*), billy buttons (*Chrysocephalum apiculatum*), saltbushes and grasses once grew along the shingly banks of a chilly Scottish river. Hayward found about forty Australian species in all, some of them sprouting in great numbers. In 1908 the tiny daisy called Bogan flea (*Calotis hispidula*) was thriving, 'looking at a distance like moss, covering as it did completely with its rich olive-green the cold gray colour of the shingle in which it grew'. Swamp dock (*Rumex brownii*) and native crowfoot (*Erodium cygnorum*) sprouted up to nine miles downstream of the mills.

One of the plants, a peppercress, could not be identified, so Druce sent it to Swiss botanist Dr Albert Thellung, who pronounced it a new species, naming it *Lepidium peregrinum* ('peppercress abroad'), guessing correctly that it came from Australia. It became extinct back home – a silent victim of cattle and rabbit grazing. It was collected in Australia only three times, in 1802, 1834 and 1884, so the last time the species was seen alive was in Scotland in 1910! Hayward and Druce feature a dried sprig of the last-known plant in their book. A strange fate to befall a small and insignificant Australian plant.

The story of the wool aliens is certainly quirky. In Middlesex and Bedfordshire during the 1950s an Australian rush (*Juncus pallidus*) was found hybridising with a native species (*J. effusus*), though the hybrids eventually died out. Australian plants sprouted near wool mills in continental Europe too – more than 170 species according to a survey published in 1949, including the saltbush (*Chenopodium auricomiforme*) first discovered and named from the Swiss town of Derendingen in 1914.

Druce was amazed that antipodean plants could set down roots so confidently at the other end of the earth.

> Who could have predicted a century ago [he wrote] that the raw material for our clothing would in great part be obtained from a country of which only a strip of coast line was then explored, but which is to-day the greatest wool-producing country in the world, whose sons have marched to the aid of the mother country and laid down their lives with men of veldt and corn-land, of rubber forest and paddy-field for their companions. And in a wonderful way to the shingle-beds of the Tweed have the ends of the earth sent their colonists, emphasising the federation of the nations and the pride of the motherland. It is the epitome of centuries of the world's discovery, invention, exploration and conquest over the realm of nature.

A former botany lecturer of mine, Professor Trevor Clifford, visited Galashiels in 1965, but he found no trace of any Australian plants. The mills had long since closed and Scotland's winters had taken their toll. The wool aliens surviving today grow mainly in milder southern England alongside ornamental Australian plants, notably eucalypts, paper daisies (*Bracteantha bracteata*), pigface (*Carpobrotus glaucescens*) and soft treefern (*Dicksonia antarctica*). None of these plants grows vigorously enough to be a pest. The tree fern is of special interest because an Australian crustacean, the amphipod *Arcitalitrus dorrieni*, lives among its fallen fronds. This tiny animal evidently travelled to England long ago on imported tree ferns and, like the peppercress, was first discovered and named in England. It probably spread from Treseder's nursery in Cornwall (which had a branch in Sydney) to the Scilly Isles, and thence to Ireland and Kew Gardens.

Other Australians resident in Britain include an earthworm in Scotland, a soil-dwelling flatworm (*Parakontikia ventrolineata*) found in a hothouse in Liverpool, which would have travelled in pot plants, a liverwort (*Chiloscyphus semiteres*) and groundsel rust (*Puccinia lagenophorae*). The rust appeared at Dungeness in 1961 as an infection on groundsel (*Senecio vulgaris*) and quickly spread across Europe, reaching the Scottish Highlands and Northern Ireland by 1963 and eventually spreading to Germany, Romania and Egypt. The earthworm (*Spenceriella minor*) was one of several taken from Mt Kosciuszko to Lephinmore in Scotland in 1975. Someone thought that the large Australian worms might do better than British worms at decomposing peat for land reclamation. They did not, but this worm lives on, slowly multiplying while doing nothing discernible to better the soil.

Were it not for England's cold snaps, two more of our animals might well be established there – the budgerigar and a frog. Over the years, English bird lovers have tried earnestly to acclimatise budgies, founding flocks in Essex, Berkshire, Hertfordshire, Huntingdonshire, Norfolk and Hampshire. A colony of budgerigars at Tresco Abbey (where the Australian amphipod was discovered) was thriving in the 1970s, when numbers peaked at a hundred or so. They were flocking with starlings, breeding in elms and sycamores, and nibbling seeds of winter grass, toad rush, sandwort and swinecress. But they disappeared after a severe winter when supplementary birdseed was no longer provided. It seems remarkable that an outback parrot could survive in England even for a few years. Florida proved more hospitable, and flocks of budgies live there today. Attempts were also made to acclimatise brush turkeys and crested pigeons, with short-lived success.

Australian frogs also lived briefly in England. Eight brown tree frogs (*Litoria ewingii*) were flown from Melbourne in 1951 and freed in a garden at St Ives in Cornwall. They settled in for some years by a pond, bleating and breeding, but the savage winter of 1963–4 snuffed them out.

Those harsh British winters keep a lid on most visitors from Australia, ensuring they never multiply into genuine pests. However, if we travel now to New Zealand, where the climate is milder and the volume of immigrants much greater, we can find Australian species misbehaving very badly – as major invaders wreaking ecological havoc.

20

Ecologically Entwined

Australians in New Zealand

> 'To-day we see a land which, in comparatively recent historical time, was practically devoid of life, looked upon now as a sportsman's paradise, and visited yearly by many thousands of people to enjoy the resources which have been created by acclimatisation.'
>
> **Walter Kingsmill, Australian politician, on New Zealand, 1918**

Nowhere on earth was acclimatisation pursued with such zeal as in New Zealand. Nowhere else did men dare to conspire with the likes of weasels and leeches. To English eyes a century ago New Zealand was an empty canvas – barren of significant animal life, with no mammals (except rats and bats) and few birds. Keen to animate the empty woods and fields, the acclimatisers unleashed a Noah's ark of exotica: stoats, chipmunks, guinea pigs, llamas, moose, nightingales, titmice, jackdaws and English toads, to name a few. The Otago society even imported 200 medicinal leeches to release in Shag River; and a fish hatchery, coveting European lobsters and

British crabs, released millions of larvae into the sea; none survived.

To New Zealand's acclimatisers, Australia was a relatively bountiful land. While our colonists were wont to denigrate our topsy-turvy animals, New Zealanders wanted and got them. Among birds, they imported and freed emus, brush turkeys, Cape barren geese, nankeen night herons, crested pigeons, bronzewings, flock pigeons, wonga pigeons, squatter pigeons, budgerigars, shrike-thrushes, magpies, finches and owls. Their journals tell us they also imported (though they might not have released) bush-curlews, grey plovers, coot, black-breasted button-quail, bellbirds, beautiful firetails, even New Guinea cassowaries – perhaps with a view to replacing the extinct moas. Some birds were freed in their hundreds (400 noisy miners in 1879 alone), but problems of supply usually limited the releases to a few pairs. Acclimatisers in Auckland, with a hankering for welcome swallows, which are impossible to keep in captivity, imported a few eggs and slipped them into the nests of a sparrow and a chaffinch. They didn't hatch, but swallows flew across the Tasman anyway and had become a feature of Auckland's skies by the 1950s. A more typical introduction was made in 1923 when the Tourist and Health Resorts Department imported twelve superb blue wrens on SS *Ulimaroa*.

Most of these ventures failed, either because too few birds were released or because the habitat was wrong, but an unmistakable legacy survives. New Zealand today is home to more than 200 000 Australian black swans, breeding in huge colonies on lakes around the country; there are plenty of Australian magpies as well, along with a few kookaburras and brown quail. Eastern and crimson rosellas, rainbow lorikeets and sulphur-crested cockatoos were spawned by later, private releases or were aviary escapes. Most of these birds keep to farmland and parks, but I have seen eastern rosellas inside rainforest in the Waitakere Ranges outside Auckland.

Australian mammals – kangaroos, wallabies, potoroos, quolls (native cats), bandicoots and possums – were brought in too, and New Zealand now lives under the curse of millions of feral possums and wallabies. The indefatigable Sir George Grey did more than anyone to Australianise New Zealand. An explorer in Western Australia, governor of South Australia and Cape Town, and prime minister of New Zealand, this remarkable man bought an island – Kawau, north of Auckland – and during the 1860s stocked it lavishly with marsupials, which he liked to hunt. He freed wallaroos, wallabies (five kinds), possums, emus, kookaburras and magpies, and kept a menagerie featuring monkeys, zebras and a wildebeest. Kawau today is wallaby heaven, home to swamp wallabies, rock wallabies, tammars and parmas. Tammars were later released near Rotorua and rock wallabies on several offshore islands.

Of Grey's wallabies, the most remarkable by far was the dainty little parma. In Australia it lived deep inside the tall forests of New South Wales, but after 1889 it was not seen again, and by 1940 it was feared extinct (the New Zealand colony was unknown at the time). For Ellis Troughton, writing its epitaph in *Furred Animals of Australia* (1941), the story represented 'yet another tragic disappearance of an animal so plentiful during early settlement that the risk of extinction and need for preservation, alive and in museums, were not realized until too late'. So in 1966 when the parma rose from the dead, in New Zealand of all places, there was jubilation across the Tasman. A batch was sent home to found a wild colony and hundreds were sent to zoos worldwide. Then, the following year, parmas were rediscovered in thick forests only sixty kilometres from Sydney, near Gosford. It seems they had survived in Australia all along, although they remain rare. Tammar and brush-tailed rock wallabies have also become rare back home, and foxes are largely to blame for this. Fox-free New Zealand is an ideal sanctuary for Australian marsupials.

A fifth wallaby, the red-necked or Bennett's wallaby, lives in New Zealand's South Island. Three freed near Waimate in 1874 proliferated prodigiously, and by the 1960s half a million were hopping free, all apparently descended from this trio. They became horrendous pests, damaging fences, driving sheep from pastures, fouling sheep feed, browsing crops and damaging rainforest regrowth. After many years of culling their numbers were reduced to below 15 000, and each year the South Canterbury Wallaby Board culls a few thousand more. Rock wallabies and tammars are also serious pests in the North Island, around Rotorua and on islands to the east.

But much worse than any of these, the worst pest in New Zealand in fact, is one of Australia's favourite animals, the brushtail possum. More than 60 million of them infest New Zealand forests and farms, where they defoliate trees, raid orchards and prey on native birds, munching their way through 20 000 tonnes of vegetation each night.

Imported from New South Wales and Victoria as fur-bearers, more than 460 were released between 1837 and 1920, and to the dismay of fruitgrowers they thrived. When farmers began culling them vigorously, acclimatisers convinced the government, in 1911, to place them under the aegis of the Animal Protection Act. An inept study found that they were not damaging forests, its author, Professor H.B. Kirk, concluding: 'Opossums may, in my opinion, with advantage be liberated in all forest districts except where the forest is fringed by orchards or has plantations of imported trees in the neighbourhood.' The government soon decided otherwise and banned further releases, though they continued illicitly for years.

Possums now infest New Zealand forests at densities of ten to twelve per hectare, three times the numbers achieved in Australia. They devour more than a hundred kinds of native plant, wreaking havoc upon rainforest. Trees such as tutu, toro, titoki and five-finger soon disappear, to be replaced by plants such as tree ferns that possums don't like. Forest succession is changing forever. In Westland, hillsides defrocked of forest are sliding into the valleys. Those trees that survive the browsing suffer more from disease, insects, wind and salt spray. And wherever possums thin the canopy, exotic deer invade, compounding the destruction. Possums take the fruits eaten by birds and even the birds themselves and their eggs.

Some very rare bird species, under threat from rats, stoats and hedgehogs, must also contend with a predatory exotic marsupial.

Possums also damage plantations, orchards and gardens, and spread tuberculosis among cattle. Bovine TB is a serious threat to New Zealand's cattle, and the possum is its main carrier, triggering severe outbreaks. Millions of dollars are spent killing possums to control disease; millions of possums are poisoned, trapped and shot each year, but the scourge persists. Curiously, brushtails in Australia don't carry this disease, perhaps because densities are much lower. New Zealand is lucky that ringtail possums are not a pest there too; the Canterbury Acclimatisation Society imported a pair in 1867 but did not establish them in the wild. This possum is smaller and its fur worth less.

As well as marsupials and birds, New Zealand now has Australian lizards, insects, worms and many plants, and the New Zealand night is alive with the grunts, growls and whistles of Australian frogs. Golden bell frogs (*Litoria aurea*), green swamp frogs (*L. raniformis*) and brown tree frogs (*L. ewingii*), released by acclimatisers, are now common on farms. Indeed, they are the only frogs known to most New Zealanders, since native species are rare. Ironically, the first two of these frogs are now endangered in New South Wales.

Acclimatisers also released Australian prawns in New Zealand waters in the 1890s and imported freshwater shrimps from Victoria as trout food. Neither effort met with success. An Australian hydroid has colonised the sea. Other creatures have slipped in with cargo. A few redbacks, huntsmen and whitetailed spiders (*Lampona cylindrata*) now live around New Zealand homes. As well, hordes of Australian insects scuttle and buzz about New Zealand today, including cockroaches, wasps, bugs, moths, mosquitoes, beetles, ants, even the lice on wallabies. New species arrive regularly. Kiwi children are stung by Australian paperwasps (*Polistes humilis*), sheep are struck by Australian blowflies, orchards are attacked by the light brown apple moth (*Epiphyas postvittana*), eucalypts and wattles by a swag of Australian insects.

The Australianisation of New Zealand becomes evident as soon as you leave Auckland International Airport and see all those eucalypts on the way into the city. These trees are planted, but at least eighteen kinds of eucalypt grow wild here, along with ten or so wattles, three she-oaks and dozens of other Australian plants including shrubs, vines and tiny herbs such as the small-flowered buttercup (*Ranunculus sessiliflorus*), established in the wild by 1852. Most were brought over as street trees, hedges or garden plants, but others, such as swamp dock, probably slipped in with cargo. The plant contingent is large enough to portend enormous problems for the future. It includes such unlikely weeds as the macadamia nut (*Macadamia tetraphylla*), bangalow palm (*Archontophoenix cunninghamiana*), coast banksia (*Banksia integrifolia*) and Geraldton wax (*Chamaelaucium uncinatum*).

Many of these plants, the bangalow palm, bleeding heart (*Omalanthus populifolius*) and silky oak (*Grevillea robusta*) among them, keep to the north of the North Island, beyond the frost line, where they are so far neither common nor troublesome. But others are behaving very badly. The lillypilly (*Acmena smithii*), under the name 'monkey apple', is one of the worst offenders (its seeds spread about by native pigeons); regional councils have banned its sale and propagation. Port Jackson fig (*Ficus rubiginosa*) is penetrating all forest types as well as clearings and islands north of Auckland. Our hakeas are proving very invasive, just as they have in South Africa. In Abel Tasman National Park near Nelson, willow-leafed hakea (*Hakea salicifolia*) and silky hakea (*H. sericea*) form dense pure stands.

But trans-Tasman trade is a two-way street and New Zealand has inflicted sweet revenge upon Australia, slipping us some remarkable pests, including weeds, a savage bird, a snail and a swag of sea creatures. Tasmania and Victoria gave New Zealand most of her possums, and these states are now burdened with New Zealand weeds. Eleven species are listed in Victoria, and three of Tasmania's twenty-two worst environmental weeds – mirror plant or taupata (*Coprosma repens*), toetoe (*Cortaderia richardii*), and New Zealand flax (*Phormium tenax*) – are garden escapes from New Zealand. Mirror plant, a delightful shrub with shiny leaves and orange berries, is weedy in New South Wales too; it thrives on cliffs around Bondi Beach, where starlings spread the seeds. Toetoe, which looks like pampas grass, has been declared noxious in Tasmania because it is taking over stream banks in the World Heritage–listed south-west.

The weka, a cheeky New Zealand rail, was freed on Macquarie Island in 1867 to feed sealers. It became a nasty predator of ground-nesting seabirds, wiping out vast colonies of prions and petrels. It probably also helped exterminate the Macquarie Island parakeet and the local flightless rail. Numbers grew to almost 500 – an untenable population – before the Tasmanian national park rangers shot the lot in the 1990s. Wekas are now rare and dwindling in New Zealand.

In south-eastern Australia a tiny New Zealand snail, *Potamopyrgus antipodarum*, has replaced our snails in disturbed rivers. New Zealand also has a presence in our seas. During the 1920s crates of New Zealand oysters were shipped to Hobart and kept alive submerged in water, allowing encrusting creatures to escape. As a result, we gained crustaceans, shellfish and a starfish – an impressive contribution to our sea life. The New Zealand screwshell (*Maoricolpus roseus*), which now dominates the sea floor in parts of Tasmania, is probably the worst of these.

New Zealand has also passed on to us many of her own foreign imports. The European wasp, cabbage white butterfly and sirex wasp reached us from across the Tasman. Some day we are also likely to get the New Zealand flatworm, the bane of Britain's earthworms. As trans-Tasman trade continues to expand, our ecologies, like our economies, will grow more and more entwined.

21

Colouring the Landscape

Our Animals Abroad

> 'They change their clime, not their frame of mind, who rush across the sea.'
>
> **Horace, Roman poet**

Over the years, many countries have expressed a wish to see koalas roaming wild in their eucalypt plantations. Australia's wildlife laws are so strict that nothing has come of this. In any event, the FAO[1] has questioned whether koalas could withstand 'the carnivorous animals of Asiatic, African and American continents'. Australian eucalypts overseas are inhabited instead by some of our smaller, less charismatic animals, notably weevils, bugs, flies and wasps.

The story of Australian animals abroad is weird and worrying in turn. The cast of characters includes many unexpected travellers and some very destructive pests. Only a few can be introduced here; the previous chapters and appendix IV list others. They include six mammals, thirteen birds, four frogs, a lizard and hordes of invertebrates. Most of them are found on islands in the Pacific region.

The most bizarre of the many introductions took place in Tahiti in the 1930s, when thousands of tropical aviary birds were set free. 'To the average person the mention of a tropical island immediately brings forth a vision of rich green foliage, brilliant flowers, and exotic birds,' wrote latter-day acclimatiser Eastham Guild in *Aviculture Magazine* in 1938. 'Tahiti is no exception to this general impression as far as foliage and flowers are concerned, but for some reason there is practically no bird life, and according to reports made by early voyagers and by ornithologists visiting the island later there have never been many kinds of birds here.' So Guild decided to remedy the shortfall by inviting in birds from almost everywhere else. In all, he released 7000 birds of fifty-nine species – an astonishing feat. No one else in the world has ever tried to acclimatise so many animals all at once and for no good reason. Tahiti briefly became home to North American mockingbirds and siskins, South American tanagers, African, Asian and Australian finches, and much else besides. The Australian contingent included star finches, long-tailed finches, diamond firetails, zebra finches, Gouldians, plum-heads, double-bars, chestnut-breasted mannikins and red-browed finches, as well as silvereyes and black swans. The bird Guild released in the greatest numbers was apparently the Gouldian finch, an exceptionally beautiful bird and now an endangered species. 'I have liberated about 700 Gouldians,' Guild admitted, 'but there are only a few that stay about the place; they seem to prefer some locality which I have been unable to locate, but I feel that it will be a matter of only a short time before they will be seen in large quantities, as they do well here and are good breeders. I have just received a new shipment of 100 and am hoping that they may attract some of the birds at liberty before they are released.'

The place Guild could not locate was a bird graveyard. Tahiti was too humid or lacked the right seeds – whatever the reason, his 800 Gouldians soon died. In fact, most of Guild's birds perished. His actions ultimately represented the acme of cruelty and folly. From his vast Australian contingent only the red-browed finches, chestnut-breasted mannikins and silvereyes survived, flying about to this day in small flocks in the grassy valleys. The silvereyes, along with introduced bulbuls, are now abetting an ecological catastrophe by spreading the seeds of miconia, Tahiti's worst weed (see chapter 30). Without Guild's silvereyes this tree would not have become so destructive.

To colour the landscape, or to control insects, Australian birds were also liberated – not by acclimatisation groups but by individuals or governments operating in a later era – in Hawaii, Fiji, New Caledonia, the Solomons and Nauru. The motivation behind the Hawaiian releases was as silly as Guild's had been in Tahiti. 'We want to fill the Islands with birds of all desirable species that will survive here', explained a former official of the Territorial Board of Agriculture and Forestry in 1937. So the Hawaiian islands briefly became home to spinifex and crested pigeons, bar-shouldered and diamond doves, bronzewings, sulphur-crested cockatoos, stubble quail and painted button-quail. Willie wagtails were brought in during the 1920s to combat hornflies on cattle, and magpie-larks to control liver flukes, but like every other Australian bird they died away. Our birds were simply not destined to succeed in Hawaii.

Fiji proved slightly more accommodating. Here an odd selection of birds was released over the years, including brolgas, kookaburras, tawny frogmouths, brown quail, grey currawongs, magpies, magpie larks and diamond firetails. Magpies, freed around the turn of the century to cull pesky stick insects, prospered on Taveuni, and they will surely reach the other islands some day. Brown quail survive on the main islands but they are rare. The others died out.

In 1916 a private collector, Richard Trent, bought two brush-tailed rock wallabies and a joey for his menagerie on Oahu in Hawaii. When, the very next day, dogs tore apart the cage (a tent), the joey was killed but its parents escaped. No more was heard from them until 1921, when up to fifty wallabies were found living in the Kalihi Valley cliffs. By 1960 the population ranged over seven kilometres along ridges and valleys. But housing developments have hemmed them in. The present colony of fewer than a hundred has nowhere to go. They live on cliffs and ledges among exotic guavas and broad-leaved pepper trees and feed on grass and 'ulei. Plans to transplant them to other islands as game animals have come to nothing. *A Field Guide to the Mammals in Hawaii* (1982) notes, 'some people feel that the species in Hawaii should be protected'. 'The main problems of survival in Hawaii centre around man's activities, for example: death by passing automobiles when venturing onto the nearby highway, indiscriminate shooting, encroaching housing developments, and harassment and predation by domestic dogs. Certainly this fascinating mammal needs more study, because only a few observers have ever seen it in Hawaii, and precious little information has been published about wallabies in their native Australia.' This last is not true!

Australia is often blamed for supplying the brown tree snakes (*Boiga irregularis*) that exterminated the birds of Guam, in the Marianas, and that often attack human babies (they try to swallow them whole). But genetic tests show that Guam's serpents came from New Guinea. Australia has in fact contributed only one small reptile to the world (so far), the eastern grass skink, a familiar little lizard of suburban gardens in Australia that somehow reached New Zealand and Hawaii. When it appeared about the turn of the century in Hawaii, the biologist Loveridge was fooled into thinking he had discovered a new native species, which he named *Leiolopisma hawaiiensis* in 1939. Australian frogs have been taken to New Zealand, New Caledonia and Vanuatu, and dwarf treefrogs somehow got to Guam.

Six Australian crustaceans have spread abroad and all have interesting tales to tell. In the bays of northern California lives an Aussie isopod, *Sphaeroma quoyanum*, whose burrowing habits are remarkable. It bores into almost anything, including mud, clay, peat, sandstone, decaying wood, live sponges, styrofoam, asphalt, even concrete. In San Francisco Bay it riddles the styrofoam blocks that buoy up floating piers in marinas, causing piers to subside.

It is thought to be eroding vast stretches of foreshore. *Sphaeroma* is a filter-feeder, and living in its burrows and on its body is another, smaller isopod, *Iais californica*, which eats detritus. Although discovered in California in 1904 and given a strongly American name, we know that this creature, like *Sphaeroma*, comes from the Australian region and must have reached the American west coast as a hull traveller, perhaps during the gold rush days.

On solid ground in San Francisco lives yet another Australian crustacean, a little amphipod (*Arcitalitrus sylvaticus*). This close relative of the amphipod that long ago reached England (see chapter 19) crawls about in leaf litter under eucalypts. Our other crustacean expatriates are the redclaw (*Cherax quadricarinatus*), a freshwater crayfish that escaped from aquaculture in Western Samoa and the tiny barnacle I discussed earlier, found in Europe and South Africa.

Some of our most accomplished travellers are flatworms – slimy little creatures found under logs and rocks that travel well in pot plants. *Parakontikia ventrolineata*, native to the Bunya Mountains and nearby forests in southern Queensland, is now found in Melbourne and South Africa, in a Liverpool greenhouse in England, in Hawaii, California, Washington, Texas, New Zealand, Norfolk Island and who knows where else. Not bad for a gooey little rainforest worm.

Our redback spider has also done well, colonising New Zealand and Japan, inciting remarkable responses from the Japanese, who regard it with great dread. Given that many of our worst ballast pests came from Japan, our mischievous little redback seems to offer little enough in the way of redress.

As for our insects, dozens of species have travelled overseas, some as pests, others as biological control agents. And since they are more significant than all of our other animal exports put together, the rest of this chapter is devoted to them.

In the 1880s the Californian citrus industry was brought to its knees by a tiny Australian bug that somehow crossed the Pacific, probably on ornamental plants. Cottony cushion scale (*Icerya purchasi*) attacked oranges so relentlessly that farmers began abandoning them as a crop. They implored their government to help, and in 1887 US entomology chief Charles Riley told growers that 'it would be very desirable to introduce from Australia such parasites as serve to keep this fluted scale in check in its native land'. Biocontrol was then unproven and no money was available for this kind of work, but Melbourne was hosting its International Exposition the following year and the canny Riley sent an entomologist, Albert Koebele, ostensibly to assist the exposition but really to bring back scale enemies.

Koebele found a ladybird called vedalia (*Rodolia cardinalis*) that saved the day. One hundred and twenty-nine were sent to California and housed under a tent on a scale-infested orange tree. Their progeny soon stripped away the scales. The tent was opened up and the conquest of *Icerya* was under way.

J. Dobbins, one of the first farmers to receive the ladybirds, was filled with delight:

> The vedalia has multiplied in numbers and spread so rapidly that every one of my 3,200 orchard trees is literally swarming with them. All of my ornamental trees, shrubs and vines which were infested with white scale, are practically cleansed by this wonderful parasite ... People are coming here daily, and by placing infested branches upon the ground beneath my trees for two hours, can secure colonies of thousands of the vedalia, which are there in countless numbers seeking food. Over 50,000 have been taken away to other orchards during the past week, and there are millions still remaining, and I have distributed a total of 63,000 since June 1.

The citrus industry was saved. In a year, orange shipments from Los Angeles County nearly tripled. Koebele and his wife were rewarded by grateful growers with a gold watch and diamond earrings. Vedalia was hailed as a miracle, the first outstanding success story of biological control, the victor in a battle fought between two Australian insects on American soil. In the long history of biological control only one other success can rival it – the conquest of prickly pear by cactoblastis, a battle fought by two Americans on Australian soil (see chapter 37).

But the scale would strike again, in 1946, after ungrateful farmers became DDT-dependent and nearly exterminated their erstwhile ally, the ladybird. By this time cottony cushion scale had spread round the world. Australian ladybirds today do duty in lands as widely dispersed as Egypt, India, Japan, South Africa, Israel, Italy, Portugal, Hawaii, Cuba and Brazil.

Other Australian insects have become pests overseas and have also required biological control. Crops elsewhere are blighted by our sugarcane planthopper (*Perkinsiella saccharicida*) (Hawaii), light brown apple moth (New Zealand), citrophilous mealybug (*Pseudococcus calceolariae*) (California) and green planthopper (*Siphanta acuta*) (Hawaii again), a pest of coffee, mangoes and sugar cane. Like cottony cushion scale, these insects have thrived on the world stage by performing two feats – juggling food plants, switching from Australian natives to foreign crops; and hitching rides overseas, in some cases travelling on ornamental Australian plants. To counter them, affected countries have imported our beetles, wasps, midges and egg-sucking bugs. Some of these insects have proved decisive, quelling pests of sugar cane, oranges and eucalypts and, more recently, controlling weedy Australian plants.

Some of their stories are remarkable. Early this century Hawaiian ferns came under assault from a most unexpected quarter, an Australian beetle. In 1900 the large fern weevil *Syagrius fulvitarsus* was noted lurking among maidenhairs in hothouses, and it soon spread into Hawaiian rainforests where it devastated native ferns, including one species that seemed doomed to extinction. But when the weevil was identified as Australian, a parasitic wasp (*Ischioganus syagrii*) was dispatched from New South Wales and Hawaii's ferns were saved.

Most of the earth has been stripped of its mantle of forest, and quick-growing eucalypts now feed much of the world's hunger for wood. Inevitably, Australian insects have travelled abroad to exploit this burgeoning resource. The eucalyptus weevil (*Gonipterus scutellatus*) and the common eucalypt longicorn (*Phoracantha semipunctata*) are proving the most destructive of the twenty-odd species. The very viability of many plantations was in doubt until a parasitic Australian wasp (*Patasson nitens*) was imported to help subdue the weevil. The longicorn has not been controlled and is now one of Australia's most widespread animals, found all over Africa, southern Europe, the Middle East, South America and California.

Australian birds and mammals, then, haven't made much of a mark on the world – except in New Zealand – but our invertebrates certainly have. Our insects are very important in some countries, and so too is the roving redback, the bay-eroding *Sphaeroma* and the harbour-encrusting barnacle *Elminius*. Biologist Frank Wilson was right to conclude that, 'Although Australia has played but a small part in the provision of useful plants for cultivation, it has been of importance as a source of pests and of beneficial insects for their control'.

1. The Food and Agriculture Organization of the United Nations.

22

Inheriting a Degraded World

Exporting Our Flora

> 'These species have polluted our vegetation and our landscapes. We may draw a parallel to industrial air pollution . . .'
>
> **South African botanists A.V. Hall and C.H. Boucher, maligning Australian wattles, 1977**

Deep in Hawaii's rainforests, concealed by mist and gloom, a new ecological menace has everyone worried because its identity seems so unlikely – it is a tree fern (*Cyathea cooperi*) from Australia. In Haleakala National Park in 1987 rangers stumbled upon thousands of these intruders, some of them four metres tall, infesting a remote mountain valley. On Kaui island hundreds of tree ferns sprout along roads sliced through the mountains, and on Oahu they have stolen into Manoa Valley, invading landslip scars. Hawaii is famed for her tree ferns, but her native species are too frail for gardens and parks so this Australian impostor, hardy and fast growing, is planted instead. Ironically it has become a symbol of native rainforest. Winds carry its spores high into the mountains. The Haleakala tree ferns appeared twelve kilometres from the nearest nurseries down on the coast. On Kaui, hundreds of young plants were apparently spawned by just five ferns planted in the 1970s.

This fern is not integrating well into the rainforest. Epiphytes don't like its trunk and smaller plants don't grow well below it. In 1992 biologists declared it to be 'an invasive, disruptive species capable of radically modifying its habitat'. They want it declared noxious and removed from sale. Now they are wondering about other Australian tree ferns grown in nurseries, especially *C. woollsiana.*

Surprisingly – or perhaps it's not surprising at all – this tree fern has popped up elsewhere as well – in South Africa, New Zealand, the Azores and even on Mauritius, where it is forging into heaths as well as rainforests, elbowing out two Mauritian tree ferns. At home it is misbehaving too, having spread, via gardens, from its native haunts in Queensland and New South Wales into forests south of Perth, where it sprouts along streams.

This troublesome tree fern is part of a remarkable phenomenon – the explosive spread of Australian shrubs and trees across the world stage. Doubt it not, our plants are now major invaders of the globe's warmer zones. Charles Darwin in *The Origin of Species* noted the propensity of our wattles to sow themselves and become naturalised in the Nilgiri Mountains of India. Now wherever you look – in southern Europe, Africa, the Middle East, southern Asia, the Americas and on islands everywhere – wattles and other Australian plants can be seen running amok, proliferating in the wild. Only where winters are harsh are Australian invaders poorly represented.

You might not expect our plants to fare well in the United States, but a recent book, *Invasive Plants: Weeds of the Global Garden*, features six of them – a quarter of the total – in its gallery of America's weediest trees. Our beloved blue gum (*Eucalyptus globulus*) is there, along with the umbrella tree (*Schefflera actinophylla*) and earpod wattle (*Acacia auriculiformis*). In South Africa, 40 per cent of environmental weeds are Australian, the figure climbing to 59 per cent for shrubs and trees in the winter rainfall zone - an astonishing statistic. I had planned to draw up a list of all these green exports but abandoned the task upon realising how many there are. Hawaii, for instance, doesn't have just tree ferns; it has eucalypts, wattles, casuarinas, tea-trees, brush box, umbrella trees, the north Queensland rainforest tree *Flindersia brayleyana* and even the staghorn fern (*Platycerium superbum*). I have seen silky oaks (*Grevillea robusta*) sprouting all over its hills. New Zealand's contingent is much, much larger, and includes such obscure plants as wonga vine (*Pandorea pandorana*), devil's twine (*Cassytha pubescens*) and the 'buttercup brush cherry' (*Syzygium paniculatum*).

In two places on earth - South Africa and Florida - Australian plants rate as the worst of all weeds. Around Cape Town, three wattles (*Acacia saligna, A. longifolia* and *A. cyclops*) and three hakeas (*Hakea sericea, H. drupacea* and *H. gibbosa*), along with three foreign pines, are the worst offenders. In vast battalions they march over hills and across plains, destroying the Cape's fynbos heathlands, probably the most diverse plant habitat on earth. I have seen them in dense thickets on the slopes around Cape Town. Many rare plants are threatened as a result. South Africans also do battle with eucalypts, a tea-tree (*Leptospermum laevigatum*) and even a banksia (*Banksia ericifolia*).

Florida's plight is as bad, if not worse. The south of the state is a festering stew of foreign pests, the exotic ingredients including South American crocodiles, Australian budgerigars, Chinese walking catfish, Cuban frogs and hundreds of alien plants. This rich concoction has brewed exuberantly in the steamy subtropical climate, helped along by America's lax controls over imports and by Miami's role as a major nursery centre and port. Of 456 million plants imported into the United States in 1993 nearly 80 per cent came through Miami. Two of the three worst weeds in this pest paradise are Australian trees.

The paperbark tea-tree (*Melaleuca quinquenervia*) is probably

Florida's worst scourge. Seen through American eyes this is a tree from hell. There are literally billions of them, occupying half a million hectares or more, and claiming another six hectares every day. A single tree can drop 20 million seeds, which are so hardy they can survive six months under water. Plants three years old can set seed, and trees can flower five times a year. Their regenerative powers are amazing, melaleuca fenceposts often sprouting into trees. They swallow up to five times the water of the sawgrass prairies they replace, thus helping drain the Everglades' water supply, already overexploited by cities and farms. They cause wildfires and raise asthma levels. According to the *Melaleuca Management Plan for Florida* (1994), 'the uncontrollable expansion of melaleuca constitutes one of the most serious ecological threats to the biological integrity of south Florida's natural systems'. They will turn the entire Everglades into a dark forest if they are not stopped. Says weed expert Dan Thayer: 'the only areas they won't colonise are those areas covered in concrete'.

Florida's other nightmarish invader is the 'Australian pine', an astonishing hybrid of the river oak (*Casuarina cunninghamiana*) and the beach she-oak (*C. equisetifolia*), growing to the height of the former and with the salt tolerance of the latter. This hellish hybrid crowds out seashores and dunes, forming thickets so dark and shedding leaf litter so toxic that nothing can grow below. By taking over mangroves and dunes it is obliterating nest sites of rare turtles and crocodiles. These 'pines' (not true conifers) often grow alongside paperbarks and Brazilian peppertrees (*Schinus terebinthifolia*), Florida's other top-ranking pest.

Like most of the world's worst weeds, these trees were originally planted for a purpose. Paperbarks were first sown in Florida in 1906 as swamp-tolerant ornamentals. Seeds were even broadcast from planes to help drain the Everglades. During the 1930s the US Army planted thousands around Lake Okeechobee for stabilisation; 65 million grow there today. As recently as 1969 they were considered 'one of Florida's very best landscape trees'. The 'pines' were valued for shade and timber and bank stability. Most of southern Florida was once a treeless plain of swampy sawgrass, and Australian trees have proved ideal colonisers of the soggy or saline soil.

Knowing Florida's plight to be extreme, in 1997 I decided to go and see for myself. I drove past dark forests of paperbarks much vaster than any growing in Australia, lining highways for tens of miles. American paperbarks grow straighter and more crowded than ours, with up to 37 000 trees per hectare, forming forests so gloomy that nothing grows inside. It was amazing to see a familiar native tree playing the role of supreme villain. These were the worst weed invasions I had ever seen.

Near West Palm Beach some American weed experts took me on a tour. They were keen to show me around; they wanted to see me squirm. We strode into a thicket where a bobcat had left claw marks up a tree and epiphytes hung from the upper branches. I peeled back a strip of bark and a native anole lizard scampered away. Around the sun-lit edges, scores of melaleuca seedlings were sprouting – the next wave of invasion. We saw Australian wattles (*Acacia auriculiformis*), 'pines', umbrella trees and tuckeroo (*Cupaniopsis anacardioides*), the latest Aussie invader.

My American colleagues are not confident of ever conquering melaleuca. For that they need nothing short of a biocontrol miracle. A leaf weevil (*Oxyops vitiosa*) was released shortly after my visit; a sawfly and a tiny psyllid bug will probably be next. The paperbark is a superior weed and eight or more agents will probably be needed. Will they work? Biological control can claim few successes against trees. What worries me most is that the paperbark behaves so much like a weed in its native Australia, totally dominating the swamps where it grows.

Most counties in southern Florida now control paperbarks and 'pines', but these trees still have their fans. Retirees from the north find the sawgrass plains dreary and they like our trees. When councils come to cull, 'Save the Pines' petitions go out. 'Many people wish to enjoy picnics and sea views from shady places', complained one resident in the *Miami Herald*. They 'aren't seeking skin cancer ... The parks people can never replace the towering and valuable beauty of the Australian pines.' Paperbark apologists claim that squirrels, blue jays and woodpeckers like these trees, and certainly paperbark forests are not the ecological deserts most biologists claim. On the once-treeless prairies they offer perches to vultures and kites, nectar to small birds and shelter to mammals, reptiles and migrating birds. I saw ample evidence of animals among the paperbarks – birds taking the nectar, road-killed raccoons and opossums, and those intriguing bobcat marks. According to one study, paperbarks are used by forty-six different birds, nine mammals, thirty-four reptiles and amphibians, and twenty-seven fishes. But areas with moderate, rather than dense, infestations provide the best habitat.

After our tour I showed my colleagues slides of paperbark forests, taken in Maroochy shire near Brisbane, where I was conducting fieldwork on another fast-disappearing habitat. I returned home to find that bulldozers had levelled much of the largest stand of paperbarks remaining, one I had listed as a high priority for conservation. I trudged across the bulldozed fields, shaking my head in dismay, thinking back to my American friends, and how they would love to see a scene like this acted out in Florida. What they want to get rid of, we want to save but can't. Paperbarks are the fastest-disappearing habitat in south-east Queensland.

That is one of the ironies of this story. There is another. My American friends grilled me about pond apple (*Annona glabra*), one of the very few trees that is native to Everglade swamps, though it is scarce nowadays. Is it true, they asked, that pond apple is now a weed in Australia's paperbark swamps? Yes, indeed. Pond apple is rated the greatest threat to the wetlands of the Wet Tropics of north Queensland, although, like tuckeroo in Florida, it is only in the early stages of spread. We have less chance of controlling it than Americans have of controlling melaleuca, because farmers use it as custard apple rootstock and biological control has been ruled out.

So much more could be said about Australian plants overseas. I might have talked about the rampant eucalypts I saw in Zimbabwe, for instance, or the worried weed experts I met in South Africa. Or I could have chosen the tiny, isolated island of Réunion, west of Mauritius, where she-oaks crowd the beaches and infestations of black wattle (*Acacia mearnsii*) cover 5000 hectares. (This plant, incidentally, has also colonised France, Spain, Portugal, Italy, Yugoslavia, Romania, South Africa, Tanzania and California.)

Australian trees and shrubs prove ideal for reclaiming degraded land. Because they evolved on the most infertile and eroded continent, they do especially well in regions damaged by centuries of agriculture, where the topsoil has washed away and rainfall is failing. Eucalypts, wattles, casuarinas and silky oak grow faster than the native alternatives, and our aid agencies promote them with scant regard for the weed problems they cause. Ethiopia was saved from economic ruin by eucalypts, which furnished it with firewood and timber, and the idea was mooted of renaming the capital 'Eucalyptopolis'.

This, then, is the key to their success overseas: Australia, barren and eroded, is the landscape of the future, with plants superbly adapted to inherit a degraded world. Our trees and shrubs thus have a special role to play in the progressive globalisation of world ecology. For just as weeds from every country are flooding into Australia, so too are our plants making inroads in every corner of the globe. We are not just the victims, but also the perpetrators, of a vast ecological unravelling affecting the whole biosphere.

23

It's Civil War

Natives can be Pests too

> 'Governments don't want to deal with it because there are no votes in killing koalas.'
>
> **Roger Martin, koala expert, saying the unsayable, 1998**

The nations suffering the most from Australian pest invaders, as we have seen, are New Zealand, the United States and South Africa. But there is one other country where the problems are formidable, where Australian animals and plants are radically rewiring the ecology, and that is Australia itself.

In earlier chapters I have mentioned the Australian rainforest tree sweet pittosporum, a truculent weed in Africa, Europe, New Zealand and the Caribbean. As it happens, this tree is a major weed here too. In most states it is now marching into forests where it never grew before, converting sunny woodlands into gloomy groves. In Jamaica it shades out dozens of rare plants, and in South Australia it is threatening a rare orchid in exactly the same way.

Coast tea-tree, umbrella tree (one of my pet hates), cootamundra wattle and silky hakea all feature prominently as weeds in Australia as well as in foreign lands. I have torn up seedling umbrella trees in both Florida and Queensland. I have driven past tea-tree invaded dunes in Western Australia and South Africa. I have seen Brisbane golden wattle creeping along roadsides in Queensland, Western Australia and Zimbabwe. Many weeds begin their careers at home, it seems. There are animals that misbehave at home and abroad as well – the redback spider, Queensland fruit fly, cottony cushion scale and brushtail possum, among others. This should not surprise us. Overseas, rabbits, rats, starlings and thistles are pests in their native lands.

A plant such as sweet pittosporum gets a chance to misbehave in Australia when people grow it in gardens far from its original home but in regions with a matching climate. A native of Australia's south-east, it has gone feral in Western Australia, South Australia and western Victoria. In some places it benefits from soil enrichment and fire control. Around Sydney and Melbourne it has crept out of gullies into dry woodlands, converting them into dark forests. Introduced blackbirds help spread its seeds.

Within Australia hundreds of native plants and animals behave at times as pests. Many of them are invaders overseas too, but the majority are pests only within Australia. They include native rats that invade cane fields, locusts and scale insects on crops, and plenty of displaced animals and plants.

Australians have long had a penchant for mixing up their fauna. Acclimatisers took kookaburras to Western Australia and Tasmania and stocked Murray cod, trout cod and blackfish in rivers where they didn't belong. Government fisheries continue this ill-considered work to this day. In Queensland, for instance, alien native fish are put into rivers that carry threatened species, with no thought for the overall ecology. Fishermen and farmers are redistributing yabbies and redclaw crayfish. Native birds have won freedom from aviaries far from home and we now have feral populations of native parrots, silvereyes, emus, owls and brush turkeys. Western Australia has exotic kookaburras, cockatoos, corellas, lorikeets and finches. In inner-city Perth, the rainbow lorikeet is a dominant bird, one that behaves for all the world like an arrogant new arrival.

Scores of native plants have followed sweet pittosporum by escaping from gardens into forests that had never sheltered them before. Wattles from eastern Australia now grow in the west, and vice versa. In Victoria more than a hundred Australian plants are listed as bushland weeds and the number keeps growing. Most of them are not troublesome, but some are big-time pests, including sweet pittosporum, bluebell creeper (*Sollya heterophylla*) and three errant wattles.

Other plants and animals have colonised parts of Australia accidentally after travelling with cargo. They include lizards, frogs, wasps, cicadas, earthworms, flatworms, slaters, shellfish and a vast assortment of plants. Outback saltbushes and grasses now sprout around stockyards and railway lines near ports, their seeds having hitchhiked on stock trains – in fleeces, fodder and dung. Palmdarts (small butterflies, *Cephrenes augiades* and *C. trichopepla*) have come south and west with cultivated palms, redback spiders have migrated east in cargo, and outback mosses now grow in at least one coastal national park (Wilsons Promontory in Victoria), their spores presumably transported there by campers.

Some animals and plants have become pests without anyone taking them anywhere. Kangaroos, benefiting from the baiting of dingoes and creation of pastures, are now troublesome on farms and in national parks, where they overgraze and threaten rare plants. A government expert in South Australia told me that virtually every national park in that state has a kangaroo problem. In Victoria's Hattah-Kulkyne National Park, and on Tasmania's Maria Island, hundreds of kangaroos have been destroyed to permit regeneration. In the absence of predators, koalas in South Australia and Victoria are multiplying too, and they are killing off so many gum trees that many biologists, conservationists and land managers want them culled – a controversial idea that nonetheless deserves support.

Australia is home to some very aggressive birds, and because we have thinned out forests or planted fruit-bearing shrubs, noisy miners, bellbirds, currawongs, butcherbirds and ravens are thriving and now pose a threat to other birds. I know of seven endangered birds that face threats from other native birds. Predatory currawongs, for example, were pushing Gould's petrel towards extinction in New South Wales until national park rangers intervened.

These problems have their parallels overseas. In Africa, elephants are destroying most of the national parks set up to save them, and North America is blighted by such native pests as parasitic cowbirds, browsing snow geese, translocated trout and catfish, urban raccoons, lake-hopping lampreys, misplaced bullfrogs and hordes of native weeds.

Australian examples are endless and include crown-of-thorn plagues on the Great Barrier Reef, woody weeds and prickly salt-bushes in the outback, tobacco blue mould, caterpillars on cotton, mosquito plagues, pandanus dieback, tree cankers – even the native weeds infesting my garden. When I began writing this book I planned to devote several chapters to these issues. But as I read and travelled more, and talked to more biologists, the scale of the problems kept growing and I realised it would be impossible to do them justice in this book. So instead I am writing a sequel, *The New Nature*, to explore, among other things, the problems posed by 'weedy' natives.

Native species are becoming pests for two reasons. Many of them we have translocated (accidentally or deliberately), providing them with new habitats far from home. We are stretching definitions when we continue to call these invaders 'native'. Australia is one of the largest countries in the world. Were it divided into smaller nations, as Europe is, wattles from eastern Australia growing around Perth would not be considered native; they would be recognised as exotic weeds. In fact, many biologists now adopt this way of thinking. The book *Western Weeds: A guide to the weeds of Western Australia* (1997) features dozens of eastern Australian plants – eucalypts, tea-trees, hibiscuses, kangaroo apples, tree ferns, grasses and, of course, sweet pittosporum. All of these plants invade forests and displace native species in the west.

We have changed the landscape and ecology so profoundly that some species have benefited enormously *in situ*, to the detriment of others. The proliferation of kangaroos, aggressive birds and native woody weeds provide examples of this. Nature abhors a vacuum, and the beneficiaries of clearing and other change are not always exotic. Our challenge today is to become more ecologically astute, to recognise that native species can be pests, too, that will sometimes need controlling (killing). Australia will have matured as a nation when we can calmly debate the merits of shooting koalas, for conservation's sake.

V

A ROGUES' GALLERY

Who are our exotic pests? Why do they often succeed so much better than the natives they supplant? Introducing some of Australia's nastier pests – before holding up a mirror to ourselves.

24

The Shuffled Pack

An Alien Who's Who

'All things bright and beautiful,
All creatures great and small ...'
C.F. Alexander, English hymn

Anything, absolutely *anything* alive can become an exotic species. Australia today is home to exotic oysters, ostriches and orchids, ferns, fungi and fish, snakes, seaweeds and shrimps. The rabbits, toads and foxes we hear so much about are just the tip of a vast and unmeasured iceberg, one so dangerous it may one day sink our economy by destroying our crops, forests and fisheries.

Most people, including most biologists, are oblivious to all this alien life around them. Even basic information on exotic species is often difficult to obtain. Take marine biology. Back in 1978, biologists judged there were about twenty exotic species in our seas. By 1990 there were known to be almost fifty. In 1996, when I began writing this book, the list had grown to about a hundred; by 1998 the number had doubled again. The list will keep growing, not so much because new species keep sneaking in (although they do) but because scientists are at last conducting proper surveys. Major studies were begun only in 1996 after governments recognised the dire threat posed by ballast invaders.

We know even less about our insects. Tim New's *Exotic Insects in Australia* avoids giving any estimate of the number of exotic insects and mites we have – because no one has any idea. The Australian Academy of Science puts the number above 2000 but bases this entirely upon one US estimate. The true figure must be much higher, since Hawaii alone has more than 2000 exotic insects and hundreds of foreign spiders and mites. Most of our introduced insects are beetles, wasps, flies and bugs. Most aphids in Australia are exotic, but we have only two foreign butterflies, the cabbage white (*Pieris rapae*) and wanderer (*Danaus plexippus*).

Animal guides always seem to underestimate the exotic component, even when the information is available. The *Slater Field Guide to Australian Birds* makes no mention of the peafowl, pheasant, junglefowl and turkey, while the Australian Museum's *Mammals of Australia* omits feral cattle and alludes only fleetingly to Asian house rats, which are not listed in the index. None of our best reptile books mentions the slider, the colourful turtle that lives around Sydney, and the mourning gecko is listed in these books as native, which it surely is not.

As for our weeds, three CSIRO scientists tried recently to draw up a complete list, which was published in 1997 as the *CSIRO Handbook of Australian Weeds*. More than 2200 exotics were included. But the following year another report, *Recent Incursions of Weeds to Australia 1971–1995*, catalogued dozens of plants not mentioned by the CSIRO. That same year I was sent a list of Wet Tropics weeds by the Wet Tropics Management Authority and found that eight of the weeds mentioned (out of fifty-two) were not listed in either of these books. I then contacted John Hosking of NSW Agriculture, who told me he had drawn up a more comprehensive list of about 2600 exotic weeds. By early 1999 his list had grown to 2700. (Biologists often rank weeds by how badly they behave.

Ecologically the most feared are the 'transformer' species, or 'ecological engineers' – those few species that can change whole habitat structures, converting grasslands into forests, or wetlands into shrublands; pond apple, mimosa bush and prickly acacia do this. 'Ruderal' weeds, by contrast, only invade very disturbed areas such as roadsides and flooded riverbanks. 'Agrestals' invade farms and gardens and often cause problems for farmers. Thornapples [*Datura* species] and thistles are typical ruderals and agrestals. The 'adventives', which are hardly weeds at all, sprout from discarded seeds but rarely reproduce in the wild; apple trees and tomatoes are good examples.)

No estimates at all exist for exotic fungi, although expert Tom May, at the Royal Botanic Gardens in Melbourne, told me there must be several hundred, possibly even thousands of foreign species. All of our crop and garden plants have their own exotic leaf spots and rusts. We also have exotic mosses, including some significant invaders of woodlands (*Pseudoscleropodium purum* and *Brachythecium albicans*), exotic liverworts, and at least two exotic lichens (*Xanthoria parietina* and *Rinodina pyrina*).

That so much uncertainty exists shows how lightly we take our problems. If we could only walk through our forests or dive through our seas and see all the aliens lit up in neon all complacency would vanish.

In appendix III, I have drawn up a list of exotic vertebrates and a very crude list of invertebrates, although these may be underestimates. Aquarium fish, for example, are difficult to monitor because so much dumping goes on, strange fish sometimes appearing in streams without necessarily breeding. A population may flourish briefly then die away, like the rosy barbs (*Puntius conchonius*) that lived in a Brisbane stream for twelve years. My list runs to thirty fish species – rather more than the seventeen recorded by the Bureau of Flora and Fauna in 1987, although some of the cichlids listed for north Queensland may not be properly established. The number of exotic birds, at twenty-six, is surprisingly high, but many of these are confined to small islands – the peacocks on Kangaroo and Rottnest islands, for example, and the redpolls, chaffinches and yellowhammers on Lord Howe – so their impact is limited. Australia's feral ostriches, also very restricted in distribution, occur only around Port Augusta and Burra, where their forebears were freed long ago from feather farms.

Among the mammals, I might have included *Homo sapiens* on the list, but I will say more about our own shoddy record later. Sheep occasionally form feral flocks, and dogs often go wild, sometimes mating with dingoes and diluting the breed (indeed, dingoes may be doomed as a pure breed of dog). In the past Australia had two kinds of wild squirrel, American grey squirrels living in Melbourne and Ballarat and Indian palm squirrels released by zoos in Sydney and Perth. I remember, when I was a child, the palm squirrels scampering wild around the grounds of Taronga Park Zoo; they died out during the 1970s.

No firm numbers can be given for marine creatures because so much uncertainty exists. Indeed, biologists can't agree whether certain species are native or not. According to the CSIRO's Chad Hewitt, twenty-five exotic seaweeds inhabit our seas and another fifty-one species *may* be exotic, but another expert scoffed at such numbers, suggesting there might be only *one* exotic seaweed.

In any case, our exotic species play a much larger role in Australia than these lists alone can suggest. We boast only one exotic amphibian, for instance, yet the cane toad probably outnumbers any of our 200 native frogs, and in eastern Queensland it may well be the most abundant vertebrate of all. Rabbits, and probably house mice, outnumber any native mammal (we had 300 million rabbits before calcivirus struck), and the mosquito fish may well be our most abundant freshwater fish.

In size, too, exotics stand out. Our largest wild mammals are all exotic, and so too are our largest bird (ostrich) and amphibian (cane toad). A toad can grow to 23 centimetres and 1.5 kilograms, dwarfing our largest frog, the white-lipped tree frog (*Litoria infrafrenata*), at 14 centimetres. The brown trout is easily the biggest fish in our mountain waters, although it is much smaller than the Murray cod of inland rivers. Great striped garden slugs (*Limax maximus*) are also massive, reaching 20 centimetres in length.

Other parts of the world have their own unique exotic problems. None of our exotics compares in size, for instance, with the Indian elephant, the world's largest feral animal. On the Andaman Islands in the Bay of Bengal, some elephants employed in forestry work escaped long ago, forming feral herds that trample the undergrowth. In Borneo biologists speculate that the wild elephants in the north are descended from domestic stock taken there hundreds of years ago by the Sultan of Sulu. This would explain their very restricted distribution on the island, the absence of fossils and the lack of indigenous names. I have hiked along their trails through the jungles while observing orang-utans. If they are truly an exotic species there, their impact – in trampled saplings, excavated salt-licks and vast muddy trails – has been remarkable.

Hawaii, with African chameleons, Indian mongooses, Chinese turtles, American antelope, Australian wallabies and lizards, South American poison arrow frogs and birds from just about everywhere, has probably the most scrambled ecology on earth. North America, especially Florida, carries a remarkable contingent of foreign birds, reptiles and fish, the legacy of a *laissez faire* pet trade. Africa has Himalayan thar (mountain goats), European fallow deer and American squirrels and coypus (giant rodents), and Patagonia in South America is suffering enormously from introduced beavers. Europe has beavers, raccoons, mink, musk ox and squirrels from North America, porcupines, mongooses, genets and monkeys (the 'apes of Gibraltar') from Africa, rats and deer from Asia, South American coypus and Australian wallabies.

The following chapters introduce some of Australia's leading exotic pests: honeybees, cats, foxes, buffalo, donkeys and that most destructive of all invaders, *Homo sapiens.*

25

Seizing the Advantage

Exotic Roads to Success

> 'All animals are equal, but some animals are more equal than others.'
>
> George Orwell, *Animal Farm*, 1951

When a honeybee (*Apis mellifera*) finds a flush of new flowers and wants to tell its hive-mates, it dances: if the flowers are nearby it moves in circles, switching direction back and forth; if the food source is distant, it dances in figures of eight, wagging its body from side to side. The direction of the dance, conducted on the combs inside the hive, gives its sisters a bearing according to the azimuth of the sun.

No other insects can communicate like this. The way honeybees live, learn and interact in busy little cities is unrivalled in the insect world. No wonder they are one of the planet's most successful animals. From their native homes in Europe, Asia and Africa honeybees have gone feral across most of the world. They are perhaps Australia's most widespread exotic invader, found in nearly all habitats, including rainforest margins, mangroves and outback plains. I see them (and hear them) almost everywhere I go. No Australian bee can begin to match them in range and numbers, and their impact upon the land has been profound – they steal nectar from native insects and birds, and take nestholes from possums and parrots.

Why do exotic pests so often fare better than native species? Shouldn't the native animals and plants that have evolved in Australia over millions of years be much better adapted than recent arrivals? Why have rabbits from cold, wet England thrived beyond belief in Australia's outback when so many of our outback marsupials have been driven to extinction? Much has been written about this question, and most of the discussion comes back to four propositions: exotic species do well because they have left behind the predators and diseases that control them back home; because they benefit from disturbance; because, like the honeybee, they are superior in some way; or because they find vacant niches to occupy. Let's consider these ideas in turn.

The waterfern salvinia (*Salvinia molesta*) was smothering dams in Queensland until brought to heel by an imported weevil. It was thriving only because it had eluded the insects that control it back home. We have reined in a great many pests by importing their predators, parasites or diseases; indeed, the theory of biological control rests on the assumption that pests can be controlled in this way. Most invaders would not be so successful if subdued by some form of biocontrol, but the option is often impractical. We would not want tigers released in Kakadu to control buffalo.

The second proposition emphasises the role of habitat disturbance. When forests are penetrated by logging operations, cattle grazing or roads, when rivers are polluted and dammed, the ecology is disrupted, native species suffer and exotics get their chance. Many ecologists argue that disturbance is a prerequisite to invasion, that pests can't succeed in its absence. They see ecosystems as closed units that exclude outsiders unless they are damaged in some way. It is certainly true that disturbed environments carry more pests.

Weeds always do better on earth broken by ploughs, along road verges and around drains, houses and grazing cattle. In the Tiwi Islands north of Darwin, ninety-five weed species grow around the Aboriginal settlements (including temporary camps) but only ten invade forests, doing best in fertile soils where buffalo trample. In the grasslands of south-eastern Australia an exotic moss (*Brachythecium albicans*) turns up wherever cows graze. Southern Australia's heathlands never suffer from weed attack (the soils are too infertile) except near roads and stormwater pipes or where nutrients have invaded first.

Botanist Jamie Kirkpatrick explains how such weeds get an edge: 'Only 1.1–1.4 days of normal individual human urine output will fertilise a square metre of soil sufficiently to raise it from being able to support only native heath to levels suitable for crop plants and most weeds. Eighty-three banana skins will do the same job. The equivalent figures for other commonly deposited substances are: steer faeces, 0.2–0.4 days worth; dog urine, 2.7–4.0 days; cat urine, 4.0 days; orange peel, 37; apple cores, 114; cigarette butts, 375. Thus, the weeds along tracks in national parks can be attributed not only to transportation of their seeds by people, but also the transformation of the soils by their effluvia.' Popular camping grounds in Tasmania, Kirkpatrick notes, support a urination ring of annual weeds. 'This growth is most prolific directly outside the doors of huts, a comment on Tasmanian weather as well as male night laziness.'

There are hundreds of disturbance-dependent exotic pests, including many farm weeds and pasture grasses. In Kakadu, para grass (*Brachiaria mutica*) is exploding into damaged wetlands now that most buffalo have been removed, and in Queensland, big-headed ants (*Pheidole megacephala*) and Amazonian earthworms (*Pontoscolex corethrurus*) invade along walking tracks. Some of these invaders are drawn to disturbance because they come from disturbed environments overseas, having evolved alongside people and livestock. Some of Europe's farm weeds, such as plantains

(*Plantago* species), may have progressed from using sediments ploughed by retreating glaciers to soil worked by hoes. In Europe's post-glacial environment they have adapted so as to grow and breed rapidly, use nutrients freely and flourish among people. As Tim Flannery says, 'These characteristics of its flora and fauna have made Europe a "weedy" place. Mobile, fertile and robust, Europe's life forms were purpose-made to inherit new lands.'

Some habitats are *naturally* disturbed, and weeds invade them too, even in the absence of people. Beach dunes whipped by wind, storm and wave suffer enormously from weeds, as do river and stream banks when floods strip back silt. In habitats that strongly resist invasion, such as intact rainforest, weeds will creep along the stream banks. Deep inside Lamington Plateau the only weeds I have seen are the South American mistflower (*Ageratina riparia*, a horrendous invader) lining the streams and the wild tobacco plants (*Solanum mauritianum*) that sprout after tree-falls. Some pests, on the other hand, can invade without any disturbance at all – pigs, for example, and buffalo, toads, trout, honeybees, pond apple, mimosa bush, mission grass and stinking passionfruit (*Passiflora foetida*), to name a few. I have seen a toad inside pristine rainforest on Hinchinbrook Island, and Ludwig Leichhardt saw buffalo in the Northern Territory in what was then untrammelled wilderness.

The third proposition emphasises the superior capabilities of the pests themselves. Some invaders grow more quickly, squirt out more eggs, ooze potent poisons, withstand harsh climates and enjoy a very broad diet. Like the honeybee they have an edge over the creatures they displace. These kinds of invaders usually do very well in their country of origin. Many weeds are thorny, toxic, fast growing, self-pollinating and bearers of long-lived seeds. Some of them, the tropical grasses especially, use very efficient C4 photosynthesis. They are often exceptionally fecund. A groundsel bush can turn out a million seeds in a season, a square metre of bulrushes (*Typha*) 17 million seeds. A Pacific seastar can release 19 million eggs in one go. Some invaders can breed asexually and found a population from one immigrant – a great asset for a stowaway. The flowerpot snake, the mourning gecko, the New Zealand snail (*Potamopyrgus antipodarum*), some flatworms and many weeds do this. Among fish, live-bearers appear to have an edge over egg-layers, there being six feral species in Australian waters. And some invaders are exceptionally good at fasting. Giant African snails, khapra beetle larvae and potato cyst nematodes can go for years without even a snack. A piece of prickly pear can sit on a shelf for years without dying.

Such feats, though remarkable, should not dazzle us into thinking that our native species are inferior and primitive. Our life forms perform very well as pests overseas, as we have seen. The Australian paperbarks in Florida provide a fine example, a single tree capable of dropping 20 million seeds all at once, a superior accomplishment by any standards.

The fourth proposition emphasises the idea of the vacant niche. Australia's giant marsupials, probably pushed along by Aborigines, died out during Pleistocene droughts, and feral livestock (along with domestic cattle and sheep) have taken their place. We have even supplied exotic dung beetles to tend their droppings, replacing native ones that vanished long ago. Trout have occupied a vacant niche in Australia's mountain streams, previously devoid of large fish. In another example, quick-growing annual weeds find it easy to displace native grasses in southern Australia because the grasses are adapted to the warmer and wetter climate that prevailed several thousand years ago. If the summer rainfall zone keeps creeping southwards, more such weeds will spread from gardens.

We have created many new habitats in Australia, and while some native species (magpies and brushtail possums, for example) have been quick to claim the new niches it is hardly surprising that exotic species, some having lived alongside people for thousands of years, have an edge. Which largely explains the success of such newcomers as sparrows, goldfinches, thistles, lumbricid earthworms, head lice, smallpox and household cockroaches.

The idea of the vacant niche (controversial to some) has also been invoked to explain why Australian shrubs and trees do so well overseas. South Africa's fynbos heaths were left treeless by climatic extremes, and Australia wattles (and exotic pines) have now moved in. The Everglades in Florida emerged from the sea only 5000 years ago and very few American trees had settled in when the marshes were sown with paperbarks.

In some pests we can see all four factors at play. Honeybees *are* superior insects, but they have also freed themselves from some of their diseases. Unlike many native bees, they do well in disturbed environments, but they are perhaps also occupying a new niche, that of a large communal bee. A close relative, the Asian honeybee (*Apis cerana*), is poised to strike Australia soon, and it may prosper here as well, for all the same reasons. The little mosquito fish that dwells in my local stream also has everything in its favour. No native fish is so tolerant of pollution, none bears its young live, and few can breed so young – at only four weeks old. It is 'one of the outstanding successful freshwater fish species of the world', say biologists, speaking more in dismay than praise. Mosquito fish thrive on disturbance, doing best in weed-infested, sunlit, slightly polluted streams denuded of their fringing forest. In Australia they are free of most of the twenty-three parasites that afflict them in their native home in North America. They have found a vacant niche in Australia, thriving in a habitat shunned by native fish. And then there is the cane toad. It is superior in size, toxicity and its ability to breed in brackish pools; it has left behind its parasites and most predators; it benefits from disturbance by entering forests along roads and tracks, and it thrives in the new niche provided by paddocks and farm dams. By understanding these advantages, by seeing how each of these exotic species prospers, we can better appreciate the many management problems that confront us.

26

A Bad Rap

Cats: Scoundrels or Scapegoats?

> 'It looked good-natured, she thought; still it had very long claws and a great many teeth, so she felt that it ought to be treated with respect.'
>
> **Lewis Carroll, *Alice's Adventures in Wonderland*, 1865**

On my very first African game drive, beside the Zambezi River, along with elephants, buffalo and giraffes, I saw a feral cat. Yes, even Africa has feral cats, roaming the savanna among lions and leopards. They consort with African wild cats, the species from which Egyptians bred the first domestic cats some 3000 years ago. Feral cats these days can turn up almost anywhere – in deserts, on remote islands, in woodlands, cities and most places in between. Once revered by the Pharaohs, cats are reviled today by many greenies, who blame them for all manner of crimes, many of which they do not commit. Indeed, they are one of the very few conservation issues that may actually be overrated.

And yet cats have committed some heinous acts. New Zealand lost an entire bird species to a single cat. The Stephen Island wren was remarkable – a mouse-sized, almost flightless creature first discovered between the jaws of the lighthouse-keeper's cat. By late 1894 it was extinct. One cat had snuffed out the lot in less than a year. Such precious meals it had! An angry newspaper scribe in 1895 argued against island lighthouse-keepers owning cats, 'even if mouse-traps have to be furnished at the cost of the state'.

Feral cats have proved very destructive on small islands. On Macquarie Island, Australia's tiny subantarctic possession, they helped exterminate the Macquarie Island parakeet and a unique flightless rail. To this day they slaughter thousands of seabirds there. But although cats are certainly damaging in these isolated habitats, where prey is often naïve towards predators and the populations are small, their impact elsewhere is harder to quantify.

Suburban cats, because they dispatch lots of birds, are often condemned as major killers, second only to their feral kin in the bush. But the evidence is not convincing. Cats kill millions of birds in gardens, true enough, but ecologically there is nothing wrong with this – predation is a fact of life. Birds are killed in forests, too, by falcons, owls, quolls, dingoes, snakes, goannas, even spiders. Pet pussies are simply the urban equivalent of these killers. Hunting by pet cats only becomes a worry if the death rate of the birds exceeds their birthrate. By and large, this doesn't seem to be the case. The birds caught by cats are usually abundant species that thrive on development. Some of them – including willie wagtails, crested pigeons and magpie-larks – are probably faring better today than ever before. This is certainly true of the common garden lizards that cats like to kill. Some studies show that leafy suburbs actually support more birds than intact forests, despite all the cats, because gardens planted with berries and nectar-rich flowers produce more food. If any species is threatening birds in suburbia it is probably the pied currawong, a vicious native bird that raids nests and devours chicks and eggs. Native noisy miners also make mischief by driving away smaller birds.

Feral cats in the bush, however, can be a serious problem, though probably not to birds, which they seldom eat. Studies of their diet confirm what cartoonists have always known: cats prefer rats, mice and other mammals. Rabbits are often their staple diet; cats may sometimes be helping the ecology by keeping bunny numbers down.

Cat expert Chris Dickman argues that cats may have exterminated many small mammals in central Australia last century. Perhaps, but does this justify a campaign against cats today? The extinctions of a century ago occurred in uninhabited woodlands where anti-cat campaigns today can have no effect. And Dickman may be wrong. 'Despite the abundance of observations linking cats to extensive losses of native species,' he himself admits, 'other evidence suggests that their impact has been minimal.' Other theories may explain the disappearing rodents, including exotic viruses brought in by house mice. Biologists Bomford, Newsome and O'Brien are also equivocal. 'It is not known', they write, 'whether feral cats are a major problem for conservation of native wildlife, even though feral cats are known to eat hundreds of thousands of native animals each year, including some rare species.'

Cat haters point to the threat posed by cats to the endangered eastern barred bandicoot, which was tragically reduced in Victoria to a single colony near Hamilton. Yet these bandicoots are doing well in Tasmania, an island that is cat infested but fox free (I have even seen one foraging on the lawn of a Hobart garden). Overall, this bandicoot has probably suffered much more from foxes than from cats, the latter becoming a problem only because the last mainland colony lay near a town. Fox-free Tasmania stands out as the only state that has not lost small mammals to extinction.

Foxes are probably the greatest blight on our smaller mammals, especially numbats, bettongs and rock wallabies. They sometimes attack platypuses, tortoises, penguins and brolga chicks as well, and they spread around the seeds of olives and blackberries. Proof of their destructive power is provided by Operation Western Shield, a remarkable pest control operation under way in Western Australia. Since 1996 the Western Australian government has baited all its national parks and forest reserves with meat laced with toxic 1080. Numbats, bettongs, wallabies, quolls and bandicoots have multiplied as fox numbers have fallen. The brush-tailed bettong has even been lifted from the endangered species list. I camped recently at Dryandra Woodland (where fox baiting began earlier, in the 1980s) and was delighted to have these marsupials bouncing up to my table at night begging for handouts. The return of all these animals is attributed entirely to fox control, since cats do not take dried meat baits. In the south of Western Australia the rare numbat, western quoll, brush-tailed bettong and red-tailed phascogale, extinct everywhere else in Australia, may owe their survival to poisonous plants (*Gastrolobium* and *Oxylobium* species) containing fluoroacetate, the same poison used in 1080. Plant-eating mammals in the west have evolved immunity to this toxin, storing it in their flesh, but foxes that eat them die. Fox numbers are thus kept low, permitting numbats and other mammals to survive.

The testimony of early pioneers supports the idea that foxes are more pernicious than cats. 'The introduced fox is to my mind the greatest pest in Australia', wrote Amy Crocker, who grew up on the Nullarbor Plain. 'People of little experience say the domestic cat gone wild is a pest – perhaps they are, in certain localities, but speaking for this district and from my own personal knowledge, wild cats were here a good 30 years before foxes came, and we still had our own little animals.'

Black rats, similarly, are rarely portrayed as killers, but as destroyers of island life they may rank higher than cats. On Lord Howe Island they knocked off five bird species, and on Christmas Island they helped exterminate Maclear's rat (*Rattus macleari*), a unique native rodent. They reached the island in 1899 by hiding in hay, and ten years later no Maclear's rats remained. Trout, too, have driven several species close to extinction, an achievement that cats, on hard evidence, cannot match. (I wonder how many cat-hating trout fishermen there are.) Rabbits nip back vegetation so severely they are probably responsible for exterminating several native herbs, such as the peppercress mentioned in chapter 19. They ringbark trees and devour their seedlings, skewing the whole structure of outback woodlands. Rabbits may have helped wipe out small outback wallabies and bandicoots by taking their food and grazing down their cover. They are possibly the worst of all our pests because of the extraordinary numbers they achieve.

Many conservationists treat cats as if they were our number one pest, but I believe foxes, rabbits, pigs, toads, trout and some weeds all pose a greater menace. Goats, donkeys, carp, mosquito fish, Pacific seastars, green crabs, honeybees, bumblebees and Amazonian earthworms concern me a great deal too. And worse than any of these is probably phytophthora, the dreaded fungal disease, along with the chytrid fungus killing our frogs. By saying this I don't wish to exonerate cats, simply to broaden the debate. Instead of heedlessly angering cat owners by vilifying their pets (and I should say that I have never owned a cat), we might look around us at all the other pests receiving less attention. The lion's share of ecological pest research goes into mammals, leaving us woefully ignorant about smaller invaders such as earthworms, ants, millipedes, bumblebees, molluscs and crabs. Green crabs, for instance, are voracious predators capable of dramatically altering seashore ecology. A recent article on big-headed ants condemned them as a 'ferocious killer of native invertebrates' that is 'quickly gaining a reputation as one of Australia's most destructive native pests'. This kind of talk is usually reserved for big furry critters such as cats. Does that tell us something?

27

Where the Deer and the Antelope Roam

Hoofed Introductions

> '... to have antelopes gladdening our plains as they do those of South Africa, and camels obviating for us as for the Arab the obstacle of the desert.'
>
> The *Age*, 2 April 1858

For many years wild antelope roamed free in Australia. They were blackbuck from India, descended from pairs set down by acclimatisers near Geraldton, Bunbury and other western towns in about 1912. Males are black with long corkscrew horns; I have seen their lively herds running wild in southern India. The antelope liked Australia, and here and there they thrived. By 1929 the Western Australian Acclimatisation Committee was worried by their numbers, noting, 'it would be extremely difficult to get rid of them'. After World War II a hundred or so were counted on one station alone, but they disappeared in the 1980s, defeated by drought and shooters.

These were not our only antelope. The acclimatiser Frederick McCoy advocated the introduction of African eland, the giants of their tribe, in the outback because of their 'delicate and nutritious flesh'. A pair of them was released in Western Australia near Kellerberrin early this century but they died within days, apparently poisoned by pea-bushes. Zebras were reportedly freed in the west about 1900, but they died away too. Remarkably, antelope were still being set free during the 1930s, when Taronga Zoo-bred blackbuck were released into Royal National Park near Sydney, but the bush there is much too thick and they were not seen again.

Asian rainforest cattle were also granted freedom in Australia, and they prospered. In 1849 twenty Bali banteng, the domesticated form of a magnificent rainforest cow, were brought to Victoria in the Northern Territory, only to run wild when the settlement was abandoned soon after. The world forgot them for nearly a century until their descendants were spotted in the wilderness of what is now Gurig National Park on Coburg Peninsula. Today they are hunted by sportsmen who pay up to $20 000 for a trophy. In Java they are an endangered subspecies found wild only in remote jungles. Bulls are black as night with clean white legs.

There are biologists today who say we should place high conservation value on these cattle. If so, we should probably conserve our dromedaries too, for Australia is the only place in the world where one-humped camels run free. Wild camels became extinct in Africa and Arabia aeons ago, their descendants surviving only as beasts of burden. Dromedary ecology is a discipline that can only be studied in Australia. Brought to Australia last century as pack animals to help tame the outback, they were freed (along with donkeys) when motor cars usurped their role, and they reverted to a wild species. Instead of taming the outback they made it wilder. Aborigines in western Queensland hunted these beasts. 'They loathed the *kamel-le-coola* man, as they called the camels', wrote pioneer Alice Duncan-Kemp. 'The Myalls speared them for meat, and because they rooted along the sandhills' edges for subterranean nuppamurras (water traps) which the desert blacks looked upon as their own secret reserves and guarded jealously at all costs. One could hardly blame them.'

In 1862 McCoy had complained about Australia's dearth of wild ruminants. Today we have camels, buffalo, banteng, cattle, horses, donkeys, goats and six kinds of deer – an impressive tally for a once-hoofless land. McCoy believed the work of importing ruminants would become 'a lasting benefit to the millions of men who will in the fullness of time inhabit this land', but although these animals have changed the land forever, Australia has not benefited. The water buffalo, according to the Australian Museum's *Mammals of Australia* (1995), 'has caused severe environmental damage, including accelerated soil erosion, channelling of flood-waters, saltwater intrusion into freshwater habitats, loss of vegetative cover, reduction in the diversity and abundance of wetland flora and fauna and disfigurement of landscapes by their wallows, trails and dung pats'. As well, water buffalo spread cattle diseases, especially tuberculosis, and wild herds are heavily culled.

Buffalo, like banteng, went feral decades before the heyday of acclimatisation. Eccentric explorer Ludwig Leichhardt, trekking from Brisbane to the Northern Territory, knew his travails were nearly over when he stumbled across buffalo dung in the wilderness. Ending their fears of starvation, his men shot a young bull. Leichhardt shared the meat with local Aborigines, who knew this animal well. He wrote in his journal: 'They called the buffalo "Anaborro": and stated that the country before us was full of them.' In Western Australia in the 1830s, another explorer, George Grey, twice saw hoofprints, probably made by buffalo wandering across from the Northern Territory. Top End buffalo roam into Western Australia to this day, sometimes reaching the hinterland of Broome.

Donkeys also roam over much of the outback. Australia's population runs to several million. Their stronghold is the Kimberley region, where Kimberley horse disease, brought on by a poisonous pea-bush, forced graziers to use them in place of horses. Descended from Nubian wild asses tamed in Egypt 6000 years ago, donkeys now reach densities of ten or more per square kilometre and severely erode the ravines where they graze. Were it not for the pea-bush the Kimberley would not be so damaged today.

McCoy would no doubt be awed by the success of our ruminants. In the Top End they reach densities equal to those in some African national parks. Australia's floodplains carry up to twenty-five buffalo per square kilometre compared to an average of one per square kilometre in Asia. Buffalo and banteng reach four and three times, respectively, the typical numbers achieved in Asia, where tigers and diseases take their toll. No wonder they so damage the land. Buffalo proliferated so quickly that 100 000 hides were exported in 1911 alone. By 1985 more than half of all the world's wild population lived in Australia. Hoofed animals now dominate the biomass of some of our reserves, with as many as five species in a national park such as Gurig (banteng, buffalo, horse, pig and sambar deer). Ecologically, they have replaced the diprotodons and other megafauna that vanished during Pleistocene droughts.

All of our ruminants, except deer and banteng, are now considered serious pests. Goats denude hills, erode slopes and compete with rock wallabies for food. Donkeys eat grass needed by cattle and kangaroos, promote weeds, precipitate erosion and aggressively keep stock away from water. Feral horses do the same things on a smaller scale.

Our deer are not usually considered a problem. We have quite a selection – Indonesian rusa, Indian chital and hog deer, South-East Asian sambar, and European fallow and red deer, all found in small herds in very scattered locations. Rusa live on islands in Torres Strait, where they nibble seaweed and sip seawater, on Groote Eylandt in the Northern Territory and in Royal National Park outside Sydney. I have seen fallow deer consorting with kangaroos in Sundown National Park in southern Queensland, reminding me of a tale told in 1861 about a fallow deer said to head a mob of 'roos in the Dandenongs.

The spirit of acclimatisation is evident in the attitudes we adopt towards deer. Governments in some states protect them, and in 1977 the red deer won a place on Queensland's coat of arms. Deer herds are even condoned in certain national parks. *The Mammals of Australia*, so damning about most exotics, offers no words against deer. Even Eric Rolls, in *They All Ran Wild*, claimed, 'Deer have done no noticeable damage in Australia'. This is nonsense. In Royal National Park rusa are degrading fragile heathlands and shrublands. They are overgrazing and trampling, baring soils to erosion, even ringbarking trees with their antlers. They eat so much grass they are blamed for the failure of grey kangaroos to survive in the park. In 1972 culling was begun, not to eliminate them but only to control their numbers. Ian Mahood, of the National Parks Service, wrote frankly about the problems in *Australian Deer*, adding that 'Valiant attempts have been made to catch poachers'. But why these efforts to protect a pest, unless double standards apply? National parks are less sympathetic today and most rangers probably want deer removed. In New Zealand, red deer, feeding alongside possums, cause severe erosion, entire slopes sometimes subsiding after their attentions.

Thanks to McCoy and others, ruminants (including cattle and sheep on farms) now dominate our land; few corners of Australia are unscarred by cloven hoofs. I once visited Prince of Wales Island in Torres Strait and, instead of the remote wilderness I was expecting, found saltpans crisscrossed by hoofprints of all sizes. This rugged island is home to feral horses, cattle, goats, deer and pigs. How disappointing to think that the red kangaroo, at 85 kilograms, is only our thirteenth largest wild mammal, behind the buffalo (at 1200 kilos), banteng, cow, camel, horse, donkey, pig and five kinds of deer.

28

The Ultimate Pest

Our Destructive Ways

> 'It is ironic to me to hear people of European ancestry accuse other organisms of being "invasive exotics, displacing native species".'
>
> **J.L Hudson, American seedsman, 1997**

If there is a world's worst pest, an exotic invader that surpasses all others, surely it is the human species. No other animal has swarmed across the globe in such numbers or displaced so many other life forms in the process. Humans enjoy the widest distribution of any mammal, ranging from snowfields to deserts, rainforests to mangrove swamps. Our ability to invade new habitats is unsurpassed. As one mammal book puts it, 'Besides his normal cosmopolitan terrestrial habitat, man has also established suitable temporary living conditions at sea, far under the oceans, on the polar icecaps, in the atmosphere, on the moon, and in the void of outer space.'

Like many invasive animals, humans are omnivores, our catholic diet incorporating seeds, tubers, fruits, leaves and stems from many plants, animals of many kinds, and even fungi and algae. A versatile diet helps us to occupy a much broader niche than any other animal. Unusually for a pest species we bear very few offspring, and these mature remarkably slowly. But a low growth rate is more than offset by our large size, great longevity and low mortality. Very few predators stalk humans. But to really appreciate the human success story we need to understand the role of disturbance in ecology. It is a key tool by which people divert nutrients to their needs. It is also the environmental trigger, in all the world's habitats, by which exotic species invade in our wake.

In a stable rainforest, specialised dominant trees control most of the nutrients. But when storms, floods or landslides strike, or when aged trees fall, soil is bared to sun and rain and nutrients are freed to a different group of plants and animals. These colonisers, gap-fillers or edge species are opportunistic plants and animals that thrive on disturbance. The plants have bigger leaves than interior species, and they grow faster so they can tap the sudden supply of sun and soil. Their leaves, produced in haste, are not as strongly constructed (or tough) as those of forest interior trees, and insects and mammals, people included, find them easier to eat and more nutritious. Since many of them grow close to the ground, they are also more accessible. Many of these edge plants bear prolific crops of small fruit, which entice edge-dwelling birds who then carry their seeds to new clearings. Animals are drawn to the clearings by all the leaves and fruit, and predators follow. Nutrients cycle freely between plants and animals. Eventually, as succession proceeds, the tall trees return, displacing this animal-rich system.

Many of the plants and animals we farm were originally colonising or edge species. Bananas, pawpaws (papayas) and passionfruit are quick-growing plants of the rainforest edge with big soft leaves. The red junglefowl, ancestor of the domestic chicken, is an edge bird of Asian rainforests. In the Thai jungles I have heard the cocks crowing at dawn, and flushed flocks of wild 'chooks' around the edges of clearings. Cows are descended from a wild ox (the aurochs) that hid in deep forests and trekked to clearings to feed.

When people grow crops they simulate natural disturbance. Slash and burn agriculture is the clearest manifestation of this, but it holds true of all farm systems. As Bill Mollison has said about permaculture, 'most certainly, increased edge makes for a more productive landscape'. Tilling soil to expose earth produces the same opportunities created when storms tear up trees, floods dump silt or glaciers retreat. Many of our vegetables are bare-soil colonists or gap-fillers.

There is almost no difference, ecologically, between farm weeds and vegetables. Both are short-lived plants that grow quickly to exploit bare soil. The most obvious difference is that vegetables taste good, though many weeds are edible too. Some of our foods were in fact bred from plants we call weeds today. Wild oats (*Avena fatua*) and artichoke thistle (*Cynara cardunculus*), ancestors of their namesakes, are both serious weeds. Chickens, cows and pigs were probably domesticated after they kept wandering into old clearings left by early farmers.

Humans reveal themselves as a colonising or edge species in many ways. Even in highly urban societies people prefer to live and holiday on beaches, riverbanks and lake edges. Hunter-gatherers find the most yams, fruits and game by searching in clearings and along edges. 'Ecotones', as ecologists call edges, are rich in resources because they support animals and plants from adjoining environments (land and water, for example) as well as specialised edge-dwellers. (Our visceral dislike of forest interiors shows through in European fairy tales, which portray forests as the dark and evil domains of witches, trolls and giants. In *The Hobbit*, Mirkwood is sinister and unknowable, the abode of giant spiders and hostile elves. Tolkien, of course, drew on medieval folk traditions.)

But humans go much further than most edge species. We don't wait for lightning strikes or landslides to create disturbances; we use firesticks and bulldozers to create our own. Humans aren't the only animals that disturb habitats. Elephants smash down woodlands, beavers build dams that silt up valleys and some of the dinosaurs must have been very destructive. But never before has earth seen an animal as destructive as us. We are systematically turning the world into one giant clearing, disturbing every habitat, pushing over forests and churning up soil, creating more and more edges, feeding off the nutrients freed by our activities.[1] Soon there will be no large intact forests left, only remnants dissected by roads and cuttings.

This process is disastrous for forest-interior species. We have pushed more animals and plants to extinction than any other 'pest'. But those animals and plants that like clearings and edges have never had it so good. There are probably more dandelions growing in the world today than ever before, and certainly more cane toads, thistles, mosquito fish, honeybees, sparrows and lantana. Humans are returning the earth to an earlier stage of succession and these species are prospering in our wake. We play a facilitating role for colonising species. As Jack Harlan wrote in *Crops and Man* (1975), 'If we confine the concept of weeds to species adapted to human disturbance, then man is by definition the first and primary weed under whose influence all other weeds have evolved.'

These 'weeds' include the cattle and sheep under our care. Domestic livestock, strictly speaking, fall outside the ambit of this book, but they do require a mention. They have done more damage to Australia than any feral animal by triggering erosion, compacting soil, trampling burrows and nests, erasing the understorey, grazing out certain plants, spreading weeds and altering forest succession over vast areas. Cattle and sheep have probably contributed more to extinctions than foxes or rats. They are generally more destructive than wild buffalo, donkeys and goats because they are fenced into paddocks at high densities. As one university biologist complained to me, those people who rant about the cat should add 'tle' to the name and pursue a worthier rogue. It is ironic that domestic cattle have done so well out of human stewardship when most wild cattle species – yak, banteng, bison, gaur, anoa, kouprey – are rare or endangered. The aurochs, or ancestral cow, has long since been driven to extinction.

Our arrival on earth may be a logical consequence of a world grown less stable. The Pleistocene ice ages plunged the world deep into chaos, fostering evolution of species that thrive on change. We are the supreme product of our times. People addicted to change now run the world, the process has come full circle, and our pest problems can only multiply as a result.

1. Each year humans shift around some 40 billion tonnes of rock and earth.

VI

WHERE ARE WE HEADED?

Our problems will keep growing as more pests escape and sleepers wake. How will the future look? The Wet Tropics offer a glimpse of tomorrow; the Homogocene, a view of the distant future.

29

Expanding and Infilling

Pests Old and New Tighten Their Grip

'Nature abhors a vacuum . . .'
Benedict Spinoza, *Ethics*, c. 1678

At night as I lie in bed I hear a new sound – the *chuk chuk chuk* of an Asian house gecko (*Hemidactylus frenatus*) calling from across the road. These lizards first appeared in Brisbane in 1983, turning up at a container terminal, then at the wharves. Since then they have advanced across the city, taking over factories, shops and houses in dozens of inner-city and riverside suburbs. Even so, I never expected them to reach my bushy outer suburb, which they accomplished early in 1998, and it surprised me even more to hear them at my parents' house a couple of months later, then at the homes of several of my friends; 1998 proved a very good year for this gecko. The sound of the Brisbane night has changed forever. The last time it changed so dramatically was in the 1950s when the cane toad arrived.

Twenty-five years ago, when I was a lizard-obsessed teenager, *Hemidactylus* was called the 'Darwin house gecko' because you had to go to Darwin to see them. No one imagined they would one day come south to Brisbane; we thought the world was more stable than that; they are supposed to be a tropical reptile anyway. When they reached Brisbane I was working at the Queensland Museum and we thought a harsh winter would snuff them out. Now we wonder whether they will invade every Brisbane home and how much further they will go. To Byron Bay? Coffs Harbour?

These invaders may not matter very much, but another newcomer, the Indian myna, is alarming bird lovers. Why these nasty birds, which thrive in Sydney, Melbourne and north Queensland, did not invade Brisbane long ago remains a great mystery. They have lived for decades just to the west and north; as a kid I saw one in Brisbane, a vagrant flocking with starlings, but mynas did not enter the city in earnest until about 1982, spreading east from the Darling Downs. They nest in tree hollows and will force out prior tenants, sometimes piling straw over the eggs of parrots. Only a couple of years ago I rarely saw them around Brisbane; now I usually spot one or two on the drive across town to see my parents. Last weekend, for the first time, and on the same road I always take, I saw a small flock. A few months ago I was surveying bushland and saw them loitering around three separate eucalypts with hollows. Yikes! I had hoped their impact might be slight, that native noisy miners would keep them in check, but now I am expecting to see fewer parrots and dollar birds and more of these clamorous interlopers. Tree hollows are scarce around town, a legacy of logging, and we can ill afford to lose more to exotic birds.

Another newcomer to Brisbane is a little black ant (*Iridomyrmex* species) that scurries around concrete pavements and nips bare-legged picnickers in city parks. Hundreds of these ants infest my front garden. Ant experts cannot properly identify this species, or say where it came from, or how it got here. Now one of Brisbane's most abundant insects, its impact on the local ecology is unknown.

Other changes are occurring as I write. Three exotic fishes, carp, tilapia and platies, are spreading in local waterways. I recorded the first platy in my local stream just a few weeks ago; the Queensland Museum wants the specimen. Tilapia appeared in Forest Lake, a few miles south of my house, early in 1999. In 1995 two major new waterweeds put in an appearance – Senegal tea plant (*Gymnocoronis spilanthoides*), an escape from an ornamental pond, and alligator weed (*Alternanthera philoxeroides*), which spread from vegetable plots tended by the Sri Lankan community, who grow it in place of mukunu-wenna (*A. sessilis*), a similar traditional vegetable. (It escaped in Melbourne in the same way). As well, plenty of established weeds are multiplying. Asparagus fern, first recorded in Australia in 1972, is now prolific. Umbrella trees have taken over some bush where I used to ramble years ago. Glycine, a pasture legume, is swallowing up vacant land in town at an alarming rate, and Singapore daisy, introduced in the 1970s, is running rampant on creek flats.

Other naturalists, in other cities, are seeing similar changes. In Canberra myna numbers are exploding, dating back to the release of some birds in around 1968. Along with starlings (also proliferating), they are taking over the nestholes of parrots. Foxes increased dramatically after Canberra's last drought and radiata pines (*Pinus radiata*) are fast invading bushland from maturing plantations. Cootamundra wattles are stealing away from rural estates, and broom, Montpellier broom (*Genista monspessulana*) and broadleaf privet are exploding. Alligator weed, one of the world's worst weeds, turned up in 1994; the oriental weatherloach, a burrowing fish, ten years earlier. Perch and weatherloaches are both increasing in numbers. Canberra now has seven exotic fish to eight natives.

Australia is in the throes of ecological upheaval, and most of this change is coming not from new invaders but from old pests tightening their grip on the land. It is important to understand that most pests in Australia have yet to occupy their full range; they are still migrating outwards or increasing in density (infilling) or both. By the time every invader has gone everywhere it can go

Australia will look very different. This chapter makes some predictions about the future.

Cane toads are travelling very fast, and by the time you read these words they may well have reached Kakadu. From the moment they were unleashed in the 1930s they showed no inclination to stay within the cane fields, scattering in all directions, helped along, no doubt, by well-meaning farmers. Southbound toads reached Brisbane by 1954 and the Gold Coast by 1974. Westbound toads got to Roma, 300 kilometres from the coast, by 1974, and toads freed at Mossman hopped north to Cape York, reaching the tip in 1994. Toads taken to Karumba in the Gulf went west, reaching the Roper River by 1995, and are now invading Arnhem Land.

I have never seen a site newly conquered by toads, but I know what it looks like. In 1988 I worked as a naturalist at the Cape York Wilderness Lodge (now called Pajinka) before toads had arrived. Quolls (native cats) scampered about the dining hall, huge amythestine pythons hunted the quolls, and a big goanna called Geoffrey scrounged around for scraps. The toads arrived in 1994 and no quolls have been seen since. Geoffrey disappeared, and while there are still a few pythons and goannas about, they are smaller than the giants of yore. Toads now hop about the lodge everywhere, tumbling regularly into the swimming pool.

Scientists have calculated the invasion rates of a number of species. In the Northern Territory the toad front is advancing at an impressive rate of about 30 kilometres a year. The virus that slaughtered pilchards in 1995 travelled at speeds varying from 12 to 43 kilometres a day, depending upon ocean currents, spreading more than 2500 kilometres east and west over three months. Starlings overseas have been clocked advancing at 200 kilometres a year, Australian barnacles (*Elminius modestus*) in Europe at 30 km/year, cabbage white butterflies at 13 to 170 km/year and bubonic plague at 400 km/year. Pests travel more quickly when people help them along. The cane toad was taken to Ballina in the 1970s and to Karumba in the 1960s, and odd toads turn up in Adelaide and Perth. Asian house geckoes did not get to

Brisbane by walking there; ships brought them south. They have yet to occupy most towns between Brisbane and Townsville, though it won't take them long.

Biologists today use a computer model, Climex, to predict how far invaders will go. By feeding in data on a pest's climatic range back home they can generate a map showing how well it should fare here. These maps make interesting reading. They tell us that mimosa bush, for instance, could spread from Kakadu down to Rockhampton, that rubber vine and pond apple will get from Queensland to Western Australia, that Japanese kelp will reach Wollongong and Western Australia and that cane toads, if global warming proceeds as expected, will hop south of Sydney to Eden. Even without this model we can predict some of the changes. The flowerpot snake has colonised Cape Town (34 degrees S), so we can expect it one day to establish in Sydney (on the same latitude, and with a comparable climate), after travelling south in pot plants. I may live to see this snake – another Asian reptile – in my garden.

Some pests, thankfully, are held back by major natural barriers. The deserts separating eastern and south-western Australia have stalled many invaders. Western Australia remains sparrow, starling and myna free – a commendable achievement – and the state quarantine department keeps a list of more than a hundred insects found in the east they want kept out. Codling moths (*Cydia pomonella*) and European red mites (*Panonychus ulmi*) are often intercepted on illegally imported fruit. European wasps (*Vespula germanica*) turn up regularly around Perth and the state has probably lost its fight against this invader. The west also has pests that the east doesn't want – the dreaded Mediterranean fruit fly (*Ceratitis capitata*), the green snail (*Helix aperta*) and the Senegal dove among others. In 1980 someone thought it would be clever to smuggle green snails into Perth to rear for the table. They escaped of course, and despite an eradication campaign costing nearly $1 million they now thrive around the city, sometimes attacking crops. These snails will go east one day in a suitcase or packing crate.

So how will the future look when every invader has filled every vacant corner in Australia? I'm not sure I really want to know, but I can say that among the groups that will expand the most will be the fungi, weeds and sea creatures. Phytophthora will continue its march across southern Australia, in time striking the heart of every national park in the south-west, rewriting the ecology there. In 1997 it appeared on Kangaroo Island, stirring panic among biologists concerned for the island's many rare and unique heathland plants. Another fungus to watch is the spectacular red-and-white fly agaric (*A. muscaria*), a mushroom that grows symbiotically on the roots of northern hemisphere trees, including beeches. Where pine plantations abut antarctic beech forests in Victoria and Tasmania, it is learning to grow with antarctic beeches (*Nothofagus*). Fly agaric may one day penetrate the darkest rainforests of Tasmania, and biologists fear it could exterminate hordes of rare relictual Gondwanan fungi. It is deemed a serious threat to forest ecology in New Zealand.

Among weeds, garden and pasture escapes will continue to dominate. Hundreds of garden plants are festering in bushland on the fringes of our cities, and we may wonder how far they will go. The future could be ghastly. Lantana, cultivated in the 1840s and weedy by 1862, is the only subtropical shrub, apart from the prickly pear, that now occupies its full range, and its impact has been incredible – it dominates gullies and rainforest margins all over eastern Australia. Dozens of subtropical shrubs began their conquest much more recently and if some of them fare even half as well our forests face an interesting future. Asparagus fern did not reach Sunshine Coast beaches until the 1980s, yet by 1998 it had become one of the four worst weeds. Each year another five weeds, mostly garden plants, invade those beaches.

A vast expansion of pasture grasses has been proposed for northern Australia. Five million hectares have been sown so far, and another 5 million hectares have spread by themselves. Apparently, a total of 47 million hectares is achievable, including an easily attainable 22 million in Queensland. That goal, put recently by beef producers and the Meat Research Corporation to the pasture workshop I attended, would spell doom for biodiversity in our tropical woodlands. Many plant and animal species would be wiped out. To add to the problems, pasture scientists keep breeding up new varieties. New lines of leucaena, bred for resistance to the tiny leucaena psyllid bug, will cross-pollinate with wild leucaena plants, spreading resistance into the wild and improving weed vigour. In an act of unbelievable shortsightedness, pasture scientists in Wagga Wagga released a new line of African lovegrass (*Eragrostis curvula*), a failed pasture grass and major weed (declared noxious in some areas), insisting it would not behave like a weed. What will happen when it cross-pollinates with weedy forms? In the past, willows and pampas grass became much weedier after newly imported material bred with the wild plants, improving their vigour. New cuttings of olives imported recently from Israel and Italy will do the same thing.

The twining legumes brought in by agronomists in recent decades worry me enormously too, for they are fast becoming rampant weeds in gullies and damp forests. Plantation pines are also thriving. Trees invade slowly and their impact tends to be underestimated. So far ten pine species, superbly adapted to Australia's infertile soils, are taking off, invading forests, swamps and heathlands. Aquatic weeds are also multiplying. Cabomba and alligator weed will spread all over eastern Australia, helped along by the dams we build and the pollutants we direct into streams. The Murray River is especially pest prone because of its vast size and degraded state. Alligator weed, found in the Murrumbidgee River (a tributary of the Murray) in 1994, and water hyacinth, which infests 10 000 hectares near Moree, will probably sail all the way to the mouth. Tilapia (found recently in a tributary stream near Toowoomba), carp and oriental weatherloaches will match their invasions underwater.

The opportunities for expansion of our marine creatures are enormous. Harbour surveys have found that Melbourne's Port Phillip

Bay has 145 exotic species while Portland, further west, has only nine. The great challenge for the Quarantine Service will be to control coastal ballast water and hull hygiene, to prevent Port Phillip Bay and Hobart harbour from becoming relay stations that spread their pests everywhere.

In the long term we can expect every exotic species to expand and fill every available niche. And if all this seems bad enough, there is worse to come. Australia's next wave of pests will be her sleepers, aroused from their slumber.

30

Sleepers Wake

The Pests that Bide Their Time

> 'One of the great mysteries of introduced species biology is why many species spread slowly, if at all, for many decades – the lag phase – then erupt suddenly with drastic consequences for the native biota.'
>
> **C.H. Daugherty, biologist, 1993**

Even if Australia closed the door on imports today our pest numbers would multiply, for thousands of aliens have already entered the country and are simply awaiting their chance to escape. These 'sleepers' are all around us, in our gardens and aviaries, on farms and plantations, in laboratories and aquaria. They are the plants we grow, the pets we keep, and the diseases and germ plasm in our laboratories.

Take miconia (*Miconia calvescens*). This Latin American garden plant has overrun Tahiti, blanketing two-thirds of the island in green gloom under leaves that grow as big as tea towels, pushing forty plants close to extinction. In Hawaii it escaped from gardens in the 1960s, prompting the declaration of a state of emergency.

Operation Miconia, with a budget of half a million dollars, was launched to contain it.

This plant already has a foothold in Australia. A decade ago Peter Stanton of the Queensland Environment Department was warned about miconia by a worried American biologist. Stanton spotted a plant growing in the Flecker Botanical Gardens in Cairns. He asked the garden's director several times to remove it, but nothing was done. When the Department of Natural Resources approached the gardens, again they refused to act until warned that an order would be issued. Miconia plants were then noticed growing in gardens in Innisfail and these too were removed. Seedlings, the spawn of some unknown cultivated tree, appeared around the edges of rainforest in Kuranda and these also were destroyed. Other miconia trees were found growing in botanic gardens in Brisbane, Sydney and Melbourne. The Queensland government is trying hard to evict this tree and we must hope they succeed. A single miconia berry can carry 200 seeds, which are spread about by birds. Miconia could invade rainforests as far south as Brisbane.

According to one federal report, *Recent Incursions of Weeds to Australia 1971–1995*, 295 new weeds have established in Australia in recent years. Many of them (65 per cent) were escaped garden plants, most of which had been in Australia for some time – the daffodil, sweet pea, cotoneaster and poinciana, for example. We can expect that most new weeds in the future will also be escaping garden plants. In other words, tomorrow's weeds are already here.

Invasion biologists talk about the lag effect. Some pests take off as soon as they can, while others remain benign for decades, needing some special event to kick them off, or needing time to adapt to their new home. Athel pine (*Tamarix aphylla*) lent a refreshing shade to outback homes for more than sixty years without causing a stir, until the exceptionally wet year of 1974, when thousands of seedlings sprouted along inland waterways, replacing ancient red gums toppled by the floods. The seeds had washed downstream from remote homesteads. Following one climatic event this tree jumped from obscurity onto Australia's worst weeds list. Mimosa bush (*Mimosa pigra*), another of our worst weeds, kept a low profile for seventy years before exploding onto the weed scene.

In a few cases plants have taken off when new breeding stock came in. Pampas grass (*Cortaderia selloana*) remained a sleeper for almost a century because all plants were white-plumed females, unable to set seed. Then someone brought in a pastel form to add variety, and this pollen-bearing hermaphrodite crossbred with the others, creating a major new weed in the 1970s. In other examples, willows took off in Victoria when male plants came in, and carp numbers exploded in the 1970s when a new strain escaped. In South America, the killer Africanised honeybee was spawned from African and European stock and soon spread to North America.

A German study has found an average lag time in that country of 131 years for shrubs and 170 years for trees. On those counts most garden shrubs and trees in Australia could be ticking time bombs. Australia's *National Weeds Strategy* acknowledges the threat, warning, 'There is a need to recognise and eliminate sleepers during their benign phase or at least identify the events that could turn them into major weeds.' Yes, but how?

Two weed experts, Steve Csurhes and R. Edwards, have tried to predict which of Australia's sleepers will take off next. Basing their work on overseas experience, they came up with almost 300 suspects found in our gardens, ponds, aquaria, or growing as localised weeds. One that caught their eye is sweet thorn (*Acacia karoo*), an acacia with spines longer than pencils, first found sprouting near Perth in 1989. That infestation was destroyed, but weeders have to return each year to quash new seedlings (acacia seeds stay viable for decades).

Noxious Weeds of Australia provides some striking examples of sleepers rising from their slumber. Take stevia (*Stevia eupatoria*), a pink-flowered herb from Mexico that somehow got to Glen Innes in about 1930: 'This small colony remained more or less static until about 1960 when it became more aggressive, moving into adjacent properties by 1965. It is still spreading slowly . . .' And tangled hypericum (*Hypericum triquetrum*), a crop pest in Greece, Turkey and nearby lands: 'Apart from these countries, it is regarded as weedy only in Victoria where it occurs as a single patch of about 2 hectares at Tarnagalla near Dunolly in the north of the state. It is believed to have occurred on this site since 1910 but was not recognised as a potential weed until 1969 when it appeared to be spreading and was proclaimed a noxious weed. Although treated on a regular basis it still persists, varying in abundance from year to year.' It says something about our nonchalant approach that these weeds, although declared noxious and featured in weed books, are allowed to fester on, year after year.

Many of our more dangerous sleepers bide their time in the grounds of our botanic gardens. Mimosa bush, the repugnant claimant of 30 000 hectares of Top End wetlands, was imported long ago by the Darwin Botanic Gardens, presumably as a novelty (its leaves fold when touched). A newer escape is cutch tree (*Acacia catechu*), an Indian weed with hooked spines found sprouting beyond the gardens in 1978. If cutch ever comes to rival its close relative, prickly acacia, the Darwin Botanic Gardens will have contributed two of Australia's worst weeds.

In 1992 botanist Greg Keighery found a strange vine, *Succowia balearica*, sprouting in banksia woodland near Perth. A second patch turned up within the city, and the source of infestation became clear. Succowia had been sown in the Kings Park and Botanic Gardens as an example of a Mediterranean plant. It had crept from the gardens down to the Old Swan Brewery site and the Kennedy Fountain gardens. This vigorous vine has prickly fruit that cling to socks, and a tourist probably carried them unknowingly from the gardens to the lookout east of Perth. Keighery warned that this vine was 'potentially a serious weed to much of the vegetation of the western side of the Swan Coastal Plain and should be eradicated before further spread occurs'. His advice was followed. The gardens were also growing sweet thorn and an

African plant, *Lachenalia mutabilis*, which were also spreading into the wild.

Other botanic gardens carry sleepers too. Australia's only Madras thorn (*Pithecellobium dulce*), a large prickly tree that is spreading in Africa, Asia, the Caribbean and Florida, stands in the Townsville gardens. Melbourne's Royal Botanic Gardens boast Australia's only buffalo thorn (*Zizyphus mucronata*). Other sleepers grow in the Sydney, Brisbane, Adelaide and Castlemaine gardens. Botanic gardens are, in effect, zoos for plants, but with no cages to confine the exhibits. Many botanic gardens are positioned beside bushland, and seeds are free to blow away, wash away in nearby streams, be borne off by birds, pocketed by visitors or carried off on their clothes. Vast assortments of weird and wonderful species are kept under minimal security. The number of plants grown can be enormous. Adelaide Botanic Gardens was growing more than 8000 species back in 1878. Many of these plants are so obscure we know nothing about their prospects as weeds.

Zoos also carry plant sleepers. When South Perth Zoo set up African displays a few years back they put in African acacias, including *Acacia giraffae* (syn. *erioloba*), *A. arabica* and fever tree (*A. xanthophloea*). Acacias are extremely high-risk trees, fever trees sprouting thorns as long as fingers. In a crowning act of folly, the zoo then sold off the elephant and giraffe droppings, acacia seeds and all, as 'Zoo Poo' garden fertiliser. A fever tree was recently found near Brisbane, in the garden of a South African immigrant who was ordered to destroy it. The zoo still has theirs, but the fertiliser is now treated.

Other reservoirs of weeds are those seed collections held by pasture scientists. The CSIRO in Brisbane has 17 000 accessions (collections) of legumes, 12 000 grasses and several thousand strains of exotic bacteria for nodulating legumes. The Australian Genetic Resource Centre in Perth holds 12 000 collections of clover. And new crops are regularly added. The University of Queensland's *Listing of Potential New Crops for Australia* (1993) documents 4500 species, many of them weeds or potential weeds – for example, pitaya (night-blooming cactus *Hylocereus undatus*) and neem tree (*Azadirachta indica*). Interest in new crops is growing among recession-hit farmers and alternative lifestylers, and many of the latter live on farms set among forests.

Queensland University is currently trialling blackbuck as livestock, and this antelope may one day reclaim the Australian outback. As ostrich and deer farms go broke we may see stock turned free to found new colonies. Aquarium fish and water plants provide a colourful pool of potential pests, as we have seen. Aviary birds pose a risk too. More than 250 kinds of foreign birds are bred in Australia, including many that are feral overseas. The monk parakeet, which is sometimes seen on the wing here, is a prime candidate for naturalisation. In its native South America it nests in eucalypts, and I have seen it feeding in Australian paperbarks in Florida, suggesting a penchant for Australian trees. Another possibility is the Sudan golden-headed sparrow (*Passer luteus*), a close relative of the house sparrow; according to Environment Australia there are 586 of these birds in captivity here. I can only imagine how many diseases are held in our laboratories, and how many exotic animals, plants and fungi are held at universities and other institutes.

An interesting exercise is to put the shoe on the other foot – to imagine which Australian plants and animals will take off next overseas. The possibilities seem endless. Our trees, wattles especially, are popular in land reclamation, promoted as part of our foreign aid with no thought given to weed problems; they will continue to go feral. More of our ornamental plants will escape as well, the heath banksia (*Banksia ericifolia*) in South Africa a very recent example. Our freshwater crayfish are proving popular in aquaculture because of their big tails and high meat-to-body ratio, and the redclaw has already escaped in Samoa; yabbies (*Cherax destructor*) may be next. Australia exports silver perch and barramundi to fish farms overseas, and our rainbowfish (*Melanotaenia*) are popular in foreign aquaria (the largest producer of Australian native fish is in Florida). Our green treefrog (*Litoria caerulea*) is a popular pet in the United States, known there as 'Whites treefrog' or the 'dumpy frog', and it seems likely to naturalise in southern Florida, where the climate is ideal. Les Edwards in *Tropical Fish Hobbyist* describes it as 'my all-time favourite' and 'among the most appealing of all frogs you could desire to keep in captivity. The charming face, with its huge "smiling" mouth, has endearing qualities that separate it from all other species.' I can imagine the response of my pest colleagues in Florida when Australian green treefrogs start hopping about in their Australian paperbark forests.

Sleepers, fortunately, are a problem that Australians can do something about. Botanic gardens can be persuaded to evict their riskier tenants or, at the very least, trim off their sinister seedpods. They should be monitoring escapes into nearby bushland as a matter of course. Gardeners can temper their enthusiasm for new garden plants, farmers for new crops. Garden shows could highlight the dangers. Governments could try that little bit harder to deport sleeper weeds such as stevia and tangled hypericum. The CSIRO could keep their seeds in their vaults. It's all a question of closing the gates before the horses bolt.

31

Knocking at the Door

The Next Wave of Invaders

> 'The threats from introduced pests are certainly more immediate than any risk of military attack on Australia.'
>
> **Professor Ian Lowe, *New Scientist,* 1997**

In 1980 someone smuggled into Taiwan some gigantic South American snails, as big as apples, called golden apple snails (*Pomacea canaliculata*). They proved very quick growing, fecund and voracious, suggesting they might be worth farming for the *escargot* trade. Over the next few years they were taken to the Philippines – in a rural livelihood project endorsed by the Department of Agriculture – China, Japan, Korea and all over South-East Asia.

But the snails did not taste good. Many people, even in very poor villages, refused to eat them. Export opportunities evaporated. Then the snails escaped and multiplied into voracious pests. Farmers in the Philippines lost up to a quarter of their rice. An economist at the Institute for International Studies at Stanford, Rosamond Naylor, estimates that Philippine rice growers lost somewhere between US$425 million and US$1200 million to this snail in 1990 alone. What it has cost Asia in total can only be imagined.

Golden apple snails are now poised to conquer Australia. They infest two of our main trading partners, Japan and Korea, and our nearest neighbour, New Guinea. Their distinctive, bright pink eggs have already been intercepted in imported cargo containers. Can we keep this 'miracle snail' out forever? I very much hope so, but I doubt it.

We have kept out another pernicious snail, almost – the dreaded giant African snail (*Achatina fulica*). An even bigger snail (it grows to 20 centimetres long), it attacks crops and gardens and is even a road hazard, causing cars to lose traction and skid out of control. It too was spread around Asia as a food – Japanese soldiers took it to New Guinea during World War II. When this snail appeared at Gordonvale near Cairns in 1977 quarantine officers leapt into action, throwing tarpaulins over infected gardens and pumping in methyl bromide. The colony was eradicated. We are lucky that Gordonvale is walled in by sugar cane. Had these snails slimed their way into the Queensland rainforests our national parks would never have recovered. AQIS, who rate this pest a top priority, often find it hidden in ship's cargo. But the recent Nairn review was concerned to find that containers from snail-infested countries are sometimes unloaded without any inspection or treatment.

If these snails are two foreigners we should be especially worried about, what else is out there? What other invaders are we likely to face in the next few years?

There is the Asian honeybee (*Apis cerana*), a small relative of our introduced honeybee. In the late 1970s it turned up in Irian Jaya, probably brought in by Javan immigrants for its honey (though the yield is meagre). It swarmed into New Guinea in 1985, reaching the south coast opposite Cape York in 1992, taking by surprise the AQIS officers who were monitoring its spread. The next year it claimed the Torres Strait islands of Saibai, Boigu and Dauan, just south of New Guinea. AQIS hopes to stop it there. The bees can't fly thirty-five kilometres to the islands in the middle of Torres Strait, though they try, swarms sometimes descending upon boats at sea. Hives have been discovered hidden inside imported machinery. They will nest in four-inch pipes, and AQIS fears they may sneak into the country by roosting in the air vents of containers. In 1996, 2000 of them (all dead) and seven honeycombs were found at the Adelaide docks inside an empty container imported from Singapore. More seriously, a thriving hive was found in Darwin, near the wharves, in June 1998. Those bees were swiftly destroyed, but AQIS fears there may be more swarming through the mangrove swamps around Darwin and that the war may already have been lost. 'We're talking about a very invasive organism here', David Banks of AQIS told me. What's more, these bees will bring in the varroa mite, a serious pest of honeybees overseas (the Adelaide bees were mite infected).

European honeybees are doing so much harm to Australian ecology that a second invader might be devastating. They would steal tree hollows and nectar from native birds, threatening parrots and honeyeaters. They would feed on some of the smaller native flowers that European bees ignore, further disrupting pollination ecology. Asian honeybees live as far north as Vladivostok, so they could readily colonise most of Australia.

Diseases also loom large among future invaders. AQIS recently lost the race against one of them, Japanese encephalitis, a virus that has been on the move since the 1970s, when it crossed from eastern Asia into India, sparking major epidemics. In Indonesia it stayed west of Wallace's Line until migrating birds brought it east. In 1995 it struck Torres Strait, killing two islanders. Then in 1998 it reached far north Queensland, where hordes of feral pigs provide ideal hosts. This virus does most harm to people, horses and pigs. Most humans suffer only mild symptoms, but some victims die or suffer brain damage. In 1997 a senior scientist at the Queensland University of Technology was suspended for handling this disease in an insecure lab. We will hear more about this 'bug' as it wends its way south to our major cities.

The next pathogen to breach our borders may well be Newcastle disease, one of the diseases AQIS fears most. This killer could devastate our poultry flocks and ignite fatal epidemics among native birds, disrupting Australian ecology and even causing extinctions. Overseas, whole flocks have been wiped out. A virulent form reached Irian Jaya in the 1970s. Sentinel flocks of chickens are kept in north Queensland to test for its spread. In 1996 a Brisbane woman was fined $8000 for smuggling in 3000 doses of Newcastle disease vaccine to treat her chickens; live virus was cultured from the vaccine and proved highly pathogenic. The poultry industry claims that cooked chicken imports could bring in this disease, but AQIS insists otherwise.

Under the Australian Veterinary Emergency Plan (Austvetplan), Australia has contingency plans to extinguish major new livestock diseases, which do get in from time to time. Avian influenza was erased from a Victorian chicken farm in 1992 by the controlled slaughter of 120 000 chickens and ducks and half a million eggs. The cost: more than $2 million. In 1996 a screw worm-infected dog was brought illegally from New Guinea to Saibai Island; it was killed and the owner charged. In 1992 a woman from Wagga Wagga came home from a visit to Brazil with New World screw worms burrowing in her neck.

Apart from these livestock diseases there is an American disease, *Aphanomyces astaci*, that could exterminate our freshwater crayfish, a fungus (*Spongiophaga communis*) that could wipe out tropical sponges, and some devastating diseases of foreign eucalypt plantations. Guava rust (*Puccinia psidii*), for instance, could defoliate our eucalypt forests. We can never take exotic diseases too seriously. No one could ever have foreseen that Australia would lose five frog species to a fungus, or millions of pilchards to a virus, or whole forests to an obscure plant disease called phytophthora.

Insects are another class of invader AQIS is especially watchful for – for example, the Asian gypsy moth (*Lymantria dispar*), a major tree-killer, Africanised honeybees, giant honeybees and the

European corn borer (*Ostrinia nubilalis*). Gypsy moths travel well on ships (eggs turned up on a vessel berthing in Brisbane in 1997). New fruit flies (*Bactrocera* species) are a sure bet as future invaders. The islands of Indonesia and the South Pacific host dozens of species. So much petty fruit-smuggling goes on through airports that one day all of these flies, having evolved in isolation on each island, will end up dispersed throughout the region. Asian flies often turn up in northern Australia, blown down by winds or travelling on fruit, though they usually don't survive.

The Northern Australia Quarantine Strategy has a list of forty-one priority weeds it wants kept out. Two of them, Siam weed (*Chromolaena odorata*) and mile-a-minute (*Mikania micrantha*), were recently discovered in north Queensland and the fight is on to extricate them.

The brunt of future attacks will be borne by northern Australia, and by Cape York in particular. For millions of years Australia has been inching inexorably towards Asia. We are not as isolated as we might think; we lie uncomfortably close to Borneo, Sumatra and west Malaysia, biologically the richest lands on earth, the home of a cornucopia of tropical plants, animals and diseases. Quarantine in South-East Asia is ineffective and new pests are appearing every year. In Indonesia more and more people are on the move, trading and farming along the long island chain that points towards Australia. Since Irian Jaya has become an Indonesian province Javan immigrants have flooded the territory, bringing in livestock and crops and diseases and weeds. New pests soon cross the land border into New Guinea and once there, Torres Strait is the only barrier. Many pests have travelled this path in recent years, including Siam weed, papaya fruit fly, banana skipper (*Erionota thrax*), varroa mite and of course the Asian honeybee and Japanese encephalitis. In future we may even get the Asian shrew (*Suncus murinus*).

Dotted with islands, which attract a constant traffic of small boats, Torres Strait presents a feeble barrier. Birds wing back and forth and insects are blown southwards. The Torres Strait Treaty guarantees free movement of indigenous people, and though they are forbidden to carry fruits, vegetables (except yams), animals, hides or soil, these regulations are impossible to enforce. As much as half the population travels between the islands each year. On any island, at any time, there are scores of visitors from other islands or from New Guinea. When I worked at Cape York in 1988 the police sergeant at Bamaga admitted to me he had no idea how many islanders were visiting. Saibai villagers were moved to this township when their island was flooded in the 1940s, providing a strong cultural link between mainland Australia and New Guinea. As well, there are fishing trawlers and private yachts, and smugglers trafficking cannabis ('PNG gold') from the New Guinea Highlands. It all adds up to a quarantine nightmare, one that Australia's Northern Australian Quarantine Strategy is expected to contain. AQIS officers educate islanders and Papuans, yachties and fishermen about Asian bees, sick animals, hidden pests and undeclared foods. Traps are set and searches undertaken. Australia's best defence against some of these threats is not the Strait itself but the belt of empty land south of Cape York – a designated livestock-free zone. Australia has even instituted biological control of new pests in New Guinea, notably Siam weed and banana skipper, as an advance shield of protection.

New pests can enter through any of our ports, however, especially new ballast pests, seed contaminants and wood-boring beetles. Japan has become a giant relay station for ballast invaders. Our neighbour and trading partner New Zealand is another major source of new pests. New Zealand gave us nodding thistle (*Carduus nutans*), the European wasp and the cabbage white butterfly. In 1992 we received their European bumblebees and more recently the Chilean mite (*Phytoseiulus persimilis*). In 1990 there were plans, which never eventuated, to import more than a dozen exotic bees as crop pollinators. We should have a say in decisions like that.

Nor should we forget all the legal immigrants. Australia still imports new garden, pasture and land-reclamation plants, many of which will certainly become weeds. Agronomists would like to bring in new earthworms to aerate soil and bumblebees to pollinate crops. New biological control agents will keep flowing in, and while most will be helpful the risk of another error is very real.

Immigration has always been a hot topic in Australia, racism and parochialism playing their part in the often heated debate. But the potential impact of human immigration pales into insignificance compared to that of non-human immigration. I for one don't want to end up living next door to a hive of Asian honeybees.

STOP PRESS: Less than a fortnight before the last editing changes to this book were made the CSIRO announced that a major new pest, the black-striped mussel (*Congeria sallei?*) of Latin America, had invaded three marinas in Darwin, presumably arriving on the hull of a visiting yacht. A close relative of the notorious zebra mussel, it is the first serious pest to breach our tropical waters, the CSIRO predicting it could cause 'massive fouling of wharves, marinas, recreational and inshore vessels ... and marine farms (e.g., pearl oysters)' as it has done in India and Hong Kong. The marinas were poisoned and all the mussels killed, and the search is now on to locate the 400 yachts that had passed through the infected marinas before the mussels were found.

32

Whither the Wet Tropics?

A Hot Wet Case Study

'The fairest things have the worst fate.'
Francois de Malherbe

Do we have our priorities wrong? A few years ago greenies waged war against the Kuranda Skyrail, a tourist cable-car near Cairns whose construction damaged a small but symbolic strip of rainforest. But conservationists have failed to confront the much larger threat posed by exotic pests, a threat that puts the whole of the Wet Tropics at risk. How many green activists have even heard of harungana, thunbergia or palm leaf beetle? Yet these invaders, along with sanchezia, turbina, phytophthora and a battalion of other pests, are poised to destroy more rainforest than a hundred cable-cars ever could.

Unlike bulldozers and construction gangs, exotic invaders are easily overlooked, and herein lies part of the problem. When I toured the Wet Tropics recently I was delighted by how few pests I saw. Earlier I had visited Adelaide, where entire hillsides have surrendered to olive trees and rabbits, and the creek lines are war zones of weeds. The contrast with the north was astounding. Most Wet Tropics weeds I saw kept obediently to roadsides and riverbanks, and even the lantana seemed well behaved. Intact rainforest stands up well against weeds. The only discordant scenes I saw were patches of rainforest soil grubbed up by pigs. But I have learned since then that the Wet Tropics, Australia's biological crown jewels, are poised for invasion, and that no other part of the country stands to lose so much in the feral future.

The evidence is compelling. When I investigated the aquarium industry for this book I realised that tropical fish are most likely to go feral in the Wet Tropics. Mystery snails and golden apple snails will probably invade the same region, according to scientists I spoke to. I read up on exotic earthworms and discovered that of forty-five species known in Australia, most of them confined to the temperate half of the country, only one is a serious invader – the Amazonian earthworm found in Wet Tropics rainforests. Then I read about miconia, papaya fruit fly, giant African snails, disappearing frogs, the sudden spread of cabomba and hymenachne, and the recent arrival of Siam weed and mile-a-minute, two of the world's worst weeds – and all of them have their sights set on north Queensland. The final straw came when Steve Goosem of the Wet Tropics Management Authority sent me a bundle of reports on a raft of pests I had never heard of before, including harungana, sanchezia, turbina, clitoria, hemigraphis and palm leaf beetle – all of them threats to the north.

During my stay in the Wet Tropics I was thrilled to come upon a cassowary and chick on the dirt road up Mt Lewis. But cassowaries, it seems, are having a tough time. Their rainforest sanctuaries are shrinking, cars are hitting them, dogs are mauling them and feral pigs are eating their food. At Mission Beach, a cassowary stronghold, pigs plough the forest floor in search of tubers, seeds and the same fruits cassowaries eat. They gobble cassowary eggs and may also pass on disease. Autopsies on road-killed cassowaries have found substantial tissue damage inflicted by mycobacteria. Wallowing pigs may be infecting waterholes where the big birds drink. One of the worst pests in the north, pigs also erode rainforest soil, dismember tree roots and almost certainly spread phytophthora and the cocoons of Amazonian earthworms. A Wet Tropics study found them guilty of 'soil compaction in wallows, trail formation, ploughing of the soil surface, deep excavations along creek banks, destruction of seedlings, bark damage on specific territorial marker trees, the occurrence of the vigorous introduced grass *Paspalum conjugatum* on frequently disturbed soil' and so on. Pigs threaten two rare palms (*Normanbya normanbyi* and *Gulubia costata*), whose fruits and seedlings they eat. Indeed, their baneful reputation has a long history, to judge by the complaint of a north Queensland Aboriginal woman who charged them with stealing traditional tucker in 1895.

The role of phytophthora in the Wet Tropics remains uncertain. When patches of dying rainforest were found at Eungella, west of Mackay, in 1975, the Queensland Department of Forestry swung into action, initiating a seven-year study by Dr Bruce Brown. Because the trees succumbed inside virgin forest along ridgelines churned by pigs, Brown guessed that pigs were spreading the spores. Dead patches turned up around pig diggings near Ingham and at Crediton along logging roads. Tractors used to farm pineapples (a very phytophthora-prone crop) may have carried the curse to Eungella. After taking soil samples from about 800 sites in Wet Tropics rainforests, Brown found phytophthora in more than 200 of them, mostly from places without dying trees. Amazingly, another nine *Phytophthora* species turned up as well, most (but not all) thought to be exotic. *P. cinnamomi* does best in waterlogged acidic soils, and it may well pose a threat to rare trees, especially the endemic species unique to Mt Lewis, where the disease is spreading, although we can't be certain of this because phytophthora in north Queensland was all but forgotten after Brown's study ended in 1982. The Wet Tropics Management Authority revived the issue at a workshop in 1998 because they fear the disease will spread with road works and with revegetation projects using infected nursery stock. Conservationists should be campaigning for more research on phytophthora.

For rainforest palms a greater threat may well be the palm leaf beetle (*Brontispa longissima*). Native to tropical Asia, this little beetle hitchhiked south to Cairns in 1992, and now attacks both wild and garden palms, costing nurseries $30 million a year. Young fronds are chewed and whole trees sometimes die. Rare palm forests could disappear. A parasitic wasp enlisted for biocontrol has yet to prove useful. Asian honeybees and Asian and Pacific fruit flies are also poised to invade. We may end up with foreign bees pollinating our rainforest blooms and exotic fruit flies infesting the fruits, leaving fewer to ripen for cassowaries (if any survive).

When World Heritage listing was conferred in 1988, twenty-five weeds were identified as threats to the Wet Tropics (which include woodlands and wetlands as well as rainforests). A decade later that number had risen to fifty-three. The list will keep growing as more plants escape and as more existing weed problems are identified. Two of the world's worst weeds were found in the region only recently – Siam weed (*Chromolaena odorata*) in 1992 and mile-a-minute (*Mikania micrantha*) in 1998. They were the first sightings of each weed in Australia. Authorities are determined to eradicate these two. I wish them luck.

In the rainforests the worst invaders are coffee and four ornamentals with curious names – thunbergia and turbina (*Turbina corymbosa*), both vines; the tree called harungana (*Harungana madagascariensis*); and the shrub sanchezia (*Sanchezia parvibracteata*). Harungana worries Steve Goosem most because it spreads with such ease. From what was probably a single tree planted in Babinda in the 1930s it has invaded rainforests high on the slopes of Mt Bellenden Ker, an important national park. Birds drop its seeds into clearings punched out by cyclones. The Wet Tropics Management Authority is determined to stamp out this tree, and teams of weeders hike high into the hills to poison each one.

The ranks of weeds in the north will keep growing. People are flooding into the region, buying up land abutting rainforest, planting exotic fruit trees and palms (most of which originate in rainforest), and practising permaculture. Hundreds of unusual tropical plants are cultivated, including shade-tolerant shrubs, and their seeds are easily imported. The Wet Tropics rainforests, notwithstanding their resistance to invasion, may one day become the weed metropolis of the north.

The weeds there today tell a familiar story. Most are escapes from cultivation. They include foods (coffee, mango, coconut, guava, prayer plant), timber trees (Caribbean pine [*Pinus caribaea*], East Indian mahogany [*Chuckrasia velutina*]), aquarium plants (cabomba, salvinia), garden plants (allamanda, mother-in-laws tongue and black-eyed susan, which make up almost half the total), and an appalling number of pasture plants (hymenachne, para grass, molasses grass, guinea grass, calopo, centro, glycine and puero). Garden plants, as usual, top the list with twenty-nine species. Only *two* of the fifty-three weeds entered Australia accidentally.

'The whole issue of horticultural plantings is a very serious one', warned Stella Humphries and Peter Stanton in a 1992 report to the Wet Tropics Management Authority: 'in rainforest areas bordering residential properties we found many species naturalised, but so far, restricted to garden perimeters'. These proved 'too numerous to systematically identify and observe'. Residential development is new to much of the region and weed problems lag a hundred years behind those in Brisbane, Sydney and Melbourne, where the fallout is so appalling. Goosem worries about the influx of palms. Native palms are a striking feature of the lowland rainforests but many hobbyists are now growing unusual exotic types. When they mature, native birds will find their fruit, and foreign palms will start sprouting in national parks. North Queensland's climate could not be more favourable for exotic palms and tropical fruits.

Unusual plants already dominate the weed list for the region. Sweet prayer plant (*Thaumastochloa danielii*), imported a few years ago as a food sweetener and now traded in local markets, has become so 'extremely aggressive' that the original importer, David Chandlee, wants it eradicated. Another invader is a club-moss called peacock fern (*Selaginella willdenovii*), with resplendent blue-green metallic-looking leaves. Vigorous vines and plants that endure deep shade are the ones to watch. Cyclones advance their plans by stripping back the forests from time to time. Two hundred years from now, will such plants as coffee, African tulip and wandering jew have probed the rainforest's depths, or will they have been held off at the ragged edges? I am optimistic that the rainforest can withstand most of these invaders. To be sure, pigs, phytophthora, the frog fungus, palm leaf beetles, African snails and a few garden plants are probably unstoppable, but intact rainforest is wonderfully resilient and most pests will need disturbance to invade. While it is true that large areas were roughed up in the past by logging, and that roads, powerlines, cattle, pigs and fires are doing much harm today, there are still intact forests on mountains where invaders may be held at bay.

I am less sanguine about the paperbark woodlands, another important Wet Tropics habitat. Pond apple (*Annona glabra*) from Florida, considered the very worst Wet Tropics weed, is the bane here. Used as a rootstock for custard apples, it's a tree few people know about but one that is fast remaking the north. Pigs, cassowaries, streams and tides spread its seeds, and seedlings are sprouting in their hundreds behind mangroves, along rivers and in paperbark groves. They mature quickly into gloomy pond apple forests (see colour plate), a desolate new habitat. Ocean currents are bearing the seeds northwards, and computer models suggest pond apple could thrive anywhere between Western Australia and Brisbane. In north Queensland it is traded at markets as 'caramel custard apple' and sold as a novelty, although the fruits are not very pleasant. Humphries and Stanton rate it a problem of national, indeed international, significance. 'If nothing is done,' they warn, 'the melaleuca communities in the wet tropics will progressively be lost . . .' Biological control is an unlikely solution because custard apples would suffer, and perhaps rare native trees as well; fire is one of the few means of control.

Pond apple is troublesome enough, but there are other weeds in those wetlands, including hymenachne, a rampant pasture grass that is one of Australia's worst weeds, the shrub *Clitoria laurifolia* and rogue pine trees. As well, African big-headed ants (*Pheidole megacephala*) now live inside the bases of epiphytic ant plants (*Myrmecodia beccarii*). Unlike native ants, which they evict, they don't pollinate the ant plant's flowers, disperse its fruits or tend the caterpillars of the apollo jewel butterfly (*Hypochrysops apollo*), a bright orange insect whose caterpillars live in ant plants. For these reasons, and because of habitat loss, the ant plant is listed as vulnerable and the butterfly is endangered.

So how will these wetlands be faring in two hundred years? I imagine waters alive with firemouths, green terrors and tilapia, toad tadpoles nibbling the cabomba, pond snails and golden apple snails browsing the fringing hymenachne and mimosa, and forests of pond apple thriving near stands of Caribbean pines with a Siam weed understorey. Aagh! This scenario, unfortunately, is all too plausible.

I hope north Queensland never ends up like Hawaii and Florida, the two most weed-infested and tropical regions of the United States. Hawaii, where exotics outnumber natives, is the extinction capital of the US and probably the most feral place on earth. An island ecosystem, it is more sensitive to invasion than Queensland, so direct comparison is not valid. But Florida *is* comparable. Its pest problems from escaped garden plants, fish and snails are horrific. After Cairns won an international airport in 1984, tourists from all over the world, many with mud on their boots and seeds clinging to their socks, began flocking into our 'pristine' rainforests. Ecotourism, when you factor in all the spores, eggs and burrs falling from shoes and tyres, is not as eco-friendly as most people think.

The Wet Tropics are in for tough times. Indeed, all national parks in the north are seriously at risk from pests. In Lakefield National Park on Cape York Peninsula, lion's tail (*Leonotis leonurus*) appeared a few years ago, the infestation exploding from a couple of hectares in 1992 to thousands of hectares in 1998. Kakadu, one of our most precious parks, will lose ground to cane toads, pasture grasses and legumes, Asian honeybees, big-headed ants, apple snails, tilapia, cabomba, pond apple and all the rest. Para grass has exploded in Kakadu since buffalo were controlled. A ranger there recently put out an Internet appeal: 'We are desperately trying to control para grass (*Brachiaria mutica*), which is spreading rapidly in Kakadu . . . One of our other major problems is *Crotalaria goreensis* [a pasture legume], which is spreading explosively in the drier parts of the Park. Any information on this one would also be appreciated . . .' Kakadu is another war zone where conservationists are battling development (a uranium mine) while ignoring the impact of pests. Do we have our priorities wrong?

33

The Homogocene

Visiting the Future

> 'The wilds of the new millennium are melting into degraded landscapes infested by exotic weeds, weakened by exotic pathogens, chewed over by exotic browsing mammals.'
>
> **Chris Bright, eco-author, 1998**

Some invasion biologists believe that the earth is about to enter a new era, one they dub, with wry humour, the Homogocene. The richness of planetary life will give way, they predict, to a simplified series of homogenous ecosystems. Ecology will take the same road as human culture, with a few dominant species, including ants, mosquito fish and leucaena, taking over from thousands of local alternatives.

Charles Elton said it first back in 1958 in his classic text *The Ecology of Invasions*: 'If we look far enough ahead, the eventual state of the biological world will become not more complex but simpler – and poorer. Instead of six continental realms of life, with all their minor components of mountain tops, islands and fresh waters, separated by barriers to dispersal, there will be only one world, with the remaining wild species dispersed up to the limits set by their genetic characteristics, not to the narrower limits set by mechanical barriers as well.' Earth will return to the state that prevailed during the early Jurassic, 200 million years ago, when one giant continent, Pangea, straddled the globe. Back then, many species must have occupied enormous tracts of land, just as house mice and big-headed ants do today.

A single supercontinent, according to a recent analysis by American biologists, could support only about half the animal species alive today, dooming the others to extinction through competition. Commenting on that study, biologist Gábor Lövei writes bluntly: 'If we cannot stop this McDonaldization of the biosphere, we stand to lose a substantial part of global biodiversity ...' Eco-writer David Quammen predicts a remarkably bleak future in what he calls 'the Planet of Weeds'. 'Having recently passed the great age of biogeography,' he writes, 'we will have entered the age *after* biogeography, in that virtually everything will live virtually everywhere, though the list of species that constitute "everything" will be small.' (Isaac Asimov has speculated on a future world in which exotic pests even infect distant space stations. We could end up with intergalactic weeds.)

San Francisco Bay may provide a foretaste of that future. More than 200 exotic creatures, including crustaceans from Australia, crowd its waters, completely dominating the local ecology. Since 1970 new species have been accumulating there at the rate of one every twenty-four weeks. The bay 'offers a special window on where the global game of mix and match is leading at its most extreme', writes William Stevens in the *New York Times*. 'Species from other oceans now dominate the bay, and more are being crammed in all the time. They cover the bay's bottom virtually wall to wall, and no part of the larger bay ecosystem has escaped their impact; in some places they appear to account for all life.

'They have, in fact, created a brand new ecosystem and perhaps, in time, will write a new chapter in evolution.'

But will the future everywhere look like this? I doubt it. Busy harbours are a special case. Melbourne's Port Phillip Bay, with its 145 recorded exotic species, is certainly hurtling in this direction, but most habitats will not suffer this much, neither will homogenisation proceed this far. Australia will never get all the big animals of Africa or all the fish in the Amazon. To be sure, plants, insects, diseases and some marine species will keep on travelling, but other life forms, especially frogs, reptiles, mammals and fish, won't find it so easy. Australia's harsh soils and climate will keep many invaders at bay. Certainly areas of native vegetation will continue to shrink and more and more disturbed habitats will emerge, many of which will take on that international look. The Mediterranean regions of the world (including southern Australia) are trending in this direction, the same birds, snails, grasses and thistles becoming dominant in widely dispersed regions. Alfred Crosby makes much of this phenomenon in his exaggerated book *Ecological Imperialism: The Biological Expansion of Europe.*

In 1996 I flew to Florida expecting to see another example of this process. Southern Florida and southern Queensland have traded dominant trees, and each now carries forests of paperbarks and slash pines. The paperbarks around Brisbane have an understorey of native blady grass (*Imperata cylindrica*) and North American groundsel bush (*Baccharis halimifolia*), two plants that thrive in Florida as well. I thought I might find the same hybrid habitat emerging in both places. But Florida's paperbarks grow together so densely that nothing, not even grass, can sprout beneath them. In future, when enough Australian insects have been brought in for effective biocontrol, the canopy may thin out, and blady grass – a hated weed there called 'cogon grass' – may invade. But as yet the only forests in Florida that reminded me of home were some stands of mixed paperbarks and pines that looked like thickets I have seen near Noosa.

Instead of homogenisation we may see the birth of entirely new habitats spawned by exotic plants, freed from their normal constraints, sprouting in plague numbers. Already there are Jerusalem thorn and mimosa thickets in the Northern Territory, pond apple stands in north Queensland and camphor laurel forests around Lismore. In *The Nature of Cleland*, a guidebook to Cleland Conservation Park near Adelaide, Anne Hardy paints an idyllic picture of the 'Olive Community' with its understorey of lavender, dog rose and blackberry and its fauna of rabbits, blackbirds, native finches and frogs. Here is a cheerful acceptance of the feral future. A true South Australian, Hardy accepts that the weeds are there to stay and gets on with the job of describing them. She grants the Olive Community the same legitimacy as the Blue Gum and Stringybark communities.

Pure forests of weeds will be one extreme. But Australia's native trees are highly competitive, and we are more likely to see mixed habitats forming, with natives and exotics jostling for dominance. Eucalypts and pines favour the same sandy soils and we may see vast mixed forests of the two, along with paperbark–pine woodlands and rainforests with a coffee understorey. Nineteenth-century naturalist the Reverend William Woolls foretold the rise of hybrid habitats over a hundred years ago when he wrote, 'it is not too visionary to predict, that, in the course of another Century, many of the native plants will exist only in enclosed or remote places; whilst a mixed Flora, adapted to the altered circumstances of the colony, will usurp the place of the past vegetation.' The new habitats that emerge will eventually find their way into our nature guides and documentaries.

Australia's native wildlife won't always suffer from these changes. Many animals have already found a niche in the new habitats that are forming. Lantana (*Lantana camara*), one of our most bountiful weeds, has become a valued resource for wildlife. Wallabies and bandicoots shelter among its thickets, whipbirds, fairywrens and scrubwrens forage among its foliage, Richmond birdwing butterflies sip the nectar, native birds take the fruits and reed bees nest in the stems. Unique among weeds, lantana has become a keystone species for many animals, providing a greater range of resources than almost any native plant. Unfortunately, lantana also has a downside, for it takes over farmers' fields and shades out native plants, such as endangered native jute (*Corchorus cunninghamii*).

All over Australia, native animals are learning to live with weeds. The marram grasslands behind beaches furnish feed for wombats, shelter for small marsupials and nest sites for oystercatchers, plovers and fairywrens. The camphor laurel forests in New South Wales sustain vast flocks of fruit pigeons; indeed, they owe their very existence to the birds' enthusiasm for the fruits and dispersal of their seeds. Most fruit- and seed-eating birds in Australia now take exotic foods, and dietary studies show that parrots and pigeons in many regions are now weed dependent. Long-billed and western corellas are extreme examples: having lost their links with natural habitats, they live today in farmland and feed upon crop seeds and weeds. They are model citizens of the feral future.

A great many rare and threatened species have also become weed dependent. The vulnerable black-breasted button-quail relies upon lantana for shelter, having lost most of its dry rainforest habitat to farming. As protection against foxes, rare southern brown bandicoots use blackberry brambles and southern barred bandicoots hide in gorse. The nearly extinct Norfolk Island parrot lives almost entirely upon olives and cherry guavas (*Psidium cattelianum*), and the endangered orange-bellied parrot and northern hairy-nosed wombat are also weed dependent. Ironically, to conserve these species, we must consider the weeds they rely upon. Conflicts have already arisen over the idea of planting gorse for barred bandicoots. In future, as more and more endangered species turn to weeds for shelter and sustenance, we may end up by planting exotic pests simply to save rare animals. I will have more to say about this trade-off in the sequel to this book, *The New Nature*.

Exotic animals have also become important in Australian food webs. Rabbits, black rats and house mice now sustain a large proportion of Australia's birds of prey. Around Mildura, young bunnies make up 60 to 90 per cent of the diet of local eagles, goshawks, harriers, kites and falcons. Most wedge-tailed eagles in Australia have sampled exotic fare and our raptors are suffering grievously now that calcivirus is removing rabbits from the menu. In central Australia house mice make up as much as 97 per cent of the diet of barn owls, and in our cities peregrine falcons, once thought to be endangered, are growing plump on street pigeons. In the feral future, natives and exotics will become more and more interdependent.

If new habitats are destined to arise in the Homogocene, so too are new species. Many exotic pests, stranded in new lands, will one day evolve into something new – a prospect Elton did not foresee. We can expect evolution to proceed rapidly wherever exotic colonies are founded by only a few individuals with a narrow genetic pool. Cane toads all over the Pacific may radiate into a suite of new species, which biologists could call *Bufo australis, Bufo novaeguineae, Bufo hawaiiensis* and so on. The Australian rabbit could be called *Oryctolagus destructor.* As soon as invaders are bequeathed new, local names they become our special responsibility, and those whose numbers dwindle in future may be the objects of conservation concern. Tomorrow's preservationists may run campaigns to save the Kosciuszko trout and the Kakadu mimosa from extinction. Australia may also become responsible for conservation of exotic species that are disappearing in their native homes, such as banteng and the pink gladiolus (*Gladiolus caryophyllaceus*), a weed in Western Australia that is rare in its native South Africa. New Zealand could accept responsibility for saving the parma wallaby, golden bell frog and green swamp frog, three rare species in decline in Australia.

The evolution of new pests is already under way. Electrophoretic studies show that Australia's house sparrows are unique. British geneticists Parkin and Cole have found 'divergence from the ancestral British stocks that can be explained as being due to an initial period of inbreeding followed by a similar rate of differentiation'. New Zealand's parma wallabies, after a century of overcrowding on

Kawau Island, have grown smaller and breed later, the females at two years compared to one year in Australia. These differences appear to be genetically fixed, for Kawau parmas fed supplements remain small. New Zealand rock wallabies and swamp wallabies may also be shrinking, but her thriving brushtails have grown. The European wasps (*Vespula germanica*) in Australia may also be winging their way towards a new horizon. Most colonies in the northern hemisphere die away in winter, but ours keep expanding, year after year, recruiting more and more queens, rearing more and more workers, growing to a hundred times the size of European colonies. This trend is significant because European wasps destroy vast numbers of native insects; overseas they sometimes kill people.

New species may also arise through hybridisation. Black ducks are mating with mallards in southern Australia (and in New Zealand), Cootamundra wattles are crossing with local wattles (after escaping from gardens), and in Western Australia a native club-rush (*Isolepis cernua*) is interbreeding with an African invader, budding club-rush (*I. prolifera*). Some of our most successful weeds – notably lantana (bred up in European hothouses hundreds of years ago) and cord grass (*Spartina townsendii*), spawned by English and American cord grasses – are themselves artificial species. An obvious example of a new species, I believe, is the 'Australian pine' invading Florida. A hybrid of two she-oaks, it has the height of one and the leaning limbs of the other, but doesn't look much like either. A few years ago the classification of she-oaks was revised and many new species were named, some of them derived by hybridisation. Many are less distinctive than this tree, so it could certainly be recognised as a new species – say, *Casuarina horribilis* or *Casuarina americana*.

Exotic species will also direct the future evolution of our wildlife. Honeybees now pollinate such vast numbers of shrubs and trees that biologists expect our native flowers to change shape to suit bee visits. Asian honeybees and bumblebees will further this trend. We will see native shrubs evolving larger spines to fend off goats and cows, snakes developing immunity to toad toxins, plants immune to phytophthora, grasshoppers that can eat pine needles, marsupials more alert to foxes, and sea creatures that love the taste of Japanese kelp. Most of these changes, and a great many others, are probably under way already. Together they will amount to a profound shift in Australian evolution. The future ecology of this continent is being altered irreversibly in a million different ways.

The Homogocene, the Planet of Weeds, or whatever we choose to call it, will be an unfamiliar place in many respects. There will be new kinds of forests containing new species and old, but many creatures we know today will be missing. This is the point to ponder: our endless shuffling of the biological pack is creating many more losers than winners. As the authors of this confusion we have a responsibility to try to manage the outcome, to play a stewardship role, to save those species most at risk, and to preserve a sample of all the natural habitats in their original state. Unfortunately, this task requires a level of vision and resources that we have so far not proved willing to provide.

34

The New Architects

Redesigning the Land

> 'They wept like anything to see
> such quantities of sand:
> "If this were only cleared away,"
> They said, "it would be grand!"'
>
> **Lewis Carroll, *Through the Looking Glass,* 1872**

All along the shores of southern Australia, weeds are redesigning our beaches. Steep hillocks are growing where once was shallow sand. Dunes are rising closer to the sea than ever before. Wherever you look European plants are hard at work, building European landscapes from Australian sand.

When Baron Von Mueller visited Warrnambool in 1875 and saw how cattle and horses had ruined the dunes, he advised that stabilising plants such as gorse be sown. Marram grass (*Ammophila arenaria*) was brought from Europe the following year and planted up and down the coast. It proved ideal. Port Fairy Borough Council extolled this 'wonderful sand stay' in an 1895 broadsheet: 'So complete has been the reclamation of the lands, and where for years, nothing but pure shifting sands prevailed, there now thrives an area of magnificent grass, growing as thick as a cornfield, to the height of four feet.' Marram was sown so widely, its living tufts carried so far by waves, that it is now a major invader of beaches in southern Australia. A recent parliamentary inquiry on weeds found it to be 'the dominant plant of most primary dunes along the Victorian coastline'.

Beach spinifex (*Spinifex sericeus*), the native counterpart to marram, is a creeper that traps blowing sand against its stems, forming wide, low dunes. Marram grass, tall and stout, instead forms hillocks up to five metres high, often steeper than the angle of repose of bare sand. Wind funnelling below it maintains a European-style landscape of steep hills and deep valleys. Sea wheat-grass (*Thinopyrum junceum*), another exotic beach weed (probably a ballast stowaway), is incredibly salt tolerant, building dunes at the water's edge where none stood before. It too was planted out to reclaim sand. On a vast spit at Waratah Bay on Wilsons Promontory it has, over thirty years, built a system of dunes 200 metres wide, six metres high and nearly a kilometre long. Two other dune-makers, European sea rocket (*Cakile maritima*) and American sea rocket (*Cakile edentula*), create communities of low, irregular dunes. The former, another ballast stowaway, is the more common, found on almost every beach in southern Australia, often growing closer to the sea than any native plant. You can recognise it by its purple-tinted, fleshy stems and leaves and lilac four-petal flowers.

These aliens not only grow faster than native plants, trapping more sand and building bigger dunes than ever before in a confusion of new shapes, they also leave less sand to blow inland for back-dune plants and animals. The resulting changes to landscape and ecology are profound. Few beaches in southern Australia are untouched by their work. Sadly, the unruly sands in Von Mueller's day needed taming only because stock were allowed to trample it and towns were built too near the sea.

Where sand gives way to muddy bays, other invaders can be seen at play. Cord grasses (*Spartina townsendii* and *S. maritima*) sprout in the tidal zone beyond the mangrove fringe where no native plant can grow. These plants, sown by farmers to convert 'unproductive' mud to pasture, are now prolific weeds. By turning mudflats into vast meadows sometimes a hundred hectares in area, they steal habitat from wading birds – a serious problem in Victoria. Sediment captured by the metre-tall stems builds terraces at a rate of several centimetres a year, killing landlocked mangroves.

Weeds and feral animals are the architects of the new Australia. Quietly but industriously they are redesigning our beaches and wetlands, our hills, valleys and sea floors. Australia, oldest of lands, is receiving an uninvited facelift from its newest immigrants.

In the Northern Territory, water buffalo have converted vast paperbark swamps into sterile marshes. The paperbarks grew in freshwater swamps along the coast; buffalo gouged out swim channels that let in the sea, poisoning hundreds of square kilometres of trees. The swamps, once a haven for magpie geese, Burdekin ducks, waterlilies and sedges, are now barren saltmarshes spiked by graveyards of bleached trunks. In an effort to hold back the salt, massive and expensive embankments have been built.

Buffalo, like other hoofed animals, trample stream banks, erode gullies, compact soils and stamp bare trails through forests. Each day, millions upon millions of hard hoofs rain down like hammer-blows upon our fragile land. Australia's two or three million goats, which are prone to overgraze and erode steep slopes, are a major destructive force on their own. On Moreton Island I have seen an ancient Aboriginal axe quarry sliding into the sea, destabilised by goats. Pigs, like buffalo, excavate wallows in which to bathe. They root in the soil for tubers and grubs, ploughing up patches half a hectare in area, altering soil structure and forest regeneration. By grubbing up roots they can kill trees, opening up the canopy and letting in light for weeds. In Forty Mile Scrub National Park, north Queensland, lantana has spread in this way and now infests 73 per cent of the park. Even deer, supposedly the most benign of introduced animals, are implicated in serious erosion, as is the tree disease phytophthora.

All through the outback the rabbit plagues of the 1880s and '90s, in tandem with massive overstocking, remade the land. Grasses and shrubs vanished overnight, trees were ringbarked, and soil bared to wind and rain was borne away. Sand dunes, shorn of their cover of cane grass, were blown asunder, some sand carried all the way to New Zealand. The north-east of South Australia was transformed into desert and never recovered.

Exotic animals are at work underwater as well as on land. In the Swan River estuary the Asian mussel (*Musculista senhousia*) forms sheets that cloak the river bottom, entrapping silt and converting sand flats to mud banks. In outback rivers, carp are accused of muddying watercourses by grubbing in mud and uprooting plants. Above the waterline, weeds are remoulding watercourses. Along streams, woody weeds dramatically slow water flow, increasing sedimentation. In central Australia athel pines (*Tamarisk* species) are designing new riverbeds, and the impact on stream life must be dramatic. Channel widths are narrowing as siltation and flooding increase. The willows (*Salix* species) that crowd creeks in southern Australia behave in much the same way, reshaping the channels, accelerating bank erosion, slowing water flow, increasing sedimentation and flooding. The Melbourne Water Corporation spends about half a million dollars each year controlling these popular trees which have recently won a place on Australia's list of twenty worst weeds (appendix II).

One common weed in southern Australia is restructuring the soil in a remarkable way. Iceplant (*Mesembryanthemum crystallinum*), from southern Africa (I have seen it growing in Namibia), once found a place in gardens, and it's easy to see why. Adorning its leaves and stems, like clusters of tiny jewels, are thousands of glistening cells. But this plant has looks that kill. The water in its cells is very saline, storing salt drawn from the soil. When an iceplant dies the salt escapes into the soil, forming a salty crust in which nothing can grow – except more iceplants. This weed restructures the soil profile to benefit its offspring.

All over Australia combustible African pasture grasses are fuelling massive fires, bringing about profound changes to woodland structure. But there are other weeds that behave in the opposite way by deflecting fire. In the tropics, where buffalo and cattle graze back grass, hyptis (*Hyptis suaveolens*) sprouts in sweetly aromatic fields, which burn so poorly that fire-wary shrubs and trees get a start, turning sunny woodlands into shady, fire-resistant groves. Ground-choking weeds change habitats by blocking seedling recruitment. In southern Australia many tea-tree thickets are doomed because bridal creeper has blanketed the land and the next generation of tea-trees can't get started. Rainforest remnants in New South Wales are challenged in the same way by creepers such as wandering jew. Other creepers destroy forests directly by dragging down trunks with their weight alone; ivy, cats-claw creeper, blue thunbergia and rubber vine can all do this.

So exotic species don't only affect people, animals and plants; sometimes they also reshape the geography, the very lie of the land. The changes to our landscape have been profound. Yet exotic invasions are so insidious that most people don't recognise the invaders themselves, much less the new landscapes they are building.

35

Cryptogenic World

Native or Not?

> 'What the eye doesn't see, the heart doesn't grieve over.'
>
> **Proverb**

On Stradbroke Island near Brisbane I was taken bush by 'Auntie' Margaret Iselin to see the wild foods and medicines her Aboriginal elders harvested back in the 1930s. Among the plants she showed me were:

Castor oil plant (*Ricinis communis*) – prescribed by the witch doctor as a purgative
Madagascar periwinkle (*Catharanthus roseus*) – a remedy for diabetes
Prickly pear (*Opuntia stricta*) – edible fruit and medicinal 'leaves'
Billygoat weed (*Ageratum houstonianum*) – crushed stems are applied to sores
Guava (*Psidium guajava*) – bears edible fruit
Inkweed (*Phytolacca octandra*) – a dye plant for baskets
Red-headed cotton bush (*Asclepias curassavica*) – the sap cures warts
White passionfruit (*Passiflora subpeltata*) - carries edible fruit

What makes this list interesting is that none of these plants is Australian; every one is a weed from Africa, Asia or South America. The Stradbroke Aborigines probably took up their use during the mission years and in some instances (castor oil, for example) under missionary influence.

All over Australia, Aboriginal communities use exotic plants and animals. In the outback, where bandicoots and small wallabies became extinct decades ago, rabbits are now important tucker, dug from their burrows with digging sticks of steel. Cats are eaten too, as were camels in the past. South American tree tobacco (*Nicotiana glauca*) was chewed as a stimulant. In Kakadu National Park, traditional owners insist on keeping a few buffalo for hunting. The same is true of banteng in Gurig National Park. More than a few exotic species are now dignified with Aboriginal names. The fox, in central Australia, is 'ningani', 'waltaki', 'tuuka' or 'puwutjuma', the camel 'kamulpa' or 'kamala'. In Arnhem Land, stinking passionfruit vine (*Passiflora foetida*), a source of tangy yellow fruit, is known as 'djalamartawk' or 'gana'; on Groote Eylandt as 'kiriba', a corruption of 'creeper'.

Aborigines know that some of these species were introduced but insist that others have always been here. The rabbit they know to be recent, but the cat and Jerusalem thorn (*Parkinsonia aculeata*), which entered the deserts somewhat earlier, are often perceived to be native. Writing about the thorn, one of Australia's worst weeds, botanist Peter Latz noted: 'Although this plant was probably only introduced into the area less than 100 years ago, I have had informants swear to me that it has always been present in their country.' At Belyuen in the Northern Territory I remember shocking one Aboriginal elder by telling her the hyptis sprouting at her feet was a South American weed. She had just explained to me that local children traditionally used it to make play spears.

All over the world as the past fades from view, exotic invaders, biding their time, are winning new status as native species. It has happened to the Scotch thistle in Scotland, to the rabbit, sparrow and fallow deer in England, to the barbary ape of Gibraltar and the pheasant in Europe. California has her 'native' eucalypts, Arizona its burros (feral donkeys), New Zealand her lupins, and 'native' wild horses gallop across the plains in many lands. In Australia this rogue-to-respectability phenomenon is pervasive. In the west, fields of bright yellow capeweed are admired as wildflowers, and in central Australia bus drivers show tourists fields of bright red 'native hops' (*Acetosa vesicaria*), though in fact these are invasive weeds from Africa. In the Flinders Ranges, purple fields of Paterson's curse (*Echium plantagineum*) are marketed as native wildflower displays (and sold in the Melbourne markets as 'Riverina bluebells'), though there is more awareness today that the purple flushes signify invasion and degradation. In national parks all over Australia, weeds are routinely assumed to be native, especially the African grasses that violate woodlands and the rampant garden vines that ruin rainforests. Indeed, highly degraded, vine-choked rainforest fits popular stereotypes of what jungle should look like. And very few beach-goers recognise European green crabs, Pacific oysters or marram grass as invaders, or realise that the daddy-long-legs spiders in the garage and the mosses in the courtyard are imported.

Weeds even infiltrate 'wilderness' photography. European sea spurge and marram grass (hard to avoid these days) often appear in seascapes along with foreign waterlilies and watsonias. I told one leading wilderness photographer that his beautiful calendar scene of Moreton Island showed clumps of American beach primrose (*Oenothera drummondii*) infesting a dune. They had been misidentified for him, he said, as native guinea flower, a creeper with similar yellow flowers. Asked if he would use the image again, he said he would – because it had already entered his 'system'.

Nurseries are implicated too. I caught one of Australia's leading native nurseries selling African Hottentot fig (*Carpobrotus edulis*), a weedy garden plant, mislabelled as native pigface (*C. glaucescens*). So I wasn't surprised when the 'native lasiandra' they sold me grew into a South American tibouchea. Nurseries regularly sell weedy

African fountain grass (*Pennisetum setaceum*) as native swamp foxtail (*P. alopecuroides*), which itself may not be a native plant.

These mistakes don't rile me too much because I know that even biologists get it wrong. During the nineteenth century some weeds raced away so fast that colonial scientists assumed they were native, never guessing they could infiltrate forests and remote stations with such ease. Eminent botanist Joseph Hooker observed in 1860 that such plants 'wander from the cultivated spots and eject the native, or, taking their places by them, appear, like them, to be truly indigenous'. Von Mueller complained: 'The lines of demarcation between truly indigenous and recently immigrated can no longer in all cases be drawn with precision', and concluded that 'the nativity of some our plants will probably remain for ever involved in doubts'. True enough. Botanists still argue over dozens of plants, including 'native poppy' (*Papaver aculeatum*) – probably an import from South Africa – 'native gooseberry' (*Physalis minima*) and 'native thornapple' (*Datura leichhardtii*). There is disagreement as well about common sowthistle (*Sonchus oleraceus*) and the blue couch (*Cynodon dactylon*) grown as a lawn. Bushland regenerators can't agree on whether or not to use waterbuttons (*Cotula coronopifolia*) to reclaim degraded saltmarshes: is it native or an exotic weed?[1] The authoritative *Noxious Weeds of Australia* (1992) even wonders about nutgrass (*Cyperus rotundus*), arguably the world's worst weed: 'There is doubt on the origin of nutgrass; some authorities claim it is Indian, while others feel it is more widespread, including northern and eastern Australia.' Don't expect me to clarify whether these plants are native or not. I'm as confused as anyone else. I only know that they grow in Australia as well as overseas, were recorded here early, and either arrived here naturally long ago (via birds or sea currents) or were borne here more recently by people.

In many fields of biology the origins of species remain clouded. Among protozoans, mosses, algae, fungi, animal diseases and marine creatures the level of uncertainty is tremendous. When I e-mailed algae expert Tim Entwisle about freshwater algae, he shot back the reply: 'Certainly there appear to be exotic or naturalised species of freshwater algae, but there is no proof for any of them. I couldn't even take a stab at what percentage I thought were native versus introduced by human activity.' Protozoans provide another realm for guesswork: some biologists claim that many are exotic, but Professor David Patterson of the University of Sydney believes all species (except parasites) are naturally found everywhere – that is, they are all cosmopolitan.

This vexing uncertainty has attracted most attention from marine biologists. In 1996 prominent American scientist James Carlton wrote in the journal *Ecology*: 'Ecologists and biogeographers routinely categorize species as "native" with a confidence that for a surprising number of taxa belies the quality of the supporting data.' Carlton coined the word 'cryptogenic' for creatures that cannot with assurance be assigned either way. 'In the absence of a fossil record, for thousands of species of shelf-dwelling (neritic, inshore) marine organisms that now occur transoceanically or interoceanically, and in habitats wherever ships have gone or oysters have been moved, their categorization as 'cryptogenic' seems inevitable.' If these species are common, we are profoundly under-estimating the scale of exotic invasions.

The idea of cryptogenic species has been embraced enthusiastically by Dr Chad Hewitt of the CSIRO in Hobart. He lists ninety-five such species from Australian seas, including crustacea, jellyfish, shellfish and even a fish (the variable triplefin). Among the brown algae he rates seven species as exotic and thirty-six as cryptogenic. In other words, Australia has somewhere between seven and forty-three exotic brown algae. Hewitt suggests that some of Australia's best-known marine dwellers may in fact be exotic, including the abundant giant kelp (*Macrocystis pyrifera*) and the edible mussel (*Mylittus edulis planulatus*). Genetic evidence may help decide. Giant kelp colonies in Australia are almost identical genetically to those in the Americas, suggesting very recent arrival here, perhaps on explorers' ships.

Many of our fungi are cryptogenic. Most of the mushrooms that sprout on lawns, including parasol mushrooms (*Macrolepiota procera*) and inkcaps (*Coprinus comatus*), are probably introduced (they shun native forests), but they are not recognised as such in reference books. Debate raged for years over cinnamon fungus, the pernicious forest killer. We now consider it an Asian import, but what about the other twenty-one *Phytophthora* species in Australia? Some at least are native, but how many?

A vast number of diseases also fall into the unknown category, greatly impeding our understanding of their impacts. I have mentioned mange on wombats, the chytrid fungus on frogs and uncertainty over fish diseases. PhD student Alastair Dove hopes to determine whether fish ailments such as whitespot and chilodinella are indigenous or brought in on aquarium fish. He will compare disease levels in streams with and without exotic fish. We need a lot more scholarship to resolve these uncertainties.

Scientists overseas have the same problems with certain species assumed to be Australian. New Caledonians can't tell if Moreton Bay chestnuts (*Castanospermum australe*) are native there (maybe the seeds floated across) or exotic weeds. The famous 'Australian flatworm' found in English gardens is not Australian and neither, it seems, is the 'Australian' tubeworm (*Ficopomatus enigmaticus*) found in the Thames. New Zealand has many plants and animals to ponder about, such as amulla (*Eremophila debile*), recently declared a weed from Australia after decades on the native plant list, and the swamp hibiscus (*Hibiscus diversifolius*). Of the spider *Aranea pustulos*, Forster and Forster wondered: 'It is not certain whether this spider found its own way across the Tasman Sea or was brought in during colonization but it has certainly been in this country for a long time.'

Some species, such as the Australian crustacean *Iais californica*, now bear inappropriate names because they were first 'discovered' far from their place of origin. Others have been discovered and scientifically named without anyone knowing where they really came from. The wood sorrel *Oxalis thompsoniae*, which grows mainly as an urban weed, was named in 1994 by two Australian botanists unable to decide if it was native or exotic. Another weedy plant named in Australia, *Verbena incompta*, is thought to be South American.

The confusion deepens when we start to consider the definition of a native species and where to draw the line. Are Australian plants that now grow outside their natural range in Australia still native plants? Many biologists, myself included, think not. Are dingoes native? Under state laws they are sometimes recognised as wildlife, sometimes as vermin. We *know* they were brought from Indonesia thousands of years ago so their origin is not in dispute, but we must still decide whether or not to embrace them as a native species. Some scientists insist that any pre-European introduction qualifies, but I would not accept cats as native fauna if it were found that they too had entered on early Indonesian boats.

Australia's leading dingo expert, Laurie Corbett, is adamant that dingoes *are* native. In a recent *Nature Australia* article, 'Are dingoes native Australians?', he rejects the dictionary definition of 'native' – 'one that is inborn, indigenous, derived from one's country' – pointing out that dingoes have interacted with Australian animals by competition and predation, exterminating Tasmanian tigers for instance. 'Thus, in a functional and evolutionary sense, Dingoes are as much a part of the Australian environment as, for example, kangaroos, which are indigenous, having evolved within Australia.' This is questionable. Corbett continues:

> Defining a native Australian animal as one that arrived in Australia before European settlement is also inappropriate because some, such as rabbits, that arrived in Australia after 1788 are now unique to Australia. The physiological and genetic differences between rabbits in Australia and the original Spanish rabbits are due to their interaction with the peculiarities of the Australian environment. Australia is unlikely to ever be free of rabbits so the implication is that we have a native Australian

> rabbit. Similarly, feral cats and foxes have caused extinctions and modified environments, and will probably be in Australia forever; so, in that sense, they can also be considered Australian mammals.
>
> So what constitutes a native animal? It is simply one that lives in Australia and has ecological and/or cultural impact, regardless of taxa, birth site, race, language, length of time in Australia etc. Accordingly, the Dingo most certainly is a native Australian.

By Corbett's definition, since every species has some sort of ecological impact, every one must be native. The word sheds its meaning. Native rabbits, native goats, native toads, native trout, native camphor laurels and native prickly pear. O brave new world!

In my view the dingo is exotic because, irrespective of when it arrived, *it was brought here by people.* It is, after all, a domestic dog gone wild. But because it has achieved ecological integration, becoming our top non-human predator, essential to ecosystem functioning, we need to conserve it as if it were a native species.

But what of the cattle egret? This graceful bird flew in from Asia during the 1940s and spread out across the country, reaching south-western Australia in 1952 and Tasmania by 1965. This immigrant behaves exactly like an exotic, foraging in paddocks on grasshoppers flushed by cattle. But because it flew here unaided, most (though not all) biologists regard it as a native bird. Again I disagree. It can survive here only because we felled forests and brought in cattle, and it was able to get here only because humans created stepping stones of cattle paddocks running through the length of Indonesia. Its passage was assisted, not directly by ships, but indirectly by agriculture. Into the same category falls the wanderer (*Danaus plexippus*), the butterfly sometimes called the monarch. During the nineteenth century it island-hopped west across the Pacific, perhaps aided by ships. I cannot accept it as a native animal knowing that its caterpillars feed only on foreign cottonbushes.

Definitions can matter less than how much we cherish a species. I envisage a future in which people are less concerned about what is native (because they no longer know) and direct their conservation dollars towards popular species, whether they are 'native' or not. It's happening already with buffalo, banteng, deer, brumbies and coconut palms.

Tour operators in the Northern Territory would love to see more buffalo galloping free and at least one biologist, Bill Freeland, is sympathetic. 'Over the past 50 or so years the Water Buffalo has become a treasured symbol of the Top End', he wrote wistfully in the Australian Museum's *Australian Natural History* magazine[2] in 1992.

> It was and is part of things Territorian, and its apparent demise something many regret. While noting that, yes, it had a negative impact on the environment, there is something sad about the current status of the buffalo population. The Top End's Water Buffalo, in a sense, provided a substitute for our lost marsupial megafauna. Nowhere else in Australia could you see vast herds of large herbivorous mammals (large horns and all) thundering across the plains – an unusual sight for Australia. Unfortunately, however, something went wrong and, as well as a megafaunal replacement, we ended up with a major environmental problem.

Freeland stops short of saying we should conserve buffalo, but another biologist, David Bowman, certainly says this of banteng. *Nature Australia* devotes a regular page to endangered species, many of which are threatened by invaders, but in 1992 they took the unusual step of devoting the column to this exotic pest. 'Obviously, exotic animals should not be introduced into areas dedicated to nature conservation,' wrote Bowman, 'but the presence of the banteng in Gurig National Park appears to be a special case.' Because they are vulnerable to extinction in Asia, Bowman wants them conserved here, even though they harm Gurig by trampling animal burrows, forcing tracks through forests and browsing back trees and grass. In the museum's book *The Mammals of Australia*, Corbett almost agrees: 'There may be an argument for preservation of the Australian population, despite its exotic nature.' Banteng conservation is a fine idea, but can't we do it outside our national parks? Our banteng, after all, are domestic stock gone wild and they are surely inbred and genetically modified.

In the United States, where anything and everything goes, even eucalypts have their apologists. In a long-winded article, 'In Defense of the Nonnative: The Case of the Eucalyptus', Achua Stein and Jacqueline Moxley praise the benefits that eucalypts bestow upon California as bird habitat, timber, windbreaks and ornamentals. 'The eucalypt with all of its many uses', they conclude, 'shows clearly that created ecosystems have a validity of their own and need to be protected.'

In Australia the coconut palm (*Cocos nucifera*) has been championed in much the same way. Because Captain Cook and other mariners (Bligh and Flinders among them) never saw coconuts growing on mainland beaches, most experts agree they are not native. Indeed, the coconut palms growing wild today (Niu vai breeds) have rounded nuts, showing their descent from cultivars. But to tourists, the feral coconuts at Cape Tribulation and on Hinchinbrook Island enhance the Wet Tropics experience. Unfortunately, they are odious weeds that need controlling. Less weather resistant than native beach plants, they wash away during storms, encouraging erosion, and the nuts attract hordes of enormous white-tailed rats, which then prey upon birds' eggs and native seeds, distorting the local ecology.

In 1996 a ranger culled a few coconut palms in Family Island and Goold Island national parks in north Queensland. Locals were enraged. A petition went round and the Chamber of Commerce set up the 'Cardwell and Districts National Park Watch' to safeguard remaining palms. The state member for Hinchinbrook, Marc Rowell, weighed into the ruckus, complaining about the 'madness', the 'senseless destruction', and the 'defacing of many once-beautiful spots along our coast and our islands'. A cowed Department of Environment promised that no more coconuts would be harmed.

Here is a destructive exotic species protected inside national parks by the Department of Environment itself, in response to public pressure. They protect deer too. Don't be surprised if, through popular demand, future generations see more buffalo in Kakadu, and maybe banteng as well. Professor McCoy's dream may finally be realised and the outback populated by herds of galloping antelope. More and more exotic pests will make the transition from rogue to respectability, reaching that final stage of invasion: they will filter into our conception of Australia and win our acceptance as native wildlife. This is the cryptogenic future.

1. Other plants that inspire uncertainty include water primrose (*Ludwigia peploides*), bulrush (*Typha latifolia*), mud dock (*Rumex bidens*), club-rush (*Isolepis marginata*), balsam pear (*Momordica charantia*), balsam apple (*M. balsamina*) and spiny sida (*Sida spinosa*).
2. Now called *Nature Australia*. I write a column for this magazine.

36

It Happens Naturally

Invasion as Natural Process

> 'Australia never has been truly isolated from the rest of the world's flora and fauna (and their associated pests and diseases).'
>
> **Malcolm Nairn et al., *Australian Quarantine: A Shared Responsibility,* 1996**

The mighty wildebeest migrations in Africa are one of the greatest shows on earth. Each year a million of the shaggy beasts lumber north to the Serengeti to crop succulent fields of rain-sweetened grass, watched over by some of Africa's best-fed lions. It's an extraordinary spectacle that owes much to a single grass, the prolific *Themeda triandra*. The Serengeti also draws vast herds of tourists, including battalions of camera-clicking Australians. To Australian eyes the snorting herds are an exotic spectacle, but the greenery at their feet, viewed closely, should be familiar, for *Themeda triandra* is a grass found in Australia. It is our well-known native 'kangaroo grass'. During the early years of white settlement it sustained millions of cattle and sheep until eaten out and replaced by poorer plants. Somehow, perhaps millions of years ago, *Themeda* spread between Africa, Asia and Australia. Just how it reached Australia we cannot say, but its journey is one that many plants have made.

In 1801 botanist Robert Brown set out with Matthew Flinders on his epic voyage around Australia. Brown had studied the collections of Sir Joseph Banks and he expected to see strange-looking banksias, grevilleas and other remarkable plants. What he didn't anticipate were plants he knew from the hedgerows of England. How astonished he must have been to encounter self-heal (*Prunella vulgaris*), one of England's traditional herbs, praised by Culpepper in 1653 as an 'especial remedy for all green wounds'; and sowthistle (*Sonchus oleraceus*), purslane (*Portulaca oleracea*), couch grass (*Cynodon dactylon*) and purple loosestrife (*Lythrum salicaria*), a well-known English wildflower. In *The Wind in the Willows*, loosestrife caresses Mole and Rat as they row up the river; in Australia it sprouts above platypus pools. Brown found over a hundred plants, including lichens, known to him from Europe, and his list, 'Natives both of Terra Australis and of Europe', was important enough to be published with Flinders' journal in 1814.

Had Brown compared Australia with Asia or Africa, as later botanists did, his list of common plants would have run to many hundreds. Of the 360-odd monsoon rainforest plants growing naturally around Darwin, more than half grow in New Guinea, nearly a third in India, 14 per cent in Africa and a few even in the Americas. Plants shared with Asia include the mung bean (*Vigna marina*, collected by Banks on the Endeavour River), the sacred lotus (*Nelumbo nucifera*) and sacred basil (*Ocimum tenuifolium*), the caper (*Capparis spinosa*, the source of pickled capers) and the melon *Cucumis melo*, ancestor of the rock and honeydew melons and a food appreciated by ailing explorers.

To early botanists this was all terribly puzzling. Joseph Hooker, writing in 1860 about 'the European Features of the Australian Flora', posited a continuous current of vegetation, marking out an ancient migratory highway, extending along mountain chains all the way from Scandinavia to Tasmania. Scandinavian genera, he noted, 'abound on the Alps and Pyrenees, pass on to the Caucasus and Himalaya, thence they extend along the Khasia mountains, and those of the peninsulas of India to those of Ceylon and the Malayan archipelago (Java and Borneo), and after a hiatus of 30°, they appear on the alps of New South Wales, Victoria, and Tasmania, and beyond these again on those of New Zealand and the Antarctic Islands, many of the species remaining unchanged throughout!' He proposed that the Indian plains flora extended through the Malay Islands 'whence it is continued in great force over the whole of tropical Australia'. A Viennese professor, Dr F. Unger, had similar ideas but argued, in an utterly bizarre essay, 'New Holland in Europe', that plants had gone the other way. He claimed to have found Australian plants fossilised in European rocks. He also believed that Atlantis once linked Europe and America; his essay 'The Sunken Island of Atlantis' appeared in the respectable *Journal of Botany* in 1865.

Other botanists believed in independent acts of creation, arguing that some species had appeared simultaneously in different corners of the globe. Darwin repudiated this idea at length in *The Origin of Species*, showing by experiment that plants (and animals) could disperse over vast distances and that some seeds survive in seawater more than a hundred days. He suggested seeds might travel in the stomachs of dead birds floating at sea; he even sprouted tiny seeds extracted from the dung of migratory locusts.

Today we know a great deal more about seed dispersal, ocean currents, high-altitude winds, migrating birds, continental drift and sunken land bridges. We know that although Asia and Australia have never properly embraced, hordes of plants and animals have migrated back and forth, by flying, floating, rafting, being blown on the wind or hitching rides on birds. Australia has been a target of invading species for millions of years – indeed, invasion is a perfectly natural process. Our continent was never as isolated as we might think. The proof is all around us in the many plants and animals we share with other lands – for example, birds such as waders, swifts, egrets, the osprey, peregrine falcon and barn owl. Australia's native rodents, fruit bats, boab tree (*Adansonia gregorii*) and some of our reptiles furnish proof of still older invasions. Keelbacks can eat toads only because they evolved from toad-eating snakes in Asia.

Australia also contributed plants and animals to other lands long ago. Our native hopbush (*Dodonaea viscosa*) has somehow managed to spread all over the globe (I have seen it in Africa, India and Florida), and we have dispersed fanflowers (*Scaevola*), flax lilies (*Dianella*), devil's twines (*Cassytha*) and many other plants to distant lands. Koa (*Acacia koa*), the tree that clothes mountains in Hawaii, is obviously descended from an Australian wattle, probably the blackwood (*A. melanoxylon*), and other wattles have colonised Madagascar, Mauritius, the Philippines, Taiwan, Vanuatu, Fiji and Samoa, predating the weedy invasions of wattles now taking place. Fossil evidence suggests that Australia may even be the ancestral home of the world's songbirds, and we are certainly the source of the bird lineage that gave rise to crows, shrikes and orioles.

If invasion is a natural event, if species have been travelling about the world for millions of years, does it matters much if people help the process along? Aren't we just contributing to a natural phenomenon when we bring in weeds? Can we stop worrying about the problems posed by this book? The answer, unfortunately, is no. Invasion events today are like nothing that ever transpired before.

In the past, most of our invaders came from South-East Asia by journeying down the long Indonesian archipelago, but now they are entering from everywhere – from Europe, Africa, North and South America, New Zealand, even from far-flung islands. Rubber vine, arguably our worst weed, arrived from southern Madagascar; gorteria (*Gorteria personata*), a noxious weed in Western Australia, from the granite hills of Little Namaqualand in South Africa. Hawaii has donated a tiny snail, the island of Madeira a choking vine, and the Canary Islands a whole raft of weeds, including tagasaste. Our invaders never came from these places before. As well, we are getting visits from groups that never travelled previously. Freshwater fish cannot cross between continents unless, like the barramundi, they can tolerate salt. Yet our streams today carry fish from Europe, Asia, Africa, and North and South America. Amphibians are very poor travellers and yet a toad from South America is now abundant here. The majority of our most vexing invaders, including cacti, rabbits, pigs, goats, cats, foxes, toads, buffalo, live-bearing fish, earthworms and pines, belong to groups with no previous history of migration into Australia, with no natural way of getting here. No wonder our native wildlife has fared so poorly against some of them.

The *rate* of invasion has also changed radically. More than 2700 new plants have colonised Australia in the last two hundred years; compared with perhaps one or two per millennium before that. The trickle is now a flood, as millions of years of natural migration are compressed into centuries. In Hawaii, invaders are entering at more than a million times the natural rate. The implications of all this are profound. Where previously there was plenty of time for ecosystems to assimilate newcomers, now they are being overwhelmed. And our natural environments are already under siege. People are felling forests, damming rivers, polluting harbours and generally turning the place upside down as never before. Wherever native species are retreating from this disruption, invaders, especially those from disturbed habitats overseas, seize the opportunity to move in. Ancient ecosystems are collapsing and new ones are taking their place. It all adds up to a potent and highly unstable prescription for change.

VII

THINKING AND ACTING

When we look at our plants and animals, do we understand what we see? How much can we hope to control bio-invasion? Can the government stop the pests? What can we do? Some final thoughts.

37

Seeking Magic Bullets

Biocontrol Often Misses the Mark

> 'Biological control should not be assumed to be the "magic bullet" that will always solve a weed problem.'
>
> **Dane Panetta and John Scott, 1995**

During the 1920s prickly pear cactus (*Opuntia stricta*) ranked as probably the worst weed on earth. In Queensland and New South Wales 25 million acres of land were infested and each year another hundred thousand acres went under. This curse was curbed by cactoblastis (*Cactoblastis cactorum*), a colourful caterpillar from Argentina, brought in by entomologist Alan Dodd in 1925. Cactoblastis bored through cactus pads, devouring them from the inside. Bushwhacker Phillip Wright recalled how 'the pear began to collapse everywhere, and at hatching time the air was thick with the moths, laying their egg sticks far and wide. By the end of the second year, the countryside for mile after mile was covered inches deep with rotting pear plants and a slimy jelly-like substance.' Defeat of the cactus still stands as one of the world's two most outstanding examples of biological control.

Cactoblastis worked remarkably well because it was a special case, involving a new association between predator and prey, the cactus having no resistance. The caterpillar comes from South America, where it eats other kinds of cacti, the prickly pear from North America, and they first met in Australia.

This campaign won international acclaim and other cactus-blighted countries were soon clamouring for cactoblastis. Eggs were sent to South Africa, Hawaii and Mauritius. In the West Indies, where native cacti had become pests, cactoblastis was unleashed in the Lesser Antilles, then taken to Grand Cayman in 1970. But the insect unexpectedly seized control of the invasion and, unaided by human hands, spread west through the Caribbean islands. In 1989 the unthinkable happened – cactoblastis invaded Florida.

There are six kinds of native prickly pear in Florida and one of them, semaphore cactus (*O. spinosissima*), is nearly extinct, just twelve plants surviving in the wild. When cactoblastis turned its appetite upon these, biologists acted just in time. Cages were lowered over the plants and cuttings of all twelve were taken to Fairchild Tropical Garden. The moths devastated Florida's other native pears instead, including the rare jumping prickly pear (*O. triacantha*).

Cactoblastis will keep on travelling. Fred Bennett and Dale Habeck of the University of Florida warn: 'There is a high probability that *C. cactorum* will spread north through Florida and west to Texas and into Mexico where the fruit and young vegetative parts of *Opuntia* spp. form part of the staple diet of humans and where chopped plants serve as cattle fodder in times of drought.' In a deliciously ironic twist they suggest biological control for cactoblastis itself, proposing a wasp (*Phycticiplex doddi*) named after Dodd.

In South Africa, too, cactoblastis proved disappointing after it was brought to Kruger National Park in 1988 to quell an explosion of prickly pear. Elephants and baboons had been eating the pear fruits and scattering the seeds. But cactoblastis fragmented the cacti into hundreds of pieces, which then sprouted, spawning larger numbers of smaller plants. Baboons hindered the caterpillars in their work by eating *them*.

These stories stand as cautionary tales. Biological control is often touted as the magic solution to all our pest problems: it's so much cleaner than chemicals and has a reputation for performing miracles (though it never entirely eliminates pests). In truth, biocontrol has a mixed record. More often than not it fails, and like all programmes of exotic introduction it has ignited plenty of disasters of its own. Its successes have created the dangerous illusion that it will solve our worst problems, but it is never a substitute for good land management.

Only about 6 per cent of the biological control agents released against weeds in Australia succeed completely. Another 18 per cent do some good, but most – 76 per cent – fail, either dying away or remaining rare and insignificant. Biocontrol works better than these figures suggest because biologists usually unleash several agents against one pest. Seldom does one insect do the whole job – the 'magic bullet' rarely exists.

Even when many shots are fired the target may escape. Lantana, St John's wort and prickly pear were the first plants targeted for biocontrol in Australia, and although that was almost a century ago they remain rampant weeds today. So far twenty-eight agents have been released against lantana, twelve against St John's wort (of which three are helpful but only in sunny situations). Prickly pear was controlled, but not on the first try; forty-eight insects were tested and twelve released, the first in 1903, long before cactoblastis saved the day. And prickly pear remains a problem in some places; in poor soils it produces a slimy sap in which cactoblastis caterpillars drown.

Biocontrol has worked well against certain weeds and insects, but its scope has always been narrow. It can claim few successes against trees, grasses, sedges, fungi and annual weeds, which include some of our most intractable pests. It has never been tried in the sea and has never delivered permanent control of a vertebrate. Sometimes a biocontrol agent exists but cannot be used. Exotic diseases, such as African swine fever for pigs and brucellosis for buffalo, could rein in some feral animals, but our livestock would suffer as well. Biologist Chris Dickman would like to see pet cats vaccinated (just as pet rabbits were inoculated against calicivirus) and cat flu and feline herpes released, but only a brave government would consider this. Control of rabbits by myxomatosis and calicivirus was only acceptable because rabbits are seldom kept as livestock or pets – though rabbiters fought long and hard against myxomatosis. Fertility control methods, although yet unproven, may offer better hope against foxes, cats and rodents.

The same conflicts of interest apply to plants. During the 1970s a group of graziers and apiarists took scientists to court to block biocontrol of Paterson's curse (*Echium plantagineum*) because they valued this weed. More recently, the Tasmanian government, on behalf of beekeepers and berry-pickers, childishly opposed release of blackberry rust because blackberries earn Tasmania's economy $300 000 each year (while costing Australia $42 million). Olive trees urgently need controlling, but Australia has a small olive industry and biocontrol using olive knot (*Pseudomonas syringae* pv. *savastanoi*) can be ruled out, as can control of pasture plants such as buffel grass and leucaena. Biocontrol has been considered for arum lilies in Western Australia; no suitable agent was found, but would such a scheme have proceeded? How would gardeners feel about an arum-killing bug? Many rampant pests – camphor laurel, ivy, privet, pampas grass and asparagus fern among them – still hold the affection of gardeners, so there is plenty of potential for discord.

When a plant is both useful and pestiferous, the uncontainable nature of biocontrol (compared with herbicides) becomes a hindrance. Lack of host specificity is another problem. Crop farmers in southern Australia are turning to minimum tillage to save their soil, but Mediterranean white snails (*Theba pisana* and *Cernuella virgata*) are thriving under the trash blankets and contaminating grain so badly that overseas shipments are sometimes rejected. (Minimum tillage is also favouring weedy annual grasses.) The CSIRO scoured

Europe to find two snail-attacking parasitic flies, but in lab tests the flies harmed native snails as well and the programme was dropped. So despite a huge amount of work, nothing was gained. In the war against blackberries a sawfly was rejected because it was not sufficiently destructive and also showed a liking for roses. The CSIRO then turned to blackberry rust (*Phragmidium violaceum*), recommending its release in 1984 even though it attacked native raspberries. Native plants evidently have a lower priority than roses.

One of the great myths about biological control is that agents are approved only if they prove harmless to wildlife. This has never been the case. During the early years no one gave much thought to wildlife; insects were trialled against crops and garden plants and nothing else. Tests were broadened during the 1980s to take in native plants and animals, and new moral quandaries are now arising. Agents such as blackberry rust are being released despite lab tests showing they will attack native plants or insects. The examples include a fly (*Trichopoda giacomelli*) brought in to control green vegetable bug (*Nezara viridula*), a mimosa-feeding moth (*Neurostrota gunniella*), and a moth and rust that attack rubber vine.

The rubber vine examples show how compromised the decision-making has become. In the war against this weed scientists released a moth (*Euclasta gigantalis*) even though they knew it would also attack a native vine (*Gymnothera oblonga*). They argued that rubber vine posed more of a threat to the vine than their moths, which is probably true. This moth, incidentally, was only tested against thirteen native plants in the rubber vine family (Asclepiadaceae), out of the eighty-six species in fifteen genera found in Australia. (The mimosa moth, by comparison, was tested against a hundred plants.) Scientists have since distributed a rust (*Maravalia cryptostegiae*) to control rubber vine despite lab tests showing it could harm a native rainforest vine (*Cryptolepis grayi*) listed as rare under the Queensland Nature Conservation Act. Decisions to release new agents are made by agreement of the Commonwealth and states under a process coordinated by the Quarantine Service. Environment

Australia, or any of the state environment departments, can veto an agent if they put forward a sound argument. In the case of the rust it was thought the native vine would not be attacked heavily enough to suffer harm. In north Queensland it grows fifty kilometres from the nearest infected rubber vines, and no infection has yet been found. Vines are monitored regularly. The other native vine is monitored for moth damage and only two attacks have been noted so far.

Rubber vine, mimosa bush and blackberry are such execrable pests that the decisions to release these agents were probably justified. But these examples show that careful judgment is always needed. Scientists play God when they unleash new species and we must hope they have the wisdom to match their power. As it turned out, the rubber vine moth proved disappointing, at first attacking rubber vines with gusto, multiplying dramatically and becoming a nuisance in north Queensland homes, until it was targeted by two parasitic wasps, one of them an exotic species previously unknown in Australia (how did *that* get in?). Blackberry rust is successfully damaging certain types of blackberry during wet years, but other blackberry strains are now taking their place. Damage to native raspberries has not yet been recorded but may be occurring. The rubber vine rust is a moderate success.

Scientists overseas have watched plenty of projects go awry. The rosy wolfsnail (*Euglandina rosea*), brought to Polynesia in 1977 to prey on giant African snails, almost wiped out seven native snails, although biologists, acting just in time, set up captive colonies of six of them; the seventh is now extinct. In Hawaii a biocontrol fly exterminated a native moth; in North America a weevil now harms endangered thistles; and in South Africa the Australian mealybug ladybird (*Cryptolaemus montrouzieri*) now eats the scale insects that suppress prickly pear.

During the early, halcyon years of biocontrol scant heed was paid to unintended consequences. Western Australia's Department of Agriculture was the epitome of recklessness, importing hordes of insects without any testing. Between 1902 and 1909, just to control red scale (*Aonidiella aurantii*), they brought in scale enemies from Spain, China, Japan, India, Jerusalem and New South Wales. Their collector, American George Compere, was employed by California at the same time. Fortunately, these insects were released in very small batches and most of them failed to survive. Testing of insects against alternative insect prey became standard only in the 1970s.

The sugar industry raised carelessness to an art form with its introductions of toads, mongooses and mynas, and plans for moles and shrews; no other industry abused biocontrol so extravagantly. During the years of Australia's rabbit plague there were calls for the release of skunks, civets, stoats, coyotes, jackals, lynxes, meercats, ferrets, weasels and South American army ants, though more in a spirit of acclimatisation than from any understanding of biocontrol. Prizes for foolishness should also go to the men who brought in blackbirds and starlings to eat insects, mosquito fish for wrigglers, and, in New Zealand, weasels, ferrets and stoats to kill rabbits, and hedgehogs to tackle snails and slugs. Vertebrates are seldom used in biocontrol any more, and never in Australia. They are known to feed too indiscriminately, and the disasters have piled up over the years.

But despite the calamities associated with biocontrol, its successes deserve our applause. (Biocontrol scientists fear that too much criticism will threaten their funding, and I don't want that.) In the United States, biologists such as Daniel Simberloff (in the article 'How Risky Is Biological Control?') have attacked the industry. But Australia, having experienced some great successes and no disasters in modern times, is now a leader in the field. By 1997 we had released 123 agents against weeds, a number exceeded only by the United States with 130. Other countries fall well behind – South Africa with 61, Canada with 53 and New Zealand with 24. Surprisingly, weed biocontrol has rarely been used in Europe.

Rampant waterweeds have proved amenable to control. Australia has done well against salvinia (*Salvinia molesta*) and water hyacinth.

Other plants under good control include groundsel bush, noogoora burr, harrisia cactus (*Harrisia* species) and, of course, prickly pear. Dung beetles feature in one of our more unusual success stories. Our native species shun cowpats, allowing bush flies to breed up in them, and the CSIRO has introduced forty-five dung beetles from Africa, Asia and Europe, of which seventeen have established. They help clean up paddocks, as many as 3000 beetles tackling one pat.

Biocontrol is usually very complicated, but sometimes the job can be done simply. In late 1993 sixteen dingoes were unleashed on Townshend Island in central Queensland to control a plague of 1700 goats. By 1997 only four goats remained, hiding on a cliff. The Defence Department, which controls the island, shot them out. The vegetation is now sprouting back and the dingoes will be deported before they exterminate native birds.

You might think that biocontrol is used only against foreign pests, but this is not so. A parasitic wasp was brought from New Zealand to attack native bush flies, and foreign insects, including an African wasp, have been deployed against Queensland fruit flies (*Bactrocera tryoni*), raising interesting ethical questions that haven't been addressed. The CSIRO wants to introduce parasitic wasps to attack native fruit-piercing moths in orchards (*Othreis, Khadira* and *Eudocima* species), though these same wasps will surely enter rainforests and kill moths inside national parks. In a very controversial example in North America, an Australian fungus (*Entomophaga* species) was unleashed upon pesky native grasshoppers. It now infects at least eight different grasshoppers in North Dakota, not all of which are pests, and it is spreading. American biologist Jeffrey Lockwood warns that rare grasshoppers may be exterminated by this Australian parasite, a claim disputed by its champions, who want to release a parasitic Australian wasp as well.

Scientists these days are striving to extend the biocontrol envelope. The CSIRO in Hobart is investigating a protozoan that sterilises Pacific seastars and a parasitic barnacle (*Sacculina carcini*) that attacks European green crabs, also rendering them sterile. But biocontrol in the sea is fraught with risk because our understanding of marine ecology is so rudimentary. Other possibilities are a virus that attacks carp and diseases that snuff out toads and tilapia. We may even find fungi that kill phytophthora.

Biocontrol would be put to more uses if it weren't so expensive. On average it costs nearly half a million dollars to test one agent. This may seem excessive, but the debt is often repaid a hundred times over. Unfortunately, some projects are closed down for want of funding long before the options have been exhausted. Testing becomes cheaper if another country has trialled the agent or if countries share the work, as Australia, New Zealand and South Africa often do. Third World countries often benefit from Australian work: the salvinia beetle we imported has worked wonders in thirteen other countries – in New Guinea it saved villages from destitution by cleansing their waterways of choking weed.

Unfortunately, the numbers of pests that need controlling keep swelling. Garden plants and twining pasture legumes are prime candidates for future work. But unless the public demands more funding the work will always fall behind. Farm pests inevitably take priority over environmental weeds, and except where a pest is both (for example, mimosa bush and rubber vine), environmental weeds tend to miss out, although attitudes are changing. Because scientists rely on external funding we find them doing work on the elm leaf beetles (*Pyrrhalta luteola*) attacking Melbourne's exotic elms while ignoring such serious weeds as cats-claw creeper and Madeira vine.

We need more biological control, but we should not raise our expectations too high. In 1997 I visited the Alan P. Dodd Tropical Weeds Research Centre at Charters Towers, a centre for biological control work. The manager, Trevor Stanley, told me to look closely at the rubber vines growing around town. Rubber vine is arguably Australia's worst weed and an urgent priority for control. Its native home in southern Madagascar has been scoured for agents but only two were found – the moth and rust mentioned above. (Other insects were rejected because they fed indiscriminately.) The moth did little good but the rust, released in 1995, is looking hopeful. Around Charters Towers I found that almost every leaf I turned over was speckled orange with infection. I thrilled to the hope that this rust might turn the tide in our war against this weed. But my hopes were deflated when I spoke to biocontrol experts a year later. The rust is periodically defoliating rubber vines and blocking them from flowering, but no one expects it to kill whole vines. This weed will not lose its hold on the land; farmers will simply have a better chance of controlling it. More grass is growing beneath the vines and landholders may be able to burn them back, using the grass and fallen leaves as fuel. Here is Australia's number one weed with only two biocontrol agents available, both of them potential threats to native plants and neither of them capable of quelling the weed, although one is helping.

It is this kind of modest success that is typical of biological control. It is no magic bullet, no salvation to our problems, but a valuable weapon nonetheless in a never-ending, not-quite-winnable war. Biocontrol is something we need, but not something we can depend upon.

38

The Quarantine Quandary

AQIS Wields a Small Sword

> 'An ounce of prevention is worth a pound of cure.'
>
> **Proverb**

The Australian Quarantine and Inspection Service has an impossible job to do: We expect it to keep all pests out of the country while at the same time allowing in, each year, about 7 million people with luggage, a million cargo containers, nearly twice as many air-freight containers and 150 million letters and parcels. It must also contend with smugglers, boat people, illegal fishermen, visiting islanders and errant yachties.

AQIS knows its job is ultimately impossible, that it can't keep every pest out. The government doesn't expect it to. Perfect quarantine would mean banning all trade and travel – in a country relying on trade and tourism. The Nairn review reiterated that 'a policy of "no risk" is not, and never has been, a viable quarantine policy option'. Quarantine's approach relies instead upon risk analysis. AQIS looks at the levels of risk posed by different imports and targets the worst. High-volume, low-risk imports don't earn much attention even though, in the long term, they are sure to introduce pests. With so many people and products flooding into Australia it's the best AQIS can do.

Quarantine's job is made that much harder by all the travellers who flout the rules. Astonishing quantities of foods, animals and plants are smuggled in with luggage each year. More than a third of passengers arriving in Australia who exit airports via the green, 'Nothing to declare' customs channel are illicitly carrying items of quarantine concern. One recent passenger who declared nothing but chocolate was found to be carrying apples, tomatoes, parsnips, five kilos of berry fruit, bean and pine seeds, nine packets of other seeds, thirty-three live plant cuttings, five strawberry runners and nearly a kilo of cheese. A passenger from Malaysia was caught with three litres of milk, 250 grams of meat, half a kilo of peanuts, two kilos of butter and ten kilos of other foods.

People smuggle in fruit, meat, cheese, live fish, crabs, turtles, edible snails, snakes, even small dogs. We know they often succeed because of the exotic creatures sometimes found wandering the streets or in streams. It was probably a 'smuggler' with a festering pawpaw or mango who ignited Australia's papaya fruit fly outbreak in 1995. About $55 million was spent eradicating the fly and $100 million was lost in fruit exports.

That pest's arrival was not unexpected (although Australia proved ill prepared), for the flies had been heading our way for some time. From their home in Indonesia they had spread to New Guinea in 1992. The following year they were caught in traps set in Torres Strait, and an eradication programme was under way when they appeared in Cairns. Thanks to our Northern Australia Quarantine Strategy the flies were detected in Torres Strait, but NAQS does not operate in Cairns. A few years earlier two august bodies, the Horticultural Policy Council and the Bureau of Rural Research, had recommended that fruit fly traps be set at ports, especially Cairns, but this did not happen. The traps were only put in place in 1996, *after* the outbreak, at a cost of $1.3 million, much less than the outbreak cost. Had they been set earlier the flies would have been eradicated much more quickly and cheaply.

New pests breeze into Australia so regularly that AQIS is often accused of incompetence and made the subject of government inquiries. The most recent (referred to above and in earlier chapters) is Professor Malcolm Nairn's independent review *Australian Quarantine: A Shared Responsibility*, which presented its findings in 1996. Nairn found plenty of weaknesses in AQIS, and I will mention a few.

Australia's mail, it seems, is poorly screened, and Nairn considered this a 'major cause of quarantine concern'. The Customs Service (concerned with drugs and other contraband but not pests) has an agreement with Australia Post that at least 85 per cent of letters and small parcels are released after customs profiling but without quarantine screening. The forty courier companies who bring in parcels attract even less quarantine scrutiny, if that were possible. Nairn recommended an 'immediate review of international mail operations to ensure that quarantine surveillance of all international mail is effective'. AQIS is bringing in more sniffer dogs and X-ray machines to mail centres to hunt for illicit foods, seeds, plants and animals. The extra effort is paying off. During one week early in 1999, 518 parcels arrived at Melbourne's International Mail Centre and illicit goods were detected in 176 of them, including soil and cacti (declared as 'needlework'), pork sausages ('candles'), narcotic khat leaves wrapped in banana leaves ('shoes and perfume'), seed propagation kits containing 2.3 kilograms of soil ('books') and so on. Internet catalogues make it very easy to import seeds and other dangerous things; thirteen separate parcels of seeds arrived in that week alone.

The trade in used farm machines, earthmovers and oil rigs, which are difficult both to clean and to inspect, is another big worry discussed earlier. Lots of dirt and nests of birds, bees and wasps come in on these. One truckload harboured three bird nests containing eggs and seeds of nineteen different plants, mostly weeds. In 1997 a severed human hand was found in a Japanese bulldozer, *after* it had passed quarantine, and that same year two dead cats were found in farm machinery from Saudi Arabia.

'Items such as inadequately cleaned mining and harvesting equipment, and containers loaded at ports where the giant African snail is present, are being off-loaded at Australian ports', the review reported. 'Recent experience has shown that in some cases such items are not inspected or treated before release. Surveillance of crews leaving ships is also a matter of concern.' The review found 'inadequate resources allocated to this area' and noted that 'there is a significant imbalance between the resources allocated to airport operations and the amount of attention being paid to wharf and seaport surveillance – for instance, the inadequate monitoring of refuse and of crews to ensure they do not take potentially high risk items of food from the vessel'. In other words, Quarantine looks better than it should because most of us see it at work only in airports, where standards are higher. Then again, there is plenty of sloppiness there too. I am sure many of us can remember times when we have breezed through quarantine late at night and no one has cared about our muddy boots.

The Nairn review called for major changes in AQIS's priorities. 'Quarantine decisions must take greater account of environmental considerations and this responsibility should be reflected in quarantine legislation.' (AQIS currently targets pests on economic grounds only – at least on paper.) As part of this change Nairn wanted AQIS removed from the Department of Primary Industries and made a statutory authority, a move the government rejected, in part because it *wants* the agricultural priority maintained.

The review also called for fundamental change to Australia's quarantine culture, with AQIS developing a partnership approach embracing the community. 'Quarantine must be a shared responsibility for the benefit of all Australians.' We can't expect AQIS to work alone; we must help by accepting responsibility for what we bring in, and by keeping alert for new pests here. Mile-a-minute was detected only because a woman grew suspicious about the weedy vine she'd been given. Country people should notify the Quarantine Service whenever they see unfamiliar animals and plants. We need more emergency response plans for new pests and more resources to detect them. A new pest is more likely to be discovered in the field than at an airport anyway.

Nairn also asked for greater funding for AQIS, which, amazingly, receives only a pittance. In the 1996–7 budget, Defence got $5297 million, Customs $391 million and Quarantine a mere $7.9 million. (Its total budget was $53 million, but $45 million was to be recovered from users.) The ironies here are exquisite. Exotic pests pose a much greater threat to Australia right now than exotic people. The weeds already here cost us $3.3 billion each year in eradication and lost production. Because we spend only a penny on prevention we end up paying a pound for cure. Quarantine's budget is minuscule, about 0.03 per cent of the worth of our rural industries, a tiny premium for a national insurance policy. As the Nairn committee noted, AQIS runs on the same budget that it costs to build a couple of kilometres of highway. The previous Labor government actually slashed the AQIS budget in 1992–3, shedding 750 staff and further hamstringing its functions. The minister responsible for that folly, Simon Crean, wants to become prime minister one day. (Governments over the years have often cut the AQIS budget.) Even though funding has improved since then, Nairn 'received the strong impression that some quarantine activities appeared to be driven more by the ability to charge for services than by the need to meet the objectives of quarantine'. The present government has responded to Nairn by spending an extra $76 million over four years. It's a welcome improvement but nowhere near enough.

Thanks to Nairn's review, AQIS is now sailing in the right direction. With better funding it should be able to plug at least some of the quarantine leaks identified by the review – indeed, in some cases it has already done so. New sniffer dogs at airports are cutting down on petty smuggling. New pests will still flow into

Australia but that won't necessarily be AQIS's fault. We can't expect better results when we insist on importing so many foreign foods, flowers and fish.

Three areas of AQIS's performance particularly impress me. On ballast invaders, AQIS has led the world in its research and proposed remedies. The current 'solution', reballasting at sea, falls well short of a real answer, but I'm not sure what more AQIS could do. The Northern Australia Quarantine Strategy also deserves praise. Thanks to NAQS the islanders of Torres Strait rank among the world's most quarantine-aware people. NAQS biologists scour our northern islands and shores for exotic invaders. Often they fail to keep out new pests but at least we are better informed about what does sneak in. I am also impressed by AQIS's open approach to its work. While preparing this book I usually found AQIS officers to be candid and helpful, even when they knew I might be critical of their work. Their monthly newsletter, the *AQIS Bulletin*, is a storehouse of information. Put together by the inimitable Carson Creagh, a former biologist and book editor, it's packed with fascinating and frightening snippets, penned in a witty up-beat style. It was the bulletin's regular 'Hit Parade' that alerted me to dead cats and nests lodged in farm machines, smuggled aquarium fish and much else besides. The AQIS web site is also helpful and clear. Australians remain very ignorant about their quarantine problems, but at least AQIS tries.

It all seems such fun! The *AQIS Bulletin*, especially its 'Hit Parade' column, is so up-beat in style that terrifying transgressions of quarantine end up sounding like student pranks. Consider the following examples:

> 'A passenger arrived in Sydney with a live snake tucked into one sock – and a live turtle tucked into the other! Unfortunately, the turtle escaped on board the aircraft and bit a fellow passenger ... Also at Sydney Airport, a passenger from Italy brought a small dog concealed in her carry-case and without a permit ... A few days later an elderly couple arrived from Vietnam with a bamboo cage

containing two live birds; a search of their baggage revealed five live fish.' (October 1996)

'Brisbane's eagle-eyed Quarantine inspector Steve de Jong discovered a container infested with small marine snails while performing a tailgate inspection on a container of toilet paper from Malaysia. The discovery raised two questions: why import toilet paper from Malaysia, and how does toilet paper become infested with *marine* snails?' (November 1997)

'Buck the detector dog continues to break his own records. On a recent visit of the *Spirit of Tasmania*, Buck detected a whopping 55 kilograms of fruit and vegetables in passengers' luggage disembarking via the gangway.' (April 1997)

'. . . a consignment of furniture from India was found to be infected with borers and termite tracks. It was also packed in straw that concealed a number of cow pats.' (November 1996)

'A passenger arriving on a flight from the Pacific declared an apple on his passenger declaration. Questioned by an AQIS officer, the passenger denied carrying any other items of quarantine concern . . . but a search of his luggage revealed another apple, a can of meat and some wood carvings. He was also carrying two prohibited propane gas canisters. Ironically, the passenger was later found to be a criminal law judge, visiting from a northern hemisphere country.' (February 1997)

'A recent inspection of personal effects involved more than Quarantine staff had bargained for when the client turned out to be a Chinese herbalist moving to Australia from New Zealand. His entire shop was presented for inspection: hundreds of varieties of powdered plant products, animal faeces, human placenta, seeds, insects and so on, all of which had to undergo detailed inspection.' (October 1997)

'While carrying out routine surveillance at a Brisbane wharf recently, Inspectors Bill Woods and Katie Flynn were approached by wharf staff and advised that a container of machinery from Houston,

> Texas, had a number of small lizards inside ... the lizards were North American anole lizards [major invaders in the tropical Americas].' (March 1999)

> 'During routine surveillance of cargo offloaded by accident at Fremantle, insects were detected coming out of a crate that was immediately sent for fumigation. The goods, which consisted of personal items, included two boxes of fresh capsicums, 50 wooden skewers of dried snails, 25 bags of gari root, 40 bottles of palm oil, dried chillies, dried fish and unidentified seeds. All the goods were infested and rotting, with large numbers of dead insects and rodent droppings.' (March 1999)

In other areas, however, I am less impressed. It surprises me that AQIS allows in so many foreign fish and fish products. As already discussed, Australia imports millions of live ornamental fish and vast quantities of fish bait and meal. Rock lobster pots in Western Australia are baited with herring heads from Holland, mackerel from Scotland, sardines from America and other fish from New Zealand. Prawn, salmon and tuna farmers rely upon imported fish food. Tuna long-line fishermen import squid, mackerel and other fish from just about anywhere. Australia even imports over a million dollars' worth of bait for recreational fishermen. And none of this is tested for disease.

The horrible pilchard die-off in 1995 shows that fish diseases threaten our fish farms and wild stocks, and that imported fish food poses a considerable risk. Two recent government reviews, 'Report of the National Task Force on Imported Fish and Fish Products' (1996), and Humphrey's review of aquatic animal quarantine (1995), issue warnings against our heedlessly importing fish, alive or dead. Local substitutes should be used, although the Task Force found problems with supply while recommending they be addressed urgently. Some progress has been made since then and AQIS is now conducting risk assessments on fish and fish imports. We must hope that some changes follow from this, but meanwhile these high-risk products keep flowing in, despite a second pilchard die-off in 1998.

In 1995 I penned an article for *Australian Natural History* magazine, 'Australia's Weed Scandal', which began with the statement 'Quarantine in Australia is a farce'. I was upset that AQIS was allowing in weeds in the guise of new pasture and ornamental plants. I sent the article to AQIS but they did not respond, either to me, or to the magazine. AQIS has since introduced the Weed Risk Assessment system (see chapter 40), a big step forward. AQIS tells me that new pasture grasses are unlikely to be allowed in. I hope this proves to be true. But many new plants still flow in at the whim of nurseries in search of the new and unusual. The new system will only keep out obvious weeds.

Quarantine's handling of flower imports is another cause for concern. Fresh flowers are a high-risk import because insects, mites, fungi and viruses like to travel with them. Cut flower imports have multiplied fourfold in the past three years. In 1998 about 20 *million* flowers were imported, most of them from Zimbabwe (more than 7 million), South Africa, India, Colombia and the Netherlands. Flowers are usually fumigated with methyl bromide[1] then inspected to ensure that pests have died. But new diseases and insects sometimes sneak in anyway, although no one is ever sure how. Azalea petal blight slipped in recently, along with western flower thrips (*Frankliniella occidentalis*), discovered in 1993, and poinsettia whitefly (*Bemisia tabaci*), in 1994.

Importers can bring in flowers under three different procedures: they can fumigate them on arrival, have them fumigated overseas by AQIS-approved companies or have them inspected by AQIS, which orders fumigation if pests are found. Because fumigation is not foolproof AQIS checks carefully to ensure the treatment has worked. It can't examine every bud and leaf for tiny aphids, thrips and disease, but it does inspect a large proportion – as determined by a mathematical formula that assumes pests will be distributed more or less evenly. This formula takes no account of one-off travellers, for example a pregnant bug hidden in a bud. Such 'hitchhikers', AQIS's Ian Barrass told me, usually end up in the bottom of the flower boxes, which are always inspected along with the flowers. I hope he's right but wonder how he can be sure.

Some local flower growers are angry about this system. Sal Russo of the Flower Growers Group of NSW condemns it as a 'hit and miss approach'. Australian growers are suffering both from the flood of cheap imports and from their associated pests. Western flower thrips (which attacks roses) and white rust on chrysanthemums have added substantially to farming costs. I spoke to one AQIS biologist who was remarkably candid about flower inspections, admitting, 'It doesn't work very well. Lots of pests do get in,' but, 'For the amount of trade that goes on it's not a complete failure'. This AQIS insider agreed that Australia should not be importing flowers.

AQIS claims the system works, that the number of pests eluding its inspectors is very small compared to the volume of flowers, and this seems to be true. Exact figures are unavailable because AQIS won't release its latest report on plant-pest incursions, saying it's inaccurate and needs redrafting. Some of its conclusions are summarised in the Nairn review, which reports that seventy-four new crop and flower pests became established in the years 1991–5. About one-third came in on fruits and vegetables and one-fifth on nursery plants and flowers. The level of risk posed by cut flowers may be no higher than the risks associated with aquarium fish and fresh foods and considerably lower than the risks posed by dirty machinery.

Like most countries, Australia values freedom of trade and travel above optimal quarantine, and AQIS's brief is only to minimise the risks. As the National Farmers' Federation observed to Nairn, 'freedom to trade is progressively becoming the "default" position: quarantine barriers or import controls must be justified. In this situation quarantine becomes more explicitly a program for managing risk.' No-risk quarantine is applied only to major economic pests such as livestock diseases. The problem with this approach is that we end up importing many products we don't really need, and because each product carries an associated risk, sooner or later pests will get in. The right to trade ends up overriding the national good. It concerns me greatly that we have no forum for questioning this approach, for matching the costs of the pests against the benefits of the trade, for arguing that Australia might be better off without those millions of imported roses and tulips. AQIS's job is to process trade, not to question its worth. Pests keep entering Australia, not because AQIS does its job badly (it is one of the best quarantine services in the world), but because, under current political priorities, its task is impracticable, and made more so each year by the ever-growing volume of trade.

1. Ironically, methyl bromide, the preferred fumigant for cut flowers, garden plants and containers, is an ozone depleting gas. The Montreal Protocol lists it as a class 1 ozone depletant, responsible for about 5 to 10 per cent of the world's global ozone loss. Although its use is still permitted for quarantine, a race is under way to find a replacement.

39

Are We Blind?

Barriers to Enlightenment

> 'Hey Mother Nature,
> Sure looks like we hate ya,
> But we're just so blind.'
> **Ruth Apelt, 'Mother Nature Blues', 1994**

If you were to heed Bill Mollison's advice in *Introduction to Permaculture* (1994) you might end up breaking the law. Among the plants he recommends you grow are two weeds, prickly pears and mesquite, so dangerous that most state governments have banned them as noxious, making them illegal to grow. Mesquite, one of Australia's worst weeds, is good for fodder, firewood, syrup and bee forage, says Mollison; prickly pears he recommends for fencing.

Other rampant plants he would have you grow include box-thorn (noxious in some states), honey locust, leucaena, tagasaste, iceplant, hawthorn (*Crataegus* species, declared noxious in Victoria), mirror plant, ivy, guavas, olives and fennel (also noxious in Victoria). Mollison sees in honey locust (*Gleditsea triacanthos*) an 'excellent stock fodder' tree and a source of durable, quality timber. But when this tree, sprouting spines as long as pencils (see colour plate), began thickening up in forests around Warwick, forming heavily armed, impenetrable groves, the Queensland government saw a weed, declaring it noxious in 1993; their weed teams have now almost evicted it from the state. Nurseries had been promoting it as a hardy ornamental and fodder tree, and one had 7000 seedlings in stock when the declaration came through. It's another plant that readers of *Introduction to Permaculture*, at least in Queensland, are forbidden to grow. Leucaena, says Mollison, is good for revegetating eroded hills in the tropics. True enough, but it's also a major weed, throwing out copious crops of long-lived, water-borne seeds. Once entrenched, this shrub is very hard to eradicate, as one permaculturist I know has discovered to his cost. Even the pasture men who promote it for fodder warn of its weediness, cautioning that its seedlings are 'remarkably persistent'. In Darwin and Rockhampton it's a domineering weed that chokes vacant allotments and gullies.

Mollison is not fond of critics who claim he promotes plants that do more harm than good. Fix up the land first and worry about weeds later, he says. His priorities run counter to those of orthodox pest control, which acts on the principle of controlling weeds early, *before* problems explode. The other co-founder of permaculture, David Holmgren, gives weeds a positive spin. In a 1997 article in *Permaculture International Journal* he put the view that co-evolution of wild systems containing weeds may be the key to sustainability. 'Wild nature is evolving new ecosystems from a mix of self-reproducing species at an ever increasing speed', he claims. 'In some areas, especially along streams, the ecosynthesis process is advanced to the point where forests of mixed native and exotic species are beginning to show systematic characteristics.' This may provide the best solution to land degradation, he says. The weeds are there for a reason. Elsewhere he has argued: 'We need to have good reasons, be serious about working the land in a sustainable as well as productive way before we have the moral right to get rid of the weeds.'

Fortunately, there are permaculturists who think differently, who abhor the notion that whatever blankets the land is best for the land, that weeds are nature's first-aid treatment. *Permaculture International Journal* admits to a 'a philosophical divide which has sparked heated debate within the permaculture community and strained relationships between groups that have otherwise much in common'. The movement is young and diverse, is still evolving and will probably outgrow the weed-tolerant philosophy of its founders. Mollison's failing has been to draw too liberally upon methods used by struggling Third World communities to reclaim eroded lands. Mesquite and other woody weeds may have a place in the wastelands of India, where firewood and goat-feed are scarce, but we do not need them here.

Mollison, unfortunately, is not the only public figure promoting pests. Don Burke, through his popular television show and magazine, endorses the use of highly questionable plants. The February 1999 issue of *Burke's Backyard* recommends one of Australia's worst environmental weeds, blue thunbergia ('blue trumpet vine'), as a great climber to grow in northern regions.[1] In fact, this rampant rainforest-invader has been banned as noxious by the northern shires of Hinchinbrook, Cook, Cardwell, Douglas, Johnstone and Mulgrave, making it illegal to grow across much of north Queensland. Other serious weeds lauded in the magazine include Spanish lavender (declared noxious in most of Victoria), and Western Australian bluebell creeper (*Sollya heterophylla*), which happens to be the most invasive weed in Arthurs Seat State Park near Melbourne. Don Burke is partly aware of the weed problems, but the occasional warnings appearing in his magazine seem inadequate. An article in January 1999 celebrating gloriosa lily (*Gloriosa superba*) warns of its weediness in north Queensland without explaining that it is even more invasive in southern Queensland and northern New South Wales (a 'major rainforest weed' according to one NSW book). This article was shown to me by an indignant friend who had filled three garbage bags full of gloriosa tubers during a morning on Moreton Island.

Whenever I ponder the feral future, the number of well-meaning people who have contributed to its creation always amazes me. Permaculturists care passionately about the health of the land. The explorers, the acclimatisers, even the toad-besotted Mungomery, like today's pasture scientists, were all trying to improve Australia. How could so much harm have flowed from such good intentions?

One problem is that people always underestimate the wilfulness of living things. Domesticated animals and plants are seen as compliant possessions when really they are living things that are still infused with wildness. We should *expect* our garden plants to want to escape and not express surprise when they do. (More on this in the next chapter.) Another difficulty is that cause and effect, in invasion biology, are often far removed from each other, separated by time, and often by space. Whoever planted that harungana tree in Babinda back in the 1930s could *never* have imagined that sixty years later teams of weeders would be tramping high into the rainforests to blitz its offspring. In the gardens of those weeders today may be plants that future generations will struggle to contain. So many of our disasters have been caused by people who lived and died innocent to the consequences of their actions. Mosquito fish, buffalo, mimosa bush, prickly pear, rubber vine, mesquite, athel pine, prickly acacia, thunbergia, iceplant - the examples go on and on. Because of the time lags we are forever looking backwards, berating past generations for their follies while ignoring our own. We pour hundreds of millions of dollars into killing weeds we already have while failing to keep new ones out.

We are seldom aware of the power we wield, of how seemingly trivial acts can ignite ecological explosions. Something as insignificant as a seed on a sock or a spore on a coat can rewrite world history. Someone took a diseased potato to Ireland last century and triggered the Great Potato Famine; a million people died and perhaps two million more emigrated to America and Australia. A maggoty fruit smuggled into Cairns a few years ago cost Australia $150 million (from papaya fruit fly invasion); and a woodchip ship from Japan disgorged algae that might have killed me. The burden of responsibility can be intolerable. We are clumsy giants on a fragile planet vested with powers we cannot comprehend; the lid has been levered from Pandora's box and the demons continue to emerge. No one is innocent. Spores may have dropped from my boots during my foreign travels – and a species somewhere may have died as a result. The biologists trying to save rare frogs from disease may in fact be helping it spread.

Actions affect different parties in different ways. The benefits bestowed by new pasture plants flow to graziers, the costs to crop farmers and national parks. At every turn the environment ends up subsidising economic mismanagement. As Geoff Carr has said: 'In general, the activities of horticulturalists, agriculturalists and professional land managers are astonishingly compartmentalized and narrowly focused, a situation which frequently permits short-term economic gains or "cost-effective" environmental management at the expense of irreversible biological devastation.'

The CSIRO's pasture men rarely get to see their plants misbehaving – they don't spend enough time in the bush. Bryan Hacker admitted to me that his concerns about pasture plants grew when glycine began invading his own garden. Most nursery owners and gardeners (my mother included) have never seen their prized foxgloves, gladioli and freesias running amok in the bush. A retired seed inspector told me that imported seed is often contaminated with strange seeds, but when an importer is pressing to import his seed, and the contaminant can't be identified, the seed is usually let in. Caution yields to expediency. In any case, should a problem emerge, it will be someone else's headache – perhaps someone who isn't even born yet. The connections are almost impossible to make because most careless actions produce no results. Most gardens do not broadcast weeds and most smuggled foods cause no harm.

Even pest experts are narrowly focused. When I was researching the aquarium industry for this book I spoke to weed experts alarmed about cabomba who had no idea that fish diseases and snails were problems too. No one had ever looked at the aquarium trade holistically, much less at the whole gamut of exotic invasions. Eric Rolls' classic book on exotics, *They All Ran Wild*, subtitled '*The animals and plants that plague Australia*', is in retrospect an incredibly narrow and backward-looking work, focusing almost entirely upon mammals and painting an extraordinarily understated picture of the problems. No mention at all is made of garden plants, pasture plants, aquarium fish or ballast invaders.

With so many hurdles to understanding we can begin to answer the question posed by Gábor Lövei in the journal *Nature*: 'Why have species introductions not been recognized as a global phenomenon?' As Lövei himself goes on to explain, 'Acknowledged components of global change, like ozone depletion or carbon dioxide enrichment (the "greenhouse effect"), occur on a very large spatial scale and simultaneously affect several different ecosystems. In contrast, invasions are mostly local, and involve a few species acting over a relatively short time frame.' But we don't have just a carp problem or a boneseed problem or a Portuguese millipede problem; we have an exotic species problem – a giant, ever-growing hydra that daily sprouts new heads. As any invasion biologist will tell you, it's a threat more ominous than the greenhouse effect, industrial pollution or ozone depletion. Exotic invasions – biological pollution – should be ranked as our second most urgent ecological problem after habitat loss.

Unfortunately, the conservation movement, which should be acting vigorously on this issue, has yet to wake up to the problem; most of my fellow greenies are asleep to pests.[2] Rarely does the movement seize the initiative on exotic intruders. On most of the vital issues – ballast invaders, trout stocking, phytophthora, and the importation of fish products and garden, pasture and reclamation plants – it has remained all but silent. Greening Australia is the only national conservation organisation that understands the problem of pests (or at least weeds), but because it was originally government funded, it chooses not to criticise government or industry or to enter political debate, although the problems are so dire that this may soon change. Pore through a pile of back issues of *Habitat*, the Australian Conservation Foundation's colour magazine, and you will rarely find a mention of pest issues. (I have twice offered to write for them about weeds, but they expressed no interest.) Most greenies cannot name any of Australia's worst environmental weeds, other than perhaps bitou bush. The agenda on pests has been left to committed scientists and bureaucrats toiling in government, and the calls for action have come from government reports such as *Plant Invasions*. The heroes of the movement are not Bob Brown and Peter Garrett but report-writing boffins such as entomologist Mark Lonsdale and Victorian weed expert Geoff Carr.

The sad truth is that many conservationists, despite their best intentions, are ecologically illiterate. Like everyone else they think pest problems are a legacy of the past about which little can be done today. Too often they act only from the heart, believing that evil always strikes in the guise of greedy businessmen and heavy machines. The reality – that some of our worst environmental problems stem from tiny fish and colourful plants – is somehow emotionally unfulfilling. Property developers and miners are easier to demonise than earthworms and seaweeds, and although gardening is harming Australia's environment more than mining, how do you run a campaign saying that?

At a national level, the movement is so preoccupied with saving wilderness it fails to understand that exotic pests show less respect for park boundaries than miners and developers. Greenies believe the Wet Tropics and south-west Tasmania are already saved, blissfully unaware of the escalating pest threats in these areas. Activists are kept so busy fighting for remote regions – Kakadu, Cape York, Fraser Island – they fail to notice that habitats nearer to hand, like the You Yangs and the New South Wales foreshores, are drowning under weeds. The weed problems are left for smaller groups to tackle – groups like the Big Scrub Rainforest Landcare Group in northern New South Wales, an organisation with no national voice.

The conservation movement served us very well in the past by teaching us about the ecological consequences of our actions. We learned that the products we consume use up natural resources and generate pollution, and as responsible consumers we now take these concerns into account. But Australians have yet to realise that the consequences of their actions include biological invasions. Permaculturists, in the brief history of their movement, are fast waking up to this; it behoves us all to learn the same lesson.

1. Don Burke later explained to me that this article accidentally appeared without an intended weed warning. That said, I don't think such plants should be promoted, warning or not. He is doing some very good weed-awareness work, however.
2. I speak from the inside, having served for eight years as a councillor for the Wildlife Preservation Society of Queensland.

40

Wild Organisms

Understanding Plants

> 'Virtually any plant can be a weed.'
> ***The National Weeds Strategy*,**
> **Commonwealth of Australia, 1997**

When Belair National Park near Adelaide was established in 1891, it became only the second national park in Australia and the tenth in the world. That the first park commissioners held an archaic view of the nature of a 'national park' is evident from their annual report of 1894, which described the trees, shrubs and creepers they were planting. Foreign poplars, willows, oaks, redwoods and horse chestnuts were among trees grown to shade roads and picnic grounds. Not until 1923 did the park adopt a natives-only policy.

Belair today is perhaps the most weed-infested national park in Australia. Suckering willows, poplars and ashes have claimed its streams, ivy and periwinkles blanket the banks, and thick tangles of buddleias creep high into the eucalypts. Privets, pines, roses, olives, horse chestnuts, hawthorn, boneseed and bridal creeper add to the colourful mess. According to the park management plan nearly 400 exotic species grow in the park, if garden plants, weeds and displaced natives are counted. The park stands as an embarrassing example of how easy it is to misunderstand plants. Because trees and shrubs are immobile and unthinking, people think they are necessarily benign and easy to control. Belair's Tree Planting Committee never imagined they were sowing heartbreaking problems for the future.

If I were to stand in my garden today and free a jar-full of exotic bugs or a cage-full of foreign birds it would seem irresponsible – they might establish in the wild. Yet hardly anyone thinks it unethical to plant an introduced tree, even though a tree is much more likely to spawn a problem than a bunch of beetles or birds. Over the span of its life – perhaps hundreds of years – it will turn out hundreds of thousands of seeds, and the chances are good that some of these, borne off by wind or birds or washed into drains, will sprout somewhere unbidden.

If we truly understood plants we would be appalled by the thousands of foreign species we grow. Ten per cent have become weeds already – what of the rest? Think about it: every plant, unless it has been genetically bred into submission, is a wild organism at heart, animated by the drive to survive and reproduce. Somewhere in the world its brothers and sisters are eking out a wild existence, surviving and reproducing under sometimes harsh conditions. When we see it sitting prettily in a pot in a nursery these are the last thoughts going through our minds. The small continent we live in embraces so many combinations of soil and climate that most plants in the world could probably find somewhere congenial to grow in the wild – which means that most foreign plants are potential weeds, almost every one a 'sleeper'. Instead of wondering why so many plants become weeds, we might ask why so few have yet done so. Is it because they are sterile hybrids or genetically inbred, or because they lack their specialised pollinators or soil fungi? Is it because they are outcompeted by native plants or eaten out by animals? Or is it just that they haven't yet found the opportunity?

In the government report *Recent Incursions of Weeds to Australia*

1971–1995, Richard Groves provides a taste of what may lie ahead. Since 1971, he notes, some 290 new weeds have become established in Australia. Two-thirds were escaped garden plants and most had been in Australia for some time. 'The most significant result of our analysis', Groves writes, 'is that there has indeed been an increase in the rate of naturalisation of plant species over the 25 years from 1971, and especially over the last 15 years'. Our sleepers are waking up. More and more garden plants are crossing the interface between residential areas and bushland. Some are benefiting from greenhouse warming or from floods and droughts brought on by El Niño. Yet plants can remain sleepers for hundreds of years; we are still in the early stages of an unfolding drama.

Another recent book, *Western Weeds*, addresses the current situation in Western Australia, and by including plenty of garden escapes it has the look of a gardening book, so gaudy are its plates. Freesias, jonquils, snowflakes, cannas, irises, arum lilies and gladioli smile out from the first few pages. The carrot, fig, asparagus, basil, weeping willow and Californian redwood are in there too. Here, in vivid colour, is confirmation that a typical weed is not a dock or a thistle but a rampaging garden plant.

Australians need to change their thinking about foreign plants. We should see in every species a potential weed; unless we have evidence to show it won't spread, a presumption of guilt should be applied. The federal government's own *National Weeds Strategy* is blunt enough: 'Virtually any plant can become a weed.' AQIS chose long ago to apply this thinking to exotic animals by severely restricting imports but failed to apply the same logic to plants. Other than with known weeds and close relatives of weeds, importers have had a free hand. This indulgent policy has proved disastrous, as new weeds have continued to flow in.

In recent years AQIS has woken up to the danger, and Craig Walton of the Plant Quarantine Policy Branch assures me they now apply the presumption of guilt to plants as well as animals. Botanist Alfred Ewart (see chapter 4) warned back in 1909 that new plants should not be allowed into the country without a report demonstrating their harmlessness, and now – ninety years (and hundreds of weeds) later – such a system is finally in place. Weed Risk Assessment (WRA), introduced in 1997, is a big step forward. Developed by Paul Pheloung of the Western Australian Department of Agriculture, it asks forty-nine questions of any new plant. Is it thorny, viney, shade tolerant, aquatic, fond of infertile soil? Is it highly suited to Australia's climate, is it weedy somewhere else? Does the plant benefit from fire, mutilation or cultivation? Is it toxic, unpalatable, parasitic or a fire hazard? Are its seeds (propagules) dispersed by birds, other animals, people, wind or water? Does it seed prolifically? If a plant answers 'yes' to any question it must be evaluated further; if it scores more than six affirmatives it is denied entry. WRA assumes that most weeds fit certain criteria, that pests can be predicted in advance. But one great weakness is that only a few of the questions need be answered properly for a plant to pass (ten out of forty-nine); questions can effectively be ignored if the answer is 'don't know'.

It was an early version of this test that upset the CSIRO scientists at the meeting I attended in 1995 (see chapter 11). They tried it out on twenty-three of their favourite pasture plants and ten were rejected, including buffel grass and leucaena, major pasture plants and major weeds. If these plants were up for assessment today they might not pass. Indeed, Walton told me that in future, new pasture grasses are unlikely to be admitted – the risk is too high. Mark Lonsdale's dramatic article (see chapter 11), although dismissed by some in CSIRO, proved influential. Since a plant automatically earns a point for being a grass (this is one of the questions asked), the new system could end the CSIRO's hopes, as put to the 1996 workshop I attended, of introducing new salt-tolerant grasses from Africa (sure-fire weeds). New legumes will still get in provided they survive evaluation. The ground rules here have yet to be decided, but AQIS will ask such questions as: Will the benefits outweigh the likely weed costs? Will the plants be containable if they escape?

Can they be bred to yield few seeds? An ideal legume, Walton says, will be one that only grows when sown with a specific exotic soil fungus. It will be fascinating to see the conflicts played out, to watch Australia meet the National Weeds Strategy goal of establishing 'a procedure for resolving sectoral conflicts on weed management'. Will the pasture industry wind back its operations, will new lines of non-weedy plants be developed, or will a bunfight erupt, with Australia's powerful grazing lobby demanding that screening be scrapped? Stay tuned.

During the first full year of WRA, 260 species were screened; 131 were accepted, 72 rejected and 57 shunted into the 'further evaluate' box. Most of the accepted plants are flowering annuals or shrubs, many of them from South Africa. Every time a new gardening book appears overseas, nursery owners here are alerted to new prospects and Australia ends up with more sleepers. The plants are coming in for no good reason except that a nursery somewhere wants something new to sell. Nurseries need not show that Australia will benefit in any way from the 131 new plant species.

I am amazed that so many plants are breezing through all the tough questions on the WRA form. (One plant thankfully rejected was mopane [*Cochlospermum mopane*]. A Queensland grazier, cooperating with a university, wanted to trial it as a browse legume. Mopane I know well as a tree that dominates clay soils in southern Africa, forming vast monocultures. It could have thrived on the black soil plains of outback Australia, just as African prickly acacia and Jerusalem thorn have done. Its only animal controls back home are elephants and mopane caterpillars.) Although Walton claims WRA now applies a presumption of guilt to plants, this is not quite true. Certain categories of plants – water plants, grasses, vigorous vines and so on – are found guilty by association, but plants that don't fit these criteria, the large numbers still flowing in, continue to win the benefit of the doubt. There is no proof they won't become weeds, just a lack of evidence that they will. Some of them undoubtedly *will* end up on our weed lists. No one seriously believes we can predict every new weed, just most of them.

Keeping new weeds out of Australia is one concern. Another is to extinguish the many sleepers already here. It seems unthinkable that a noxious weed such as stevia, known from only one site near Glen Innes, could have been allowed to fester there, year after year, since 1933, with no serious attempt to destroy it. Craig Walton agrees that this is incongruous and wants the National Weeds Strategy Executive Committee to respond. As it stands, weed removal is a state responsibility, even though most weeds are national concerns. There are probably a couple of hundred feral sleepers out there – most of them obscure plants marking time at only a few sites, many as yet undiscovered. The Northern Australia Quarantine Strategy works on the assumption that many of the target weeds it wants kept out of Australia may already be here, and Siam weed and mile-a-minute have proved them right.

The National Weeds Strategy committee is also uneasy about the vast seed stores held by pasture scientists, which contain enough germ plasm to ignite a millennium's worth of problems. As research projects bear fruit new 'weeds' are regularly released. The CSIRO may already have the salt-tolerant grasses it wants, stored as unassessed seeds in its collections. It certainly has supplies of high-risk mimosa and leucaena species, none of which need to be assessed by WRA because they are already in the country.

Botanic gardens, as we've discussed, are another source of risk for sleepers. Nowhere else in the land are so many potential new weeds – obscure plants from distant corners of the globe – grown together with no real security, often on the edges of bushland. Perth's botanic gardens could not have behaved more dangerously than by installing a Mediterranean garden and stocking it with the very plants from overseas best suited to the local climate; no wonder *Succowia* took off. Brisbane's botanic gardens, also on the edge of forest, feature an impressive collection of African savanna trees superbly adapted to Australia's outback. They include masasa

(*Brachystegia spiciformes*) and mfuti (*B. boehmii*), two trees that dominate Africa's miombo woodlands, along with Jerusalem thorn, harungana (fruiting freely) and several ferociously armed plants.

Nurseries are certainly a problem. They are sometimes caught selling banned plants, recent examples including cabomba and lagarosiphon. A Victorian parliamentary inquiry into weeds found in 1998 that 'many serious agricultural and environmental weeds are sold throughout Victoria under horticultural or trade names'. I know of a lawyer living in north Queensland who runs an exotic nursery as a sideline. He flies around the globe seeking out strange new plants to sell, any of which could find a new home in the Wet Tropics rainforests. Most plants in the Amazon, central Africa, Sumatra and Borneo would grow happily in tropical Queensland.

When I ponder these possibilities, I keep coming back to one big question: How many of our garden plants, in the long term, are fated to become weeds? Will more and more keep escaping until there comes a day when most of them have founded wild populations? Or are most species hamstrung by some biological limitation? Many plants, I know, are hampered by a need for special pollinators. They set no seed here because their flowers are designed to admit special insects that Australia doesn't have. When those insects reach Australia, of course, it may be open season. Geoff Carr suggests that some plants are stymied by a need for bumblebees. If so, we are in for troubling times, for bumblebees (*Bombus terrestris*) recently invaded Australia. In 1992 they appeared in Hobart after crossing the Tasman from New Zealand. One day soon they will reach the mainland.

Bumblebees are hailed overseas as 'one of the most efficient pollinating insects in the world'. They are massive bees with big tongues that probe deeper into flowers than any other bee in southern Australia. In North America their preferred flowers (notably pea flowers) are called 'bumblebee flowers'. New Zealand has four bumblebee species, introduced around the turn of the century to pollinate red clover (*Trifolium pratense*). (Similar introductions in

Australia, in 1884, 1885, 1891, 1892, 1909, 1927 and 1930, mercifully failed.) The species that appeared in Tasmania is known to visit 400 different exotic flowers in New Zealand alone. The advance of bumblebees across Tasmania is a source of delight to some and worry to others. Because bumblebees are excellent pollinators of crops, especially tomatoes, farmers and horticulturalists regularly press for their introduction in Australia, appeals AQIS has so far rejected. A disgruntled tomato farmer may well have smuggled in Tasmania's bumblebees. Their arrival is certainly raising smiles among tomato growers on the mainland.

Bumblebees are poised for a major career as environmental pests. They readily enter forests and heathlands to harvest nectar from scores of native plants, including banksias, bottlebrushes, eucalypts, pea-bushes and tea-trees. They visit many flowers that honeybees avoid. Native bees and honeyeaters, especially the eastern spinebill, are thought to be suffering from the loss of nectar. The leatherwood honey industry is likely to suffer too. Bumblebees also visit more than 150 exotic plants; they now forage on the subalpine slopes of Mt Wellington (1100 metres); and they may one day invade most of Tasmania's World Heritage areas, a matter of concern to the Tasmanian Department of Environment.

When the New South Wales Agriculture Department applied recently to release bumblebees on the mainland, Environment Australia (a section of the federal Department of the Environment) received plenty of protests from state government departments concerned about wildlife. But none of them raised the point that concerns me here: *sleepers will turn into weeds upon the arrival of a new and superior pollinator.* Bumblebees will probably reach the mainland one day by island-hopping across Bass Strait, and their impact is likely to be profound.

American leafcutting bees (*Megachile rotundata*) were introduced recently to South Australia to pollinate lucerne (*Medicago sativa*). These bees visit many other flowers as well, especially legumes and daisies, two families containing hordes of weeds. Lucerne itself is a vigorous orchard weed, and this bee will increase the seed set of the weedy populations. Several concerns were raised before leaf-cutting bees were released, but pollination of new weeds was not one of them. Asian honeybees are poised to invade Australia, and we may one day be invaded by the other three bumblebees in New Zealand. There are another dozen bees, including Mexican carpenter bees and more bumblebees, that New Zealand's Department of Scientific and Industrial Research would like to import to pollinate pumpkins, onions, sunflowers, blueberries, passionfruit and fruit trees. Some of these, if introduced, will one day sneak over to Australia, just as bumblebees, European wasps and sirex wasps have done. Once again, those 'experts' applying to import them did not consider pollination of weeds. Australia, through the Australian and New Zealand Environment and Conservation Committee, should be protesting vigorously over these introductions.

Australia, the most nectar-rich land in the world, with more nectar-feeding birds than any other country, is poised for invasion by waves of foreign bees. Symbiotic relationships between flowers, birds and insects that flourished for millions of years will be disrupted forever. That much is understood. But it shows how poorly we understand our plants that neither those scientists seeking to release new bees nor those opposing them have even considered another possibility – that new bees will help pollinate existing weeds and turn garden sleepers into destructive new pests. Here is yet another example of how ignorant we are of the intricacies of invasion. We need to wake up fast.

41

What to Do?

Embracing a New Ethos

> 'You're either part of the solution or you're part of the problem.'
>
> **Eldridge Cleaver, 1968**

Cane toads will explode into Kakadu soon and nothing we can do will stop them. Biologists have exhausted their options, science and human ingenuity have failed, and no fall-back plan exists – Mungomery's toads will invade. A poisonous intruder from South America will violate a sacred site and the ecology of the wetlands will be despoiled forever. How will the traditional owners of Kakadu judge us?

Our helplessness in the face of this invasion is particularly distressing. Indeed, we can do very little about most of the pests now invading Australia. But that does not make us powerless. We can't undo damage already done, close the gates through which the toads have bolted, but we can certainly prevent new gates being opened. We have the power to prevent future disasters.

Ten new weeds, most of them escaped garden plants, establish themselves in Australia each year. That is something we can act on. We need to embrace a new gardening ethos that accepts that gardening near bushland entails an ecological responsibility. Gardeners should choose carefully what they grow (no local weeds, nothing with berries) and dispose of waste wisely, never dumping cuttings in gullies or bushland. Nurseries should reconsider what they sell. And we should all take more care of our local bushland, which usually bears the brunt of new weeds. Those who maintain gardens alongside national parks carry enormous responsibilities and should be especially vigilant about errant plants.

Fortunately, the choice of plants to grow near bushland is greater than the many cautionary tales in this book might suggest, because most plants only misbehave in certain regions. Spanish lavender, freesias and spearmint are weedy around Perth but completely safe in Brisbane, where the climate and soils suit them less. I grow buddleias to attract butterflies but wouldn't contemplate doing so if I lived in Victoria or South Australia, where they often escape. The worst of Australia's weeds should not be grown anywhere, but with lesser pests some flexibility is warranted. Native plants are one option, so long as we recognise that some of these, when taken to new regions, can also take off as weeds. Benign azaleas and roses are better than weedy wattles.

Below are some basic principles for caring for local bushland.

Eight Principles of Eco-friendly Living

1. If you live anywhere near bushland, don't grow problem plants (obtain a list from your local council or from Greening Australia).

2. Avoid planting exotics that bear fruits and berries.

3. Never dump garden wastes in bushland or gullies.

4. Don't allow your pets to fertilise bushland with their droppings.

5. Don't let soapsuds, fertiliser, food scraps or other wastes find their way into bushland.

6. Watch over your local bushland and remove garden escapes. Weed-killers containing glyphosate are fairly safe.

7. Join your local bushcare group or Greening Australia and learn to identify local weeds.

8. Teach your children to care.

There are many other ways in which we as individuals can help. Because pests thrive on disturbance we should, wherever possible, reduce our impact on nature – tread lightly upon the earth. Soil is the skin of the land that keeps out infection (weeds), and mechanical damage to soil should be avoided. Rural land should be managed with care to avoid scarring the earth. Pole houses are better than cut-and-fill. Soil that escapes down slopes and into streams fosters pest problems wherever it goes.

Don't dump nutrients in the bush. Urine and food scraps left along walking trails provide opportunities for weeds (a major problem in some parks), as do dog droppings and horse dung (which also contains weed seeds). Never ride horses through heathlands growing on infertile soil. I was once hired as a consultant to study management issues of a Stradbroke Island beach and found the main problem was soil enrichment leading to weed invasion. Campers were tossing away food scraps and dishwashing water and using the back-dunes as a toilet, and weeds such as gloriosa lily were beginning to proliferate. Weeds are only kept out of many habitats by the infertility of the soil.

Never put exotic fish, plants or snails into ponds or dams that can overflow after rain. Just last week I saw a crop of water lettuce

(*Pistia stratioites*), a declared noxious weed, growing illegally in a garden pond close to my local stream. Fish can swim remarkable distances down 'dry' slopes after rain – I have seen it for myself. Give away or kill unwanted aquarium life, never 'liberate' it.

Clean your car of encrusted mud (full of seeds and spores) before you enter a national park. And clean your boots before returning from overseas, as well as before departure. Keep your boat free of waterweeds and encrusting animals (many waterweeds are spread about by dinghies). Think carefully before you replant natural areas (especially in phytophthora-prone regions) using nursery stock. Are exotic spores (and earthworms) lurking in the soil? The best nurseries practise phytophthora hygiene – ask them about it.

Understanding is the key to change. Most problems are born of ignorance and misunderstanding. The environment movement taught us to consider the ecological consequences of our actions in terms of energy use, resources and pollution, but biological invasions were not factored in. Because of that oversight, nurseries, gardening shows, fish stockers, seed importers, pet shops and reclamation projects have escaped the scrutiny they warrant. We can't expect governments to address unsafe activities when public support is not there.

With more public awareness we have a better chance of convincing the federal government to act in an integrated way. We urgently need a high-level government committee on exotic pests, preferably a ministerial council, which could be modelled loosely on the Invasive Species Council set up in a major initiative by US president Bill Clinton in February 1999. Co-chairing his council are the Secretary of the Interior (responsible for the environment), the Secretary of Agriculture and the Secretary of Commerce. Other council members are the secretaries of Defense, State, Treasury and Transportation, and the head of the Environmental Protection Agency. Their role is to provide 'national leadership regarding invasive species', and their brief is to develop, within eighteen months, an Invasive Species Management Plan, which among other things will recommend a 'process to identify, monitor, and interdict pathways that may be involved in the introduction of invasive species'. Australia needs an initiative like this. We have a National Weeds Strategy but nothing to address the whole gamut of pests and no body that rates such high-level representation.

State and local governments, and the media, could do so much more. As part of a broad public campaign, governments could distribute more weed posters and pamphlets and do more to promote weed awareness days. The media could take the issues more seriously. When Northern Pacific seastars first invaded Tasmania, 'Sixty Minutes' ran a television segment that posed the inflammatory question 'Should we make the Japanese pay?' without ever addressing the real issue of ballast water. (In fact, the starfish may have come from Korea or northern China.) Gardening shows and magazines have a key role to play in warning about weeds and promoting better ethics, a duty they have yet to accept.

Environmentalism has surged in popularity in recent years but most Australians remain ecologically ignorant; nature is something we love but don't understand. Yet we can't hope to care for it properly until we have all raised our ecological IQs. The study of ecology, although immensely rewarding, is complicated; our kids should begin learning about it when they are young. Good field naturalists, like good musicians, begin early, usually before their teens. We desperately need more fieldworkers to interpret the changes taking place, to spot new pests and new problems, to recognise that first bitou seedling sprouting on the beach. We need naturalists who can identify weeds, not just wildflowers. The Nairn review called for a partnership approach to quarantine involving the whole community, but that won't work without greater ecological awareness. We need pest-aware people to argue the issues. Moves are afoot to phase out tributyl tin – the poison that keeps sea creatures off hulls – because it is killing off life in our harbours. But biological pollution, because it spreads and multiplies, is arguably more harmful than industrial pollution, and a toxic harbour may be preferable to one festering with pests. We need people who can weigh these issues wisely and make the best decisions for the long-term health of the nation.

One useful thing we can do with some of our pests is to harvest them. We could be eating more rabbits and blackberries and wild olives, and wearing fox-skin coats. Ralph Waldo Emerson described a weed as 'a plant whose virtues have not been discovered', and though this overstates the case many weeds are useful. Camphor laurel wood is strongly scented and we should support those turners who use it. Plenty of urban weeds are good to eat and others serve well as medicines. Early pioneers boiled up sowthistles, nettles and amaranths and harvested centaury (*Centaurium erythraea*), asthma plant (*Euphorbia hirta*), sida-retusa and prickly pear as medicines. Blackberries make wonderful pies, and Von Mueller records that the leaves were also harvested 'as part of material for bouquets, wreaths, garlands and grave-crosses'. Dining upon weeds won't solve our weed problems but it can make for an interesting change of diet. In soups and spanakopita, for instance, docks, amaranths, nettles, and fat hen substitute well for spinach. Stems of curled dock (*Rumex crispus*) make an interesting alternative to rhubarb.

I will never forget the day, after my neighbour Jill and I decided to set up a local bushcare group, when that Brisbane City Council truck pulled up at Jill's kerb. Workmen were soon unloading shovels, forks, mattocks, gardening gloves, a wheelbarrow, hose, first-aid kit and the *pièce de résistance*, two rolls of colour film to record our efforts – all of this free of charge to a little group of neighbours boasting scarcely a dozen members. Down at the local park we have converted a tiny patch of weed-choked trees into a small but thriving rainforest. Three more groups have formed along our creek. Weeding mornings provide an inspiring spell out of doors, some good honest exercise, and a chance to feel useful and positive in a pleasant forest setting close to home.

Most Australians live in cities or towns within a sparrow's hop of remnant bushland. Wherever you live right now you can be fairly sure the forest nearest you is ailing from weeds. Around Australia there are hundreds of bushcare groups tending patches of urban bushland, keeping local habitats alive, encouraging natives and fending off weeds. Such groups help not only the patches of bushland they work upon but the larger and more important blocks farther afield, those that stand next in line for invasion from seeds spread by birds or streams. Most bushcare groups are supported by local councils or by Greening Australia, Landcare or the National Trust. These groups build upon the pioneering work begun in the 1960s by the Bradley sisters, working in weed-infested bushland near Sydney. Joan and Eileen shunned herbicides, but today's volunteers, in deference to the scale of the problem, use plenty of Roundup, an eco-friendly weedicide.

If you prefer not to join a group there are other ways to help. Some years ago, while living in a different suburb, I became so exasperated by all the umbrella trees in nearby bushland that I took to them with a jungle knife, hacking down the lot. I didn't poison the stumps and I guessed they would reshoot. Seven years passed, during which I sometimes wondered how that bush was faring. Two weeks ago I went back. The umbrella trees were hard to find. Most had resprouted, but the new shoots, which I soon snapped off, were small, and no new umbrella trees had appeared. One person *can* make a difference.

42

Life Goes On

'... a cockroach world of compromise ...'
Angela Carter, *Wise Children*, 1991

I am sitting by a stream under the shade of one of Brisbane's oldest trees, thinking about this book. Walls of weeds surround me but I like this place. I can, after all, look up to this ancient blue gum, a breathing monument to the past, a tree I know by its size was growing here well before Captain Cook sailed by (with exotic creatures encrusting the *Endeavour*'s hull). Today, South American glycine hugs its feet, a Chinese elm brushes its trunk and, at night, cane toads trill below, but I know there was a time when this tree stood here, tall and proud, with nothing exotic (other than dingoes) within hundreds of miles. It has seen all the invaders come, one after another – the guinea grass introduced by the Queensland Acclimatisation Society in the 1860s, additional weeds brought in around the turn of the century, the cane toad in the 1950s, glycine, proclaimed a wonderful pasture plant in the 1960s, the garden creeper Singapore daisy in the 1970s and the Indian mynas that winged in just a few years ago.

By looking about me I am reminded of the themes that animate this book. Most of the weeds I can see are plants that were brought to Australia deliberately, with great optimism – for pasture (glycine, para grass, guinea grass) or food (watercress, taro, mulberry), as garden ornament (Chinese elm, tipuana, mistflower), medicine (castor oil plant) or even as an aquarium plant (parrot's feather). Accidental introductions are represented too – cobblers pegs and petty spurge among others. Downstream I can see a stand of graceful river oaks, native trees that remind me of Florida, where their kinfolk are invading sand dunes and beaches.

My young son and I discovered this spot last year. We came down to scoop up some fish with our net. He asked if the taro lining the banks was a weed, then proposed a stone-throwing contest. We bashed down dozens of plants, not out of hatred for weeds but just having fun. This place is special, for it is one of the few stretches of stream in Brisbane where plenty of native fish still survive. In the depths I can see small shoals of smelt and occasionally a hardyhead, blue eye or gudgeon. Exotics outnumber them of course, but I am well used to that. Sometimes I scoop up three foreign species in one go – mosquito fish, swordtails and platies. I know from earlier surveys that no platies swam here in the 1980s, so I am also reminded how fast things change. I hope the native fish always survive here but I don't want to think too far ahead.

Many Chinese elms sprout along the banks, all of them seedlings or saplings; tipuanas too, and that too portends change. I tear out a few but their roots are strong. When planning this book I proposed to write at length about Chinese elms, Brisbane's worst weed, but there was so much else to say I never quite got round to it. If we drove on for half an hour we could find another stream with fewer weeds and more native fish, but I like it here, so close to home. By learning to like this place I am coming to terms with all these weeds and foreign fish, and with the feral future they imply.

People have been moving plants and animals around for thousands of years. It seems to be something fundamental we do, an intrinsic part of being human. We dominate the earth, we command its resources and we redistribute species as a matter of course. We are aware enough today to comprehend at least some of the consequences of this behaviour, but that doesn't mean we can change it. It's not the worst thing we do. Habitat destruction and overpopulation are bigger concerns, although I would place exotic species, or 'biological pollution', well before industrial pollution and greenhouse warming. But I can live with all of this. I won't become upset by the inevitable, even though I do expect us to behave more wisely in future.

We can't blame the animals and plants themselves for becoming pests. It makes no sense to be cruel to toads – they didn't ask to be brought here. It helps me to accept these creatures to have seen so many of them in their proper homes, where they look right because they belong. I have seen a cane toad in Costa Rica, starlings and foxes in Europe, wild boars wallowing in the Thai rainforest, and guinea grass, grazed down by antelope, on the African savanna. I once camped in a slash pine forest in Florida and counted all the birds among the trees – red-bellied sapsuckers, blue-grey gnatcatchers, red-shouldered hawks, palm warblers and so on. There was even a raccoon stealing about. Now when I look at the slash pine stands near home I feel that missing dimension – the context they lost when they were transported here.

In Australia, too, I have enjoyed many of my encounters with exotics. How exciting it was to see seven camels galloping through the searing Simpson Desert, and another time to watch a fox, just metres away, plucking blackberries. Once I camped on Moreton Island, alone and without a tent on a moonlit night, and a wild stallion, stamping on the silver sand nearby, stirred me from my dreams. It was a powerful experience, yet I am glad to know that all the island brumbies have since been shot out. Another time, sleeping under the stars near Longreach, a feral cat, obviously a dumped pet, startled me when it crawled inside my sleeping bag. It slept the night behind my knees then woke me again at dawn when it crept away.

And yet what can compare with the immense joy I felt on a visit to Wilkie Creek on the Darling Downs when rare rains thundered down and all the frogs came out! Up from nowhere came enormous waterholding frogs, pobblebonks, greenstripe frogs, salmon-striped frogs, even a few holy cross frogs, bright yellow, shiny and warty – I listed fourteen species in all. That time I was part of a survey team and in one line of pitfall traps one morning we counted out 342 ornate burrowing frogs. I saw over 700 frogs that week, more than I usually see in half a decade. It took me a couple of days to realise that this experience was made so special, not only by the range of frogs I saw, but by the absence of toads. Our campsite was a short drive south of the toad line and I could walk miles along sandy tracks at night hearing sugar gliders yapping and frogs chorusing without having to sidestep the dark knobbly shapes of toads.

So there it is. I love what is natural and I don't hate the pests. I pray we find the courage to change what we can and the wisdom to accept what we can't. We will never evict all the pests we already have, but we can try much harder to keep new ones out. A new ecology is emerging, one we don't yet understand, but one that will debase the marvellously rich diversity of life on earth unless we manage it well. That is the challenge of our feral future.

APPENDICES

Appendix I: Australia's Worst Environmental Weeds

The list compiled by Humphries, Grove & Mitchell (1991):

Prickly acacia (*Acacia nilotica*) – shade and fodder tree
Bridal creeper[1] (*Asparagus asparagoides*) – ornamental vine
Para grass (*Brachiara mutica*) – pasture grass
Buffel grass (*Cenchrus ciliaris*) – pasture grass
Bitou bush/boneseed[2] (*Chrysanthemoides monilifera*) – ornamental shrub/ballast
Rubber vine (*Cryptostegia grandiflora*) – ornamental vine
Water hyacinth (*Eichhornia crassipes*) – ornamental pond plant
Aleman grass (*Echinochloa polystachya*) – pasture grass
Reed sweetgrass (*Glyceria maxima*) – pasture grass
Hymenachne (*Hymenachne amplexicaulis*) – pasture grass
Mimosa bush (*Mimosa pigra*) – botanic garden escape
Jerusalem thorn (*Parkinsonia aculeata*) – shade tree
Mission grass (*Pennisetum polystachion*) – pasture grass
Mesquite (*Prosopis* species) – fodder and shelter tree
Salvinia (*Salvinia molesta*) – aquarium plant[3]
Athel pine (*Tamarix aphylla*) – ornamental tree
Blue thunbergia (*Thunbergia grandiflora*) – ornamental vine
Japanese kelp (*Undaria pinnatifida*) – accidentally introduced seaweed

1. Known until recently as *Myrsiphyllum asparagoides*.
2. Bitou bush exists in two forms, bitou bush (*C. monilifera rotundata*), introduced accidentally in dry ballast (becoming weedy on beaches after it was later sown for land reclamation), and boneseed (*C. monilifera monilifera*), imported in the 1850s as a garden plant and now invading woodland.
3. Experts assume salvinia was introduced as a pond and aquarium plant, as it was overseas, but proof is lacking.

Appendix II: Weeds of National Significance (WONS)

This list of Australia's worst weeds was released by the federal government on 1 June 1999. The twenty weeds were chosen on both socioeconomic and environmental grounds, and most of them are both environmental weeds and pests of agriculture. All but four of them were introduced into the country deliberately. The list was released after this book had been written and during the final editing stages, so it was not possible to discuss it within the text (but see the footnote on page xix).

Alligator weed (*Alternanthera philoxeroides*)
Athel pine (*Tamarix aphylla*)
Bitou bush/boneseed (*Chrysanthemoides monilifera*)
Blackberry (*Rubus fruticosus agg.*)
Bridal creeper (*Asparagus asparagoides*)
Cabomba (*Cabomba caroliniana*)
Chilean needle grass (*Nassella neesiana*)
Gorse (*Ulex europaeus*)
Hymenachne (*Hymenachne amplexicaulis*)
Lantana (*Lantana camara*)
Mesquite (*Prosopis* species)
Mimosa (*Mimosa pigra*)
Parkinsonia (*Parkinsonia aculeata*)
Parthenium weed (*Parthenium hysterophorus*)
Pond apple (*Annona glabra*)
Prickly acacia (*Acacia nilotica* ssp. *indica*)
Rubber vine (*Cryptostegia grandiflora*)
Salvinia (*Salvinia molesta*)
Serrated tussock (*Nassella trichotoma*)
Willows (*Salix* species), except weeping willows and pussy willow

Appendix III: Introduced Fauna in Australia

VERTEBRATES

Mammals (24)

House mouse (*Mus musculus*) – widespread
Pacific rat (*Rattus exulans*) – WA, Qld, Norfolk Island
Brown rat (*Rattus norvegicus*) – widespread
Black rat (*Rattus rattus*) – widespread
Asian house rat (*Rattus tanezumi*) – Qld
Dingo (*Canis lupus dingo*) – widespread
Fox (*Vulpes vulpes*) – widespread
Cat (*Felis catus*) – widespread
Rabbit (*Oryctolagus cuniculus*) – widespread
Brown hare (*Lepus capensis*) – widespread
Brumby (*Equus caballus*) – widespread
Donkey (*Equus asinus*) – widespread
Pig (*Sus scrofa*) – widespread
One-humped camel (*Camelus dromedarius*) – widespread
Water buffalo (*Bubalus bubalis*) – NT
Banteng (*Bos javanicus*) – NT
Cow (*Bos taurus*) – widespread
Goat (*Capra hircus*) – widespread
Fallow deer (*Dama dama*) – Qld, NSW, Vic., Tas., SA
Red deer (*Cervus elaphus*) – Qld, NSW, Vic.
Rusa deer (*Cervus timorensis*) – NT, Qld, NSW
Sambar deer (*Cervus unicolor*) – NSW, Vic.
Chital deer (*Axis axis*) – Qld
Hog deer (*Axis porcinus*) – Vic.

Birds (26)

Ostrich (*Struthio camelus*) – SA
Red junglefowl (*Gallus gallus*) – Qld
Common pheasant (*Phasianus colchicus*) – NSW, Tas., SA, WA
Indian peafowl (*Pavo cristatus*) – SA, WA
Wild turkey (*Meleagris gallopavo*) – Tas.
California quail (*Callipepla californica*) – Tas. (King Island)
Mute swan (*Cygnus olor*) – Tas., WA
Mallard (*Anas platyrhynchos*) – widespread
Rock dove (*Columba livia*) – widespread
Laughing turtledove (*Streptopelia senegalensis*) – WA
Spotted turtledove (*Streptopelia chinensis*) – widespread
Skylark (*Alauda arvensis*) – NSW, Vic., Tas., SA
House sparrow (*Passer domesticus*) – widespread
Eurasian tree sparrow (*Passer montanus*) – NSW, Vic.

Nutmeg mannikin (*Lonchura punctulata*) - Qld, NSW
Java sparrow (*Lonchura oryzivora*) - Christmas Island
Common chaffinch (*Fringilla coelebs*) - Lord Howe Island
European greenfinch (*Carduelis chloris*) - NSW, Vic., Tas., SA
European goldfinch (*Carduelis carduelis*) - widespread
Common redpoll (*Carduelis flammea*) - Lord Howe, Macquarie Islands
Yellowhammer (*Emberiza citrinella*) - Lord Howe Island
Red-whiskered bulbul (*Pycnonotus jocosus*) - NSW, Vic.
Common blackbird (*Turdus merula*) - NSW, Vic., Tas., SA
Song thrush (*Turdus philomelos*) - Vic.
Common starling (*Sturnus vulgaris*) - widespread
Common myna (*Acridotheres tristis*) - widespread

Reptiles (6)

Slider (*Chrysemys scripta*) - NSW
Asian house gecko (*Hemidactylus frenatus*) - NT, Qld
Mourning gecko (*Lepidodactylus lugubris*) - Qld
Skink (*Lygosoma bowringii*) - Christmas Island
Flowerpot snake (*Ramphotyphlops braminus*) - NT, Qld
Wolf snake (*Lycodon aulicus*) - Christmas Island

Amphibians (1)

Cane toad (*Bufo marinus*) - NT, Qld, NSW

Fish (31)

Marine

Yellowfin goby (*Acanthogobius flavimanus*) - NSW, Vic.
Goby (*Acentrogobius flaumi*) - Vic.
Variable triplefin (*Forsterygion lapillum*) - Vic.
Striped goby (*Tridentiger trigonocephalus*) - NSW, Vic., WA

Freshwater

Blue acara (*Aequidens pulcher*) - Qld
Green terror (*Aequidens rivulatus*) - Qld
Midas cichlid (*Amphilophus citrinellum*) - Qld
Oscar (*Astronotus ocellatus*) - Qld
Goldfish (*Carassius auratus*) - widespread
Convict cichlid (*Cichlasoma nigrofasciatum*) - Vic.
Carp (*Cyprinus carpio*) - widespread
Mosquito fish (*Gambusia holbrooki*) - widespread
Pearl cichlid (*Geophagus brasiliensis*) - Qld
Burton's haplochromus (*Haplochromus burtoni*) - Qld
Banded cichlid (*Heros severus*) - Qld
Jewel cichlid (*Hemichromus guttatus*) - Qld
Firemouth (*Herichthys meeki*) - Qld
Oriental weatherloach (*Misgurnus anguillicaudatus*) - Qld, NSW, Vic.
Rainbow trout (*Oncorhynchus gairdneri*) - Qld, NSW, Vic., Tas., SA

Mozambique tilapia (*Oreochromus mossambicus*) – WA, Qld
Redfin perch (*Perca fluviatilis*) – SA
One-spot livebearer (*Phalloceros caudimaculatus*) – WA
Sailfin molly (*Poecilia latipinna*) – Qld
Guppy (*Poecilia reticulata*) – Qld
Roach (*Rutilis rutilis*) – NSW, Vic.
Brown trout (*Salmo trutta*) – NSW, Vic., Tas., SA, WA
Brook trout (*Salvelinus fontinalis*) – Tas.
Niger cichlid (*Tilapia mariae*) – Qld, Vic.
Tench (*Tinca tinca*) – NSW, Vic., Tas., SA
Swordtail (*Xiphophorus helleri*) – Qld, NSW
Platy (*Xiphophorus maculatus*) – Qld, NSW

INVERTEBRATES (MOST NUMBERS ARE PROBABLY UNDERESTIMATES)

Ascidians

8 seasquirts

Echinoderms

3 starfish

Entoproctans

2–3 species

Lace corals

25–30 species

Annelid worms

13 (marine) polychaetes
45 earthworms

Molluscs

29–30 marine species (shellfish, slugs, chitons, nudibranchs)
About 50 land snails and slugs
Several freshwater snails

Arthropods

28 marine crustaceans (crabs, barnacles, shrimps, slaters)
10 land crustaceans (slaters)
35 or more spiders
6 millipedes
Vast numbers of insects and mites

Nematode worms

Many soil worms and some parasitic worms

Platyhelminth worms

2 marine worms
1 freshwater flatworm
9 terrestrial flatworms
Several parasitic worms

Cnidarians

17–26 sea anemones and hydroids

Sponges

7–9 species

Others

Unknown numbers of protozoans and other invertebrates

References

Birds – Christidis & Boles (1994): Earthworms – G.H. Baker, pers. comm.: Flatworms – Leigh Winsor, James Cook University, pers. comm.: Freshwater fish – Allen (1989): Jeff Johnson, Queensland Museum, and Alan Webb, James Cook University, pers. comm.: Mammals – Strahan (1995): Marine fish – Matthew Lockett, University of Technology, pers. comm.
See also Pollard & Hutchings (1990a) (though it is now well out of date), and http://www.ml.csiro.au/hewitt/CRIMP/ispp.html#body (Sobaity sea bream [*Sparidentex hasta*] and Japanese sea bass [*Lateolabrax japonicus*] are often listed as established but both species are probably extinct in Australia.): Millipedes – Dyne & Walton (1987): Reptiles – Cogger (1996), and Griffiths (1997) for slider: Slaters – Glen Ingram, ex-Queensland Museum, pers. comm.: Spiders – Robert Raven, Queensland Museum, pers. comm.

Appendix IV: Australian Animals Abroad

(The list for invertebrates is incomplete owing to lack of reliable data.)

Mammals

Brushtail possum (*Trichosurus vulpecula*) – New Zealand
Tammar wallaby (*Macropus eugenii*) – New Zealand
Parma wallaby (*Macropus parma*) – New Zealand
Red-necked wallaby (*Macropus rufogriseus*) – England, Scotland, New Zealand
Brush-tailed rock wallaby (*Petrogale penicillata*) – Hawaii, New Zealand
Swamp wallaby (*Wallabia bicolor*) – New Zealand

Birds

Brown quail (*Coturnix ypsilophora*) – Fiji, New Zealand
Black swan (*Cygnus atratus*) – New Zealand, semi-feral elsewhere
Peaceful dove (*Geopelia striata*) – St Helena Island
Galah (*Cacatua roseicapilla*) – New Zealand
Sulphur-crested cockatoo (*Cacatua galerita*) – New Zealand
Rainbow lorikeet (*Trichoglossus haematodus*) – New Zealand
Crimson rosella (*Platycercus elegans*) – New Zealand
Eastern rosella (*Platycercus eximius*) – New Zealand
Budgerigar (*Melopsittacus undulatus*) – Florida, escape elsewhere
Laughing kookaburra (*Dacelo novaeguineae*) – New Zealand
Noisy miner (*Manorhina melanocephala*) – Solomon Islands
Australian magpie (*Gymnorhina tibicens*) – Fiji, New Zealand, Solomon Islands
Red-browed finch (*Neochmia temporalis*) – Tahiti
Chestnut-breasted mannikin (*Lonchura castaneothorax*) – Society Islands, New Caledonia
Silvereye (*Zosterops lateralis*) – Tahiti

Reptiles

Eastern grass skink (*Lampropholis delicata*) – New Zealand, Hawaii

Amphibians

Golden bell frog (*Litoria aurea*) – New Caledonia, Vanuatu, New Zealand
Brown tree frog (*Litoria ewingii*) – New Zealand
Dwarf treefrog (*Litoria fallax*) – Guam
Green swamp frog (*Litoria raniformis*) – New Zealand

Molluscs

Land snail (*Pseudopeas tuckeri*) – New Caledonia

Crustaceans

Barnacle (*Elminius modestus*) – Britain, France, Holland, Belgium, Germany, Denmark, South Africa
Marine ispopod (*Sphaeroma quoyanum*) – US (California)
Marine isopod (*Iais californica*) – US (California), Singapore
Land amphipod (*Arcitalitrus dorrieni*) – Britain, Ireland, New Zealand

Land amphipod (*Arcitalitrus sylvaticus*) - US (California)

Redclaw (*Cherax quadricarinatus*) - Western Samoa

Spiders

Redback (*Latrodectus hasseltii*) - Japan, New Zealand, Tristan da Cunha

(Other species in New Zealand)

Insects

Too many to mention but most species in New Zealand: cockroaches, beetles, bugs, flies, mosquitoes, moths, ants, wasps etc.

Flatworms

Flatworm (*Parakontikia ventrolineata*) - England, South Africa, Hawaii, US, Mexico

Flatworm (*Caenoplana coerulea*) - South Africa, US, New Zealand

Flatworm (*Caenoplana sulphurea*) - New Zealand

Annelid worms

Earthworm (*Spenceriella minor*) - Scotland

References

Additional references for these species are provided in other chapters, wherever they are mentioned in more detail. Mammals - Lever (1985), King (1990): Birds - Long (1991), Pernetta & Watling (1978): Grass skink - McKeown (1978), Robb (1986): Frogs - Tyler (1976), McCoid (1993): Snail - Solem (1964): Crustaceans, spiders - see references for chapters 18–21: Flatworms - Leigh Winsor, James Cook University, pers. comm.: Earthworm - Sims & Gerard (1985).

Appendix V: Some Recent Quarantine Highlights

Foreign pests are continually invading Australia. The *AQIS Bulletin* is the source of most of the following reports.

1993

- Papaya fruit fly (*Batrocera papayae*) spreads from New Guinea into five islands in the Torres Strait. An eradication programme is begun.
- Citrus canker, a devastating disease that could destroy our citrus industries, is discovered in Darwin. The outbreak is contained, the disease destroyed, as were earlier Australian outbreaks in 1991 and 1984.
- Sydney experiences its first red tide, caused by toxic dinoflagellates.
- A $200 000 eradication campaign is begun against kochia (*Kochia scoparia*), a plant introduced deliberately to Western Australia three years previously to reclaim land and feed stock, despite its status as a major weed overseas. More than 6000 hectares are found to be infested.
- Western flower thrips (*Frankliniella occidentalis*), a threat to many plants, including native species, appears on flower farms around Perth, the first record from Australia. Eradication is considered impossible, but quarantine barriers are imposed.
- A South American weed, *Praxelis clematidea*, is detected in Queensland for the first time. It is found to be spreading rapidly along roadsides and railway lines in the Wet Tropics, where it is expected to become a serious pest of sugar cane.
- Melon thrips (*Thrips palmi*) are detected on capsicums in Brisbane, the first records from outside the Northern Territory. No action is taken to eradicate them, and they spread through southern Queensland.
- A disease new to Australia, *Fusarium oxysporum* fsp *vasinfectum*, is discovered infecting cotton in southern Queensland.

1994

- Siam weed (*Chromolaena odorata*), one of the world's worst weeds, is discovered growing near Tully in north Queensland. It has probably been there for twenty years. An eradication programme is undertaken.
- Avian influenza (fowl plague) breaks out near Brisbane. To contain it, 20 000 chickens are slaughtered in an operation costing about $420 000.
- Poinsettia whitefly (*Bemisia tabaci*), of a strain able to attack 500 different crops, is discovered in Darwin. It has the potential to cost Australian agriculture $300 million.
- Leafspot disease of strawberries (*Xanthomonas fragariae*) appears near Adelaide and is suppressed. A previous outbreak near Gosford was eradicated in the 1970s.
- Varroa mite, a tiny pest that kills whole bee hives, is detected on Dauan Island in Torres Strait, the first record from Australia.
- The New Guinea mango fly (*Bactrocera frauenfeldi*), which entered Cape York in 1974, reaches Cairns. It attacks bananas, mangoes and citrus fruits.
- Equine morbilli virus, a completely new disease, kills several horses and their trainer in Brisbane. This may well be a mutated native virus rather than an exotic introduction.

1995

- A new virus, apparently imported in fish food, kills millions of pilchards in southern Australia, in the seas extending from Geraldton to Noosa.
- Papaya fruit fly, a very serious fruit pest, appears in north Queensland. More than $50 million is spent eliminating it by setting up roadblocks, and baiting and destroying male flies.
- Two Torres Strait islanders die from Japanese encephalitis, a new disease from Asia.
- Formosan termites (*Coptotemes formosanus*), a major pest overseas, are detected on a boat in Sydney Harbour. AQIS flies into action and fumigates the vessel.
- Golden dodder (*Cuscuta campestris*), a serious crop weed in southern Australia, appears in Western Australia for the first time. Because it is likely to cost farmers $5 million to $7 million each year, an eradication campaign is begun.
- Melon fly (*Bactrocera cucurbitae*), a serious vegetable pest, is recorded in Australia for the first time at Perth, but it does not become established.
- A grass from Paraguay, *Nassella charruana*, closely related to serrated tussock (*N. trichotoma*), one of Victoria's worst weeds, appears at Thomastown, and 'shows every indication of becoming an extremely serious agricultural and environmental weed'.

1996

- The fruit fly *Bactrocera trivialis*, a serious pest of fruit and vegetables overseas, is detected in Torres Strait. This species has been appearing on and off since 1983. Melon fly (*Bactrocera cucurbitae*) is also recorded, but neither becomes established.
- A dog illegally imported from New Guinea to Saibai Island in Torres Strait is found to be infected with screw-worm fly. It is killed and the owner charged.
- Japanese kelp is detected in Port Phillip Bay, the first record of this serious pest from mainland Australia. Genetic tests suggest it came via New Zealand.
- The red-banded mango caterpillar (*Deanolis sublimbalis*) and a sugar cane borer (*Chilo terrenellus*) are detected in the Torres Strait islands.
- Thousands of timber-boring beetles of two species not known from Australia (*Sinoxylon conigerum* and *Xylothrips flavipes*) are found infesting imported timber at Sydney's Darling Harbour. They are fumigated.
- A new exotic ergot, a fungal disease, is found infecting sorghum just west of Brisbane. Sorghum at three research sites is placed under quarantine.
- The African lily, *Wachendorfia thyrsiflora*, recently introduced as an ornamental plant, is found spreading 'abundantly' along a drainage line in Victoria.
- Several frogs and toads from Hong Kong and Indonesia are intercepted at Sydney Airport, having arrived accidentally in air cargo.

1997

- Fireblight, a serious orchard disease caused by the bacterium *Erwinia amylovora*, appears in Melbourne.
- Anthrax breaks out at a dairy in Victoria – it is the first outbreak in Australia since 1914. A quarantine zone is imposed and the dead cattle are burned and deep buried.
- The Philippine fruit fly (*Bactrocera philippinensis*) strikes Darwin. A 50-kilometre quarantine zone is imposed.

- The mango hopper (*Idioscopus nitidulus*) reaches mainland Australia from Asia for the first time. It is recorded in both Darwin and Weipa.
- Siam weed is detected at Mt Garnet, well away from the site of first establishment.
- A senior scientist at the Queensland University of Technology is suspended after breaching safety protocols covering Japanese encephalitis. Forty-five people are exposed; fortunately none become infected.
- Eggs of the Asian gypsy moth, a pest that destroys trees in Asia and North America, are found on a passenger liner berthing in Brisbane.
- Avian influenza breaks out at Tamworth. It is eradicated.
- Panama disease, a scourge of banana crops overseas, is accidentally inoculated into plants in a Brisbane government laboratory. Security guidelines are upgraded.
- A rhinoceros beetle (*Oryctes rhinoceros*), a serious pest of palms, is captured on a plane flying from Indonesia to Australia.

Appendix VI

Animals and Plants: A Checklist of Scientific Names

acacia, prickly (*Acacia nilotica*)
agaves (*Agave* species)
allamanda (*Allamanda cathartica*)
amaranth (*Amaranthus* species)
ant, big-headed (*Pheidole megacephala*)
apple, pond (*Annona glabra*)
ash, (*desert*) (*Fraxinus angustifolia*)
banksia (*Banksia* species)
barb, rosy (*Puntius conchonius*)
barramundi (*Lates calcarifer*)
beetle, khapra (*Trogoderma granarium*)
blackberry (*Rubus fruticosus* or *R. discolor* and related species)
blackbuck (*Antilope cervicapra*)
blackfish (*Gadopsis marmoratus*)
boneseed (*Chrysanthemoides monilifera* ssp. *monilifera*)
borer, European corn (*Ostrinia nubilalis*)
bottlebrush (*Callistemon* species)
box, brush (*Lophostemon confertus*)
boxthorn (*Lycium ferocissimum*)
broom, English (*Cytisus scoparius*)
buddleia (*Buddleja madagascariensis*)
bug, leucaena psyllid (*Heteropsylla cubana*)
burr, noogoora (*Xanthium strumarium* = *X. pungens*)
bush, bitou (*Chrysanthemoides monilifera* ssp. *rotundata*)
bush, groundsel (*Baccharis halimifolia*)
bush, mimosa (*Mimosa pigra*)
butterfly, cabbage white (*Pieris rapae*)
cactoblastis (*Cactoblastis cactorum*)
calopo (*Calopogonium mucunoides*)
canna (*Canna indica*)
casuarina (*Casuarina* species)
centro (*Centrosema pubescens*)
cherry-plum (*Prunus cerasifera*)
chestnut, horse (*Aesculus* species)
clover (*Trifolium* species)
cod, Murray (*Maccullochella peelii*)
cod, trout (*Maccullochella macquariensis*)
coffee (*Coffea arabica*)
cotoneaster (*Cotoneaster glaucophyllus*)
crab, green (*Carcinus maenas*)
crayfish, redclaw (*Cherax quadricarinatus*)
creeper, bridal (*Asparagus asparagoides*)
creeper, cats-claw (*Macfadyena unguis-cati*)
crown-of-thorns (*Acanthaster planci*)
daisy, Singapore (*Wedelia trilobata*)
dandelion (*Taraxacum officinale*)
dock (*Rumex* species)
dodder (*Cuscuta* species)
earthworm, Amazonian (*Pontoscolex corethrurus*)
elm, Chinese (*Celtis sinensis*)
emex, spiny (*Emex australis*)
eye, blue (*Pseudomugil signifer*)
fennel (*Foeniculum vulgare*)
fern, asparagus (*Asparagus aethiopicus* = *Protasparagus densiflorus*)
foxglove (*Digitalis purpurea*)
freesia (*Freesia leichtinii*)
frog, ornate burrowing (*Limnodynastes ornatus*)
fruit fly, papaya (*Bactrocera papayae*)
fruit fly, Queensland (*Bactrocera tryoni*)
garlic, three-cornered (*Allium triquetrum*)
gladiolus (*Gladiolus* species)
glycine (*Neonotonia wightii*)
gorse (*Ulex europaeus*)
gourami (*Trichogaster trichopterus*)
grass, buffel (*Cenchrus ciliaris*)
grass, canary (*Phalaris* species)
grass, guinea (*Panicum maximum* var. *maximum*)
grass, kangaroo (*Themeda triandra*)
grass, marram (*Ammophila arenaria*)
grass, mission (*Pennisetum polystachion*)
grass, molasses (*Melinis minutiflora*)
grass, pampas (*Cortaderia selloana*)

grass, para (*Brachiaria mutica*)
grass, Rhodes (*Chloris gayana*)
grasstree (*Xanthorrhoea* species)
guava (*Psidium guajava*)
gum, Tasmanian blue (*Eucalyptus globulus*)
hardyhead (*Craterocephalus stercusmuscarum*)
harungana (*Harungana madagascariensis*)
heath, Spanish (*Erica lusitanica*)
hemigraphis (*Hemigraphis colorata*)
hemlock (*Conium maculatum*)
hen, fat (*Chenopodium album*)
holly, English (*Ilex aquifolium*)
honeybee (*Apis mellifera*)
honeybee, giant (*Apis dorsata*)
hyacinth, water (*Eichhornia crassipes*)
hymenachne (*Hymenachne amplexicaulis*)
iceplant (*Mesembryanthemum crystallinum*)
iris (*Gynandriris setifolia*)
ivy, (*English*) (*Hedera helix*)
jacaranda (*Jacaranda mimosaefolia*)
jew, wandering (*Tradescantia albiflora*)
jonquil (*Narcissus tazetta*)
keelback (*Tropidonophis mairii*)
kelp, Japanese (*Undaria pinnatifida*)
kochia (*Kochia scoparia* var. *scoparia*)
lagarosiphon (*Lagarosiphon major*)
lantana (*Lantana camara*)
lantana, creeping (*Lantana montevidensis*)
laurel, camphor (*Cinnamomum camphora*)
lavender, Spanish (*Lavandula stoechas*)
leucaena (*Leucaena leucocephala*)
lily, arum (*Zantedeschia aethiopica*)
lily, gloriosa (*Gloriosa superba*)
lovegrass (*Eragrostis curvula*)
lupin (*Lupinus* species)
medic (*Medicago* species)
mesquite (*Prosopis* species)
miconia (*Miconia calvescens*)
mile-a-minute (*Mikania micrantha*)
millipede, Portuguese (*Ommatoiulus moreleti*)
mother-in-laws tongue (*Sansevieria trifasciata*)
mother-of-millions (*Bryophyllum tubiflorum*)
mould, tobacco blue (*Peronospora tabacina*)
mulberry (*Morus alba*)
mullein (*Verbascum thapsus*)
nasturtium (*Tropaeolum majus*)
nettle (*Urtica* species)
oak (*Quercus* species)
oak, river (*Casuarina cunninghamiana*)
olive (*Olea europaea*)
palm, date (*Phoenix dactylifera*)
panic, green (*Panicum maximum* var. *trichoglume*)
paperbark (*Melaleuca quinquenervia*)
parakeet, monk (*Myiopsitta monarchus*)
parrot's feather (*Myriophyllum aquaticum*)
paspalum (*Paspalum dilatatum*)
pegs, cobblers (*Bidens pilosa*)
perch, silver (*Bidyanus bidyanus*)
perchlet, olive (*Ambassis agassizii*)
periwinkle, blue (*Vinca major*)
periwinkle, pink (*Catharanthus roseus*)
phytophthora (*Phytophthora cinnamomi*)
pilchard (*Sardinops sagax*)
pine (*Pinus* species)
pine, athel (*Tamarix aphylla*)
pine, slash (*Pinus elliottii*)
pipi (*Donax deltoides*)
pittosporum, sweet (*Pittosporum undulatum*)
plant, castor oil (*Ricinis communis*)
plant, mirror (*Coprosma repens*)
poplar (*Populus* species)
prickly pear (*Opuntia stricta*)
privet, broadleaf (*Ligustrum lucidum*)
privet, Chinese (*Ligustrum sinense*)
puero (*Pueraria phaseoloides*)
rat, bush (*Rattus fuscipes*)
rat, canefield (*Rattus sordidus*)
redclaw (*Cherax quadricarinatus*)
redwood, Californian (*Sequoia sempervirens*)
rose, dog (*Rosa canina*)
salvinia (*Salvinia molesta*)
scale, cottony cushion (*Icerya purchasi*)
setaria (*Setaria sphacelata*)

sida-retusa (*Sida rhombifolia*)
siratro (*Macroptilium atropurpureum*)
smelt (*Retropinna semoni*)
snail, apple (*Pomacea canaliculata*)
snail, giant African (*Achatina fulica*)
snake, flowerpot (*Ramphotyphlops braminus*)
snowflake (*Leucojum aestivum*)
sour-sob (*Oxalis pes-caprae*)
sowthistle (*Sonchus asper* and/or *S. oleraceus*)
spearmint (*Mentha spicata*)
spider, daddy-long-legs (*Pholcus phalangiodes*)
spider, redback (*Latrodectus hasseltii*)
spurge, petty (*Euphorbia peplus*)
spurge, sea (*Euphorbia paralias*)
spurrey (*Spergularia* species)
stevia (*Stevia eupatoria*)
susan, black-eyed (*Thunbergia alata*)
sweetbriar (*Rosa rubiginosa*)
sweetgrass, reed (*Glyceria maxima*)
swordtail (*Xiphophorus helleri*)
sycamore (*Acer pseudoplatanus*)
tagasaste (*Chamaecytisus palmensis*)
taro (*Colocasia esculenta*)
tea-tree, coast (*Leptospermum laevigatum*)
thorn, Jerusalem (*Parkinsonia aculeata*)
thorn, sweet (*Acacia karoo*)
thunbergia, blue (*Thunbergia grandiflora*)
tilapia (*Oreochromis mossambicus*)
tipuana (*Tipuana tipu*)
toad, cane (*Bufo marinus*)
tree, cutch (*Acacia catechu*)
tree, umbrella (*Schefflera actinophylla*)
treefrog, white-lipped (*Litoria infrafrenata*)
tulip, African (*Spathodea campanulata*)
vine, Madeira (*Anredera cordifolia*)
vine, rubber (*Cryptostegia grandiflora*)
wasp, European (*Vespula germanica*)
watercress (*Rorippa nasturtiumaquaticum*)
waterlily (*Nymphaea* species)
watsonia (*Watsonia species,* especially *W. bulbifera*)
wattle (*Acacia* species)
wattle, Brisbane golden (*Acacia podalyriifolia*)
wattle, cootamundra (*Acacia baileyana*)
weed, alligator (*Alternanthera philoxeroides*)
weed, Siam (*Chromolaena odorata*)
weka (*Gallirallus australis*)
willow, weeping (*Salix babylonica*)
willows (*Salix* species)
worm, New World screw (*Cochliomyia hominivorax*)
worm, Old World screw (*Chrysomya bezziana*)
wren, Stephen Island (*Xenicus lyalli*)
yabby (*Cherax destructor*)
yarrow (*Achillea millefolium*)

GLOSSARY

Acclimatisation – The deliberate establishment of 'useful' animals and plants in new places, either in the wild or under husbandry or cultivation.
Alien – Foreign, exotic.
AQIS – Australian Quarantine and Inspection Service, part of the Federal Department of Primary Industries and Energy.
Ballast water – Water carried by ships to keep them stable at sea, when their cargo holds are largely empty. Usually seawater held temporarily in ballast tanks.
Biocontrol – Biological control.
Biological control – Pest control using predators or pathogens released into the wild.
CSIRO – Commonwealth Scientific and Industrial Research Organisation.
Cuscus – One of several species of tropical possum with small ears.
Dinoflagellate – A single-celled, free-swimming aquatic alga often able to form a cyst.
Environmental weed – An exotic plant that invades natural or largely natural habitats.
Exotic – Introduced from another country or region. It does not mean colourful or tropical.
Feral – Established in the wild after escape from human custody.
Invasion biology – The scientific study of exotic invasions.
Invertebrate – An animal without a backbone, for example an insect, worm or shellfish.
Legume – A member of the pea family (Fabaceae) or related families capable of increasing soil nitrogen.
Nairn review – Refers to the federal government's recent review of AQIS, chaired by Professor Malcolm Nairn, published as *Australian Quarantine: A Shared Responsibility* in 1996.
NAQS – Northern Australia Quarantine Strategy.
Naturalised – Maintains exotic populations in the wild.
Noxious – A legal term for weeds declared noxious by state law, which usually means landholders must control them. Some states now say 'declared weeds' instead.
Pathogen – An agent causing disease.
Pest – A wild organism (or virus) that is troublesome, annoying or destructive.
Permaculture – A system of intense cultivation that uses mixed plantings of multi-purpose trees, shrubs and herbs, usually incorporating livestock and recycling.
Ruminant – A hoofed animal with a rumen, a stomach compartment for digesting plant fibre.
Sleeper – A dormant pest that has not yet spread according to its potential.
Weed – A plant (tree, shrub, vine, fern etc.) growing in the wrong place. A plant displacing desired plants or, in a natural habitat, displacing native plants.

SOURCE NOTES

Introduction

Antarctica – Smith (1996): **Redbacks** – Wace (1968): **Port Phillip Bay** – Environment & Natural Resources Committee (1997): **McDonaldization** – Lövei (1997), Holmes (1998): **Cootamundra wattle** – Cronk & Fuller (1995): **Kochia** – Groves (1998), Dodd (1996), Panetta (1993): **Hymenachne** – see appendix II: **Singapore daisy** – Batianoff (1992), Batianoff & Franks (1998): **2700 weeds** – John Hosking, NSW Agriculture, Armidale, pers. comm.: 16 per cent – Australia has about 17 000 native plants: **$3 billion** – Commonwealth of Australia (1997). Some estimates say $3.5 billion: **200 marine species** – Paterson & Colgan (1998): Humphries, Groves & Mitchell (1991): Their list also appears in Alexander (1996). Mimosa bush is also called 'giant sensitive plant': **Weeds of National Significance** – available at http://www.weeds.org.au/natsig.htm: ***The National Weeds Strategy*** – Commonwealth of Australia (1997): **Camphor laurel & fish** – Bishop (1993). Needs further experimentation to prove the case: Soil poisons are called allelopathic substances: **Honeybees** – Pyke (1990), Wood & Wallis (1998). Also, *Victorian Naturalist*, 1997, 114(1), has several bee articles: **Foreign oysters evict** – Alexander (1996).

1. 'Cursed is the Ground'

Biblical weeds – Moldenke & Moldenke (1952): **Theophrastes** – Hort (1968, p. 195): Cato (1934): Varro (1934): **Pliny** – Pliny (1940–63): **Pollen records** – Godwin (1975): **Roman grain** – Helbaek (1964): **Rubbish pit** – Osborne (1971): **North American cattle, pigs, horses** – Lever (1985): **Balof 2** – Allen, Gosden & White (1989), Flannery & White (1991): **Exotic mammals on islands** – Flannery (1995), Heinsohn (1998): **Cassowaries on Seram** – Coates (1997): **Admiralty possum** – Flannery (1995), Heinsohn (1998): **Embalmed cats** – Attenborough (1987): **Lapitans & rats** – Roberts (1991): **Sparrows** – Ericson et al. (1997), Summers-Smith (1988): **Pacific rats reach Norfolk** – Roberts (1991), White & Anderson (1999): **Rabbits in England** – Lever (1985): **Dodo** – Quammen (1996): **Cockroaches** – Both the common cockroach (*Blatta orientalis*) and the German cockroach (*Blatella germanica*) were introduced to England, Marshall (1973): **Rats** – Lever (1985), Strahan (1995), Watts & Aslin (1981), Lekagul & McNeely (1977): **Rabbits** – Lever (1985), Pliny (1940–63).

2. First In

Tassie Tiger? – *The Examiner*, 20 May 1991: **Devil 500 years** – Strahan (1995): **Native-hen** – Boles (1999), Corbett (1995): **Dingo scientific name** – Strahan (1995): **Wolves carried worms** – Beveridge & Spratt (1996). Goldsmid (1984) suggests dingoes introduced other parasites such as *Toxocara canis* and *Dirofilaria immitis*: **Wallabies died recently** –

Johnson, Speare & Beveridge (1998): La Billardière (1800), Goldsmid (1984). Goldsmid also discusses human diseases that may predate European contact: **Norfolk Island** – Meredith, Specht & Rich (1985), Roberts (1991), White & Anderson (1999): Pacific rats are very poor at swimming and rafting. They occupy some New Zealand islands but not others, even those close by, implying that as colonisers they need a helping hand: **Rats on other islands** – Strahan (1995): **Macassans** – Macknight (1972), a detailed study of Macassan impact: **Tamarind** – Mulvaney (1975): **Ivy gourd** – Barbara Waterhouse, NAQS, pers. comm.: **Rice traded** – Macknight (1972): **Five-leaf yam** – Low (1989), Telford (1986): **Ginger** – Bailey (1900), Low (1989): **Haddon's dagger** – Lawrence (1994): Jequirity bean has extended its range southwards during the last century, strengthening the idea that it is exotic. But Leichhardt found it in remote central Queensland, implying a long residency: **Taro** – Coates, Yen & Gaffey (1988): **Torres Strait trade** – Lawrence (1994): **Cats** – Burbidge et al. (1988): **Hewitt** – pers. comm.: **Cats** – Wagner, pers. comm.

3. 'A Seasonable Gift'

Vancouver also planted lemons, pumpkins and watercress: **Cook** – Reed (1969, p. 170): **Cook in NZ** – Thomson (1922): Bligh (1936): **Chickens on the reef** – Long (1981): La Billardière (1800, vol. 1, p. 197; vol. 2, pp. 44, 81, 84): Flinders (1814): **Baudin** – Lever (1985): Grey (1841, vol. 1, pp. 108, 236–7): Giles (1875): Bunce, D. (1979, pp. viii, ix): **Hill** – Johnstone (1984, pp. 148, 184): Other men who planted seeds included Charles Fraser in Western Australia (bananas etc.), Captain Charles Sturt in central Australia (wheat, mustard, barley) and Captain Arthur Phillip in Sydney Cove. As well, pigs were released by Joseph Jukes, the naturalist on HMS *Fly*: **Coffee a threat** – Humphries & Stanton (1992): **Olives etc. noxious** – Parsons & Cuthbertson (1992): ***CSIRO Handbook*** – Lazarides, Cowley & Hohnen (1997): **WA weeds** – Keighery (1991): **Potential environmental weeds** – Csurhes & Edwards (1998): **Phillip Island** – Ziesing (1996): **Ancient date palms** – Gough (1994): **Genetically engineered foods** – Gledhill & McGrath (1997), Shaner (1996)

4. Ecological Insurrection

North (1980): **First Fleet** – Information available from many sources, e.g. Cobley (1980), Rolls (1969): **Brown's weed list** – Kloot (1985b): **Black stem rust** – White (1981): **Smallpox** – Butlin (1983): **Deer, rabbits** – Rolls (1969): Mitchell (1848): McKell (1924?): **Harvey** – Ducker (1988): **Tenison-Woods (1881)**: **Out of spite** – Liddle (1986): McCoy (1890): North (1980): Ewart (1909)

5. 'To Our Heart's Content'

Gillbank (1986), Rolls (1969) and Jenkins (1977) are useful sources on acclimatisation: Bennett (1864): **Barkly** – Rolls (1969): **Saint-Hillaire** – Gillbank (1986): Wilson (1858): Francis (1862): McCoy (1862): Von Mueller (1870): **Mrs Fraser** – Tindale (1959): **Dandelion** – Von Mueller (1885): Von Mueller (1870): **Select extra-tropical plants** – Von Mueller (1885). Earlier editions have different text: **Beneficial birds** – South Australian Acclimatisation Society (1881): Francis (1862): **Legacy today, zoos** – see Jenkins (1977)

6. By Design

18 worst – Humphries, Groves & Mitchell (1991): **Mesquite** – Anonymous (1914): **In Victoria** – Carr (1993): **Wet Tropics** – see chapter 39: **Noxious weeds** – Parsons & Cuthbertson (1992): **Desire for pretty gardens** – Many are listed in Parsons & Cuthberton (1992): **Pennyroyal etc.** – Parsons & Cuthberton (1992): **Spiny emex quote** – Helms (1898): **Axolotls** – Mark Lintermans, ACT Parks and Conservation Service; pers. comm.: **Ferrets, lions** – Long (1988): Flannery (1994)

7. Ode to the Toad

Mungomery (1936): **Dexter** – Mungomery (1935a): **Bell** – Mungomery (1935a): **102 toads** – Mungomery (1935b): **Breeding cage** – Mungomery (1935b): **41 800** – Mungomery (1936): **Many kinds of control** – Mungomery & Buzacott (1935): **This latest importation** – Mungomery (1935a): **Cane beetles** – Mungomery (1936): Mungomery & Buzacott (1935): **Fears about toads** – Mungomery (1936): Mungomery (1936): **Nigger in woodpile** – Mungomery (1935a): **Bees** – Goodacre (1947), Mungomery (1935a, 1936), Freeland (1985): **Expected high density** – Mungomery (1936): Mungomery (1936): Mungomery (1949): **Puerto Rico** – Freeland (1985): **Toads to Taiwan etc.** – Easteal (1981): **Ancestry of Australian toads** – Easteal (1981): **Toad habitat & distribution** – Covacevich & Archer (1975), Bennett (1996): **Brackish water** – Covacevich & Archer (1975), Freeland (1986): **30 km a year** – Bennett (1996): **Eats mammals etc.** – Covacevich & Archer (1975): Ormsby (1955): **Along roads, tracks** – Seabrook & Dettmann (1996): **Digitalis-like steroids** – Freeland (1985): **Keelback** – Covacevich & Archer (1975), Freeland (1986): **Goannas etc. disappear** – Burnett (1997), Covacevich & Archer (1975), Freeland (1985): **Take hiding places** – Covacevich & Archer (1975), Freeland (1985) Parasite free – Freeland (1986): **Viruses** – Bennett (1996): **European toads, hedgehogs etc.** – Jenkins (1946): **Moles** – Boyd (1902): **Moles, shrews** – Illingworth (1921): **European toads** – Wilson (1960): **Indian myna, mongoose** – Chisholm (1919): **Owls** – Anonymous (1920)

8. 'Peopling a Barren River'

Fish numbers – modified from Allen (1989): McCoy (1862) Macinnis (1996): Kingsmill (1918): **Fish eggs** – Macinnis (1996): Nicols (1882): **Shag culls** – Jenkins (1977): **Galaxias suffer** – Crook & Sanger (1997), Cadwallader (1996) and references therein: **Pedder galaxias** – Crook & Sanger (1997), Hamr (1992), Hamr (1994): **Swan galaxias** – Crook & Sanger (1997): **Said one scientist** – Cooling (1923): **Adelaide premises, Adelaide hatchery** – Borthwick (1923): **Mosquito fish** – Arthington & Lloyd (1989). The mosquito fish was previously called *Gambusia affinis*: **Para grass** – Arthington, Milton & McKay (1983): Eat eggs, steal food – Arthington, Milton & McKay (1983): **Nip fins** – Morgan, Gill & Potter (1998): **Mound springs** – Ponder (1986): **Bell frogs** – Webb & Joss (1997), Morgan & Buttemer (1996): **Carp** – Brumley (1991), Shearer & Mulley (1978)

9. Wet Pets

Kailola (1990): **Fish policy, fish farms, 14 snails** – Humphrey (1995): **Flukes at Perth** – Ponder (1975): **Mystery snail** – Robert Ingram, AQIS, pers. comm., Baker (1998):

Waterweeds – Parsons & Cuthbertson (1992): Cabomba – Mackey (1996): **Lagarosiphon** – Parsons & Cuthbertson (1992): **Lagarosiphon, etc. on sale** – Andrew Bishop, Tasmanian Department of Primary Industries, e-mail to ENVIROWEEDS@nre.vic.gov.au on 8 February 1999: **Approved list of 50** – Department of Natural Resources (1997): PIJAC – Birkhill (1997): **Caulerpa** – Paterson & Colgan (1998): **Flatworm** – Sluys, Joffe & Cannon (1995): **Webb** – pers. comm.: **Tilapia** – Arthington et al. (1984): **Weatherloach** – Lintermans, Rutzou & Kukolic (1990): **Approved fish list** – McKay (1984), Kailola (1990): **Piranhas intercepted** – Carson Creagh, AQIS, pers. comm.: **Fined $5000** – *AQIS Bulletin*, Aug. 1997: **Illegal catfish** – *AQIS Bulletin*, Mar. 1996: Humphrey (1995): **Turkeys** – *AQIS Bulletin*, June 1996: **Diseases** – Humphrey (1995), Langdon (1990), Ashburner (1976): **Goldfish ulcers** – Humphrey & Ashburner (1993), Whittington & Cullis (1988): **Tapeworm** – Dove et al. (1997), Dove (1998) Chytrid fungus – Laurance, McDonald & Speare (1996): **Fish farms** – Department of Primary Industries & Energy (1996): **Carp** – Morison & Hume (1990)

10. When Beauty is the Beast

Thirty per cent – Panetta (1993): **Seven of eighteen** – Humphries, Groves & Mitchell (1991): **Two-thirds are garden escapes** – Groves (1998): **In Victoria alone** – Carr (1993): **Four thousand** – Batianoff & Franks (1997): **Rubber vine etc.** – Humphries, Groves & Mitchell (1991): **Ferntree** – Department of Conservation & Environment (1991): Wombats & lyrebirds threatened – Adair (1995): Hamilton (1937): Clements (1983): **Sydney weeds book** – Buchanan (1981): **Melbourne 90 per cent** – Milburn (1996): Kirkpatrick (1974): Dixon (1892): **Athel pine** – Griffin et al. (1989), Humphries, Groves & Mitchell (1991): Batianoff & Franks (1997): **Wingham Scrub** – Stockard & Hoye (1990): **Dutchman's pipe** – Sands & Scott (1996–97): **Alpine** – Sainty, Hosking & Jacobs (1998): Tenison-Woods (1881): **Cooke** – e-mail dated 25 Nov. 1997 to NURSERY-WEEDS@majordomo.nre.vic.gov.au: Environment and Natural Resources Committee (1998): ***The Good Plant Guide*** – Craw (1996): Hamilton (1892)

11. Seeking Greener Pastures

Allen (1961). His paper is called 'Glycine proves its value': **Wombat** – Alan Horshup, Queensland Department of Environment & Heritage, pers. comm. Also see Low (1997c): **200 000 ha of buffel** – Walker & Weston (1990): **Africanising the world** – see Parsons (1972): **Plant Invasions** – Humphries, Groves & Mitchell (1991): **Package at fault** – Minutes of meeting held on 27 Jan. 1995, p. 1: **Meeting lively** – Held at CSIRO Cunningham Laboratory, Brisbane, 19 June 1995: ***The Weeds of Queensland*** – Kleinschmidt & Johnson (1977): **Weedy in Arizona** – Vitousek et al. (1997): Mott (1986): Lonsdale (1994): **My paper** – Low (1997c): **Mission grass** – Panton (1993): **Scorching in many parks** – Joe Edair, Queensland Department of Environment & Heritage, pers. comm.: **Veldt grass** – Pigott & Armstrong (1997): **Simpson's Gap** – Peter Latz, Northern Territory Conservation Commission, pers. comm.: **Para grass** – Low (1997c), Humphries, Groves & Mitchell (1991): **Native fish shun para grass** – Arthington, Milton & McKay (1983): Vines – Low (1997c): **NT weed book** – Smith (1995): **Legumes & grasses dominate** – e.g., Kloot (1987a): Myers & Robbins (1991): Cavaye et al. (1989): Other articles warning about run-down and unrealistic expectations include Silcock & Johnston (1993) and Burrows (1991). Silcock & Johnston say: 'If sown pastures prove difficult to grow, "wonder" plants are sought. However, such perceived needs are often due to unrealistic expectations of the land, either environmentally or economically.': Emmery (1997): Lonsdale (1994): Lowry (1991): Silcock & Johnston (1993): **Hundreds of wonderful plants** - R. Reid, Department of Primary Industries & Fisheries, Tasmania. He projected a slide of pack camels in the desert to emphasise that pasture scientists should become more adventurous in pursuit of new plants: McIvor & McIntyre (1997): Papers appeared in *Tropical Grasslands*, vol. 31 (1997): Hacker, Date & Pengelly (1997): **Digitaria an eco-disaster?** - Rod Randall, Weed Science Group, Agriculture Western Australia, pers. comm.: ***Indigofera*** a weed - Dr Gordon Guymer, Queensland Herbarium, pers. comm.

12. 'Every Creeping Thing'

Hull foulers - Evans (1981): **Lace coral** - Allen (1953): **Port Phillip Bay** - Environment & Natural Resources Committee (1997): **Tributyl tin** - Environment & Natural Resources Committee (1997), Alexander (1996, pp. 8–39): La Billardière (1800): **Rats** - Rolls (1969): **Lord Howe** - Hutton (1990): **Rats & plague** - Cumpston (1923), Longman (1923), Holmes (1927): Goldsmid (1984) discusses other disease introductions (typhus, tuberculosis etc.): **Crows** - Hylton (1927). See also Long (1981) and references therein: **Spurrey, canary grass** - Spicer (1878): **Golden dodder** - Groves (1998): Tenison-Woods (1881): **South Australian weeds** - Kloot (1987b): **Drought weeds** - Davidson (1984–85): Wace (1977): **Kakadu** - Lonsdale & Lane (1990): **Ayers Rock weeds** - Buckley (1981): **Barrier Reef weeds** - Chaloupka & Domm (1986): **Lady Elliot Island** - Batianoff (1998): **Zeppelin** - Russell et al. (1984): **Malaria in Europe** - Smith & Carter (1984): **5500 dead insects** - Russell et al. (1984): Passenger numbers - From AQIS in Review 1997–8 and earlier AQIS reports

13. Soil Travellers

Darwin (1881): **NZ flatworm** - Blackshaw & Stewart (1992), Jones & Boag (1996): Haria (1995): **Ecoclimatic study** - Boag et al. (1995): **Phytophthora in a nursery** - Hardy & Sivasithamparam (1988): **Nutgrass** - Parsons & Cuthbertson (1992): **European heath rush** - Groves (1998): **CSIRO Earthworms** - Baker et al. (1997): **350 native earthworms** - G.H. Baker, pers. comm.: **Pontoscolex** - Dyne & Wallace (1994): **Slaters** - G.J. Ingram, formerly Queensland Museum, pers. comm.: **Flatworms** - Leigh Winsor, James Cook University, pers. comm.: **Flowerpot snake** - Cogger (1996): **Texan oil rig** - *AQIS Bulletin*, Nov. 1996. The contamination included live spiders and wasp and bee nests: **New Guinea bulldozer** - *AQIS Bulletin*, June 1996: **Locomotives to New Zealand** - Wellington *Evening Post*, 13 June 1997: **Muddy tyres** - *AQIS Bulletin*, Aug. 1997: **Balinese carvings** - Carson Creagh, AQIS, pers. comm.: **Plants in speaker box** - *AQIS Bulletin*, Sept. 1996: **Plant in tampon** - *AQIS Bulletin*, Apr. 1997: **Deep burial** - *AQIS Bulletin*, Oct. 1996

14. Ballast Blues

For general introductions to ballast invasions see Bonny (1994), Paterson & Colgan (1998),

Carlton & Geller (1993) and Jones (1991 – now outdated). The various AQIS reports and the Environment & Natural Resources Committee 1997 report cover the issues in more depth. Reports date quickly as the number of pests keeps growing: **98 per cent of trade** – Paterson & Colgan (1998): **150 million tonnes etc.** – Paterson & Colgan (1998): **Captain Cook's ballast** – Emery & Bryan (1989): **43 weeds in dry ballast** – Kloot (1987b): **Green crabs** – Environment & Natural Resources Committee (1997): **Bitou & alligator weed** – Parsons & Cuthbertson (1992): **Ballast surveys** – Jones (1991), Williams et al. (1988): **Survey of 31 ships** – Williams et al. (1988): **300 million cysts** – Hallegraeff & Bolch (1991, 1992): **More than 50 animals & plants** – Chad Hewitt, CSIRO, Hobart, pers. comm.: **Kelp** – Sanderson (1990), AQIS (1994): **Seastar** – Buttermore, Turner & Morrice (1994), McLoughlin & Thresher (1994), AQIS (1994): **Spotted handfish** – Last & Bruce (1996–97): **Dinoflagellates** – Hallegraeff et al. (1991), Jones (1991), ACIL (1994): **AQIS warns, whales & tourists will die** – ACIL (1994): **Giant sea worm** – Bonny (1995), Environment & Natural Resources Committee (1997): **Asian mussel** – Willan (1987): **Sea bass, goby** – Jones (1991): **Zebra mussel** – Ludyanskiy, McDonald & MacNeill (1993): **148 pathogens** – AQIS (1994): **Alan Taylor** – Ballast Water Symposium (1994: Canberra, ACT) (1994): **Economic analyses** – ACIL (1994): **10 billion tonnes, 3000 species** – Paterson & Colgan (1998): **Reballasting far from foolproof** – Environment & Natural Resources Committee (1997, pp. 117, 118) noted that 'it is not an ideal measure' and 'should not . . . be viewed as a long term solution . . .': **Sediment in 14** – Hallegraeff & Bolch (1992): **Voluntary guidelines** – AQIS web site: http://www.aqis.gov.au/pubs/index.htm* – Hallegraeff & Bolch (1991), Jones (1991): **Port Phillip Bay** – Environment & Natural Resources Committee (1997)

15. Where Have All the Flowers Gone?

Weste & Marks (1987): Much of the information for this chapter has come from Newhook & Podger (1972), Weste & Marks (1987), and from a series of papers that appeared in the *Journal of the Royal Society of Western Australia* in 1994 (vol. 77). A useful introduction to the topic is Weste (1993): **Discovered in Sumatra** – Newhook & Podger (1972): **Brisbane Ranges** – Weste & Ashton (1994): **Photos from bomber** – Hill, Tippett & Shearer (1994): **Stirling Ranges** – Wills (1992): Kennedy & Weste (1986): **Rare Grampians plants** – Kennedy & Weste (1986): **Rare WA plants** – Keighery, Coates & Gibson (1994): **Animals suffering** – Wilson et al. (1994): **Resprouting in Brisbane Ranges** – Weste & Ashton (1994): **Phosphonate** – Greg Keighery, pers. comm.: **Other *Phytophthora* species** – Keighery, Coates & Gibson (1994): **Kangaroo apples, native grapes affected** – White (1981): **Blackberry rust** – see chapter 37: **Native cankers** – Keighery, Coates & Gibson (1994)

16. The Sick and Dying

Pilchard die-off – Whittington et al. (1997), Hyatt et al. (1997), Griffin et al. (1997): Some government experts are claiming this virus is indigenous and not the cause of death, blaming cold water upwellings instead, but this is repudiated by Griffin et al. (1997) and unsupported by any published research. CSIRO scientists have isolated the virus, on behalf of the Joint Pilchard Scientific Task Force, which has Commonwealth and state representation, showing how seriously the disease threat is taken: **Disappearing frogs** –

Laurance, McDonald & Speare (1996), Berger, Speare & Hyatt (in press). See also http://www.jcu.edu.au/dept/PHTM/frogs/ampdis.htm. Freeland (1993). But see Oakwood & Pritchard (1999): Blomfield (1978): **Crocker** – Richards & Short (1996): Le Souef (1923): **Rinderpest** – Henning (1956): **Hunting dogs** – Creel et al. (1995), Macdonald (1996): **Lions and distemper** – Roelke-Parker et al. (1996), Macdonald (1996): Toxoplasma – Freeland (1993), Oakwood & Pritchard (1999): **Tapeworm** – Johnson, Speare & Beveridge (1998): **Saville-Kent** – Cadwallader (1996): **Macquarie perch** – McDowall (1996): **Gouldian finch** – Tidemann et al. (1992): Garnett (1993): **Second pilchard die-off** – see media releases by the Department of Primary Industries & Resources, South Australia, at http://www.pir.sa.gov.au/MediaReleases/index.html: **Cassowary** – Steve Goosem, pers. comm.: **Mange** – Skerratt, Martin & Handasyde (1998), Martin, Handasyde & Skerratt (1998): **Newcastle disease** – Carson Creagh, AQIS, pers. comm.: **Crustacean disease** – *AQIS Bulletin*, Oct. 1996: **Macaws** – Macwhirter & Ambrose (1997)

17. The Price of Free Trade

Smith (1994) – foreword to Runge (1994): Runge (1994): Yu (1994): Jenkins (1996): **Yu replies** – Yu (1996): **Western flower thrips** – Australian Academy of Science (1996): **Pilchard die-off** – Whittington et al. (1997): **$5 billion** – Ludyanskiy, McDonald & MacNeill (1993): **Comb jelly** – Travis (1993): **$97 billion** – Bright (1998, p. 175): ACIL (1994): Australian Academy of Science (1996): **WTO** – Nairn et al. (1996): The WTO decision can be accessed from the WTO homepage at http://www.wto.org/wto/about/about.htm: **Embarrassing blow** – *Financial Review*, 6 May 1998: **Holmes** – *Financial Review*, 6 May 1998

18. A Source of Perverse Pride

Le Souef (1923): **To foreign lands we have contributed** – References are provided in appendix V and in the following chapters, wherever these species are mentioned in more detail: **California has eucalypts, wattles** – Hickman (1993): **Hawaii has** – see following chapters: **Silky oaks in Hawaii** – Wriggley & Fagg (1989): **Florida budgerigars** – Anonymous (1963), Long (1981): **Peaceful doves, St Helena** – Long (1981): **Redback** – Wace (1968): **Cape Town water** – Le Maitre et al. (1996): **Florida paperbark fires** – Laroche (1994): **Brushtail is NZ's worst** – Jack Craw, pers. comm.: ***Sphaeroma*** – Cohen & Carlton (1995). *Sphaeroma* is native to New Zealand as well as Australia and the Californian population may have come from there, although Australia is the more likely candidate because of our greater volume of shipping: Australian beetle – Gerard (1994): **Ross River fever** – Miles (1984), Tesh et al. (1981): **Tobacco blue mould** – Spencer (1981): **Louse** – Hopkins (1949): **Blackwood, pittosporum** – Cronk & Fuller (1995): **500 fungi** – Tom May, pers. comm.: **Earthballs in China** – Low (1995a)

19. Colonial Revenge

Claridge – Hayward & Druce (1919): **Wallabies** – Corbet & Harris (1991), Yalden (1988), Weir, McLeod & Adams (1995): **Stonecrop** – Dawson & Warman (1987), Pain (1987): **Barnacle** – Crisp & Chipperfield (1948): Elton (1958): **Carpet beetle** – Gerard (1994):

Wool aliens – Stace (1991), Hayward & Druce (1919): **Extinct peppercress** – Leigh, Boden & Briggs (1984): **Juncus hybrids** – Stace (1975): **Continental Europe** – Kloot (1985a): **Paper daisies, pigface etc.** – Stace (1991): **Amphipod** – Richardson (1980): **Earthworm** – Sims & Gerard (1985): **Flatworm** – Leigh Winsor, pers. comm.: **Groundsel rust** – Wilson, Walshaw & Walker (1965): **Budgerigars** – Sharrock (1978), Long (1981): **Frogs** – Raff (1956)

20. Ecologically Entwined

Kingsmill (1918): **Hedgehogs, ferrets etc.** – Thomson (1922), a key reference on early acclimatisation in New Zealand. See also McDowall (1994): **Emus, brush turkeys etc.** – Thomson (1922): **Kangaroos, wallabies etc.** – Thomson (1922): **George Grey** – McDowall (1994): Troughton (1941): **Parma** – Wodzicki & Flux (1967), King (1990): **Red-necked wallaby** – King (1990): **Brushtail** – The many references include King (1990), Brown, Innes & Shorten (1993), Leutert (1988), Campbell (1990): **TB** – Barlow (1991): **Frogs** – Robb (1986): **Prawns, shrimps** – Thomson (1922): **Spiders** – Forster & Forster (1973): **Insects** – Somerfield (1977) and articles regularly appear in the *New Zealand Entomologist*: **Plants** – Webb, Sykes & Garnock-Jones (1988), Webb et al. (1995): **Hakeas** – Williams (1992): **11 in Victoria** – Carr (1993): **Tasmania's worst weeds** – Humphries, Groves & Mitchell (1991): **Weka** – Brothers (1984): **With oysters** – Dartnall (1969): **NZ screwshell** – Allmon et al. (1994)

21. Colouring the Landscape

Koalas – Metro (1976): Guild (1938): **Gouldian finches** – Guild (1940): There were releases of red-browed finches and chestnut-breasted mannikins in the Society Islands before Guild's time. His releases of these species may in fact have failed – see Long (1991): **Bird releases** – Long (1981): **Hawaii** – Fisher (1948): **Fiji** – Pernetta & Watling (1978): **Rock wallabies** – Tomich (1986), Lever (1985): **Brown tree snake** – Rodda, Fritts & Conry (1992), McCoid (1991): **Grass skink** – McKeown (1978), Robb (1986): **Frogs** – Tyler (1976): **Got to Guam** – McCoid (1993): **Sphaeroma & Iais** – Rostramel (1972): **Arcitalitrus** – Lazo-Wasem (1983): **Redclaw** – Lowery (1996): **Flatworms** – Leigh Winsor, James Cook University, pers. comm.: **Redback** – Main (1984): **Cottony cushion scale** – DeBach (1964): **Dobbins** – DeBach (1964): Sugarcane hopper etc. (Swezey 1928), Wilson (1963): Wilson (1963) was the main reference used for insects: **Fern weevil** – Pemberton (1921), Swezey (1926), Wilson (1963): **Eucalypt pests** – Ohmart & Edwards (1991), Metro (1976), Low (1996–97): **19 insects imported** – Wilson (1963): Wilson (1963)

22. Inheriting a Degraded World

Hall & Boucher (1977): **Tree fern** – Medeiros et al. (1992), Medeiros, Loope & Anderson (1993): **Major invaders** – Cronk & Fuller (1995) list countries invaded by Australia's worst exports. See also Kloot (1985a): **Invasive plants in US** – Randall & Marinelli (1996): **South Africa** – Macdonald (1985), Low (1997b): **Hawaii** – 'Alien Species in Hawaii' available at http://www.hear.org/AlienSpeciesInHawaii/index.html: New Zealand – Webb, Sykes & Garnock-Jones (1988), Webb et al. (1995): **Florida's pests** – Frank & McCoy (1995):

456 million plants – Center, Frank & Dray (1994): **Paperbark** – Laroche (1994), Davis & Ogden (1994), Austin (1978), Cronk & Fuller (1995): **Melaleuca management plant** – Laroche (1994): **Australian pine** – Cronk & Fuller (1995), Randall & Marinelli (1996), Low (1999): *Miami Herald*, 8 Mar. 1994: **46 birds etc.** – O'Hare & Dalrymple (1997): **Pond apple** – Swarbrick (1993a, 1993b): **Réunion** – Jacque Tassin, Forestry Department, Centre de Coopération Internationale en Recherche Agronomique pour le Développement, pers. comm.: **Eucalypts** – Low (1996–97): **Eucalyptopolis** – Bright (1998)

23. It's Civil War

These topics will be fully referenced in my next book, *The New Nature*: **Pittosporum** – Cronk & Fuller (1995), Brown et al. (1991): **Pittosporum in WA** – Hussey et al. (1997): **Acclimatisers** – Jenkins (1977): **Native birds** – Long (1981): **Weeds in Victoria** – Carr (1993). For South Australia see Kloot (1985a): **Outback plants around stockyards** – Gray & Michael (1986): **Kangaroo expert** – Peter Alexander: **Hattah-Kulkyne** – Cheal (1986): **Koalas** – Tyndale-Biscoe (1997b) etc: **Aggressive birds** – Low (1994) etc: **Seven endangered birds** – Gould's petrel, Lord Howe woodhen, golden-shouldered parrot, Norfolk Island green parrot, forty-spotted pardalote, black-eared miner, helmeted honeyeater: **Gould's petrel** – Priddel & Carlile (1995): **Elephants** – Laws (1969) etc: **Cankers** – Keighery, Coates & Gibson (1994): ***Western Weeds*** – Hussey et al. (1997)

24. The Shuffled Pack

In 1978 twenty exotic species – Williams et al. (1978): **Fifty exotics** – Pollard & Hutchings (1990a, 1990b). I exclude those species they suggest are misidentifications: **About a hundred** – CSIRO list posted at http://www.ml.csiro.au/~hewitt/CRIMP/ispp.html#body, dated 19 May 1996. (The list runs to 115 species with 31 listed as cryptogenic.): **Doubled again** – Paterson & Colgan (1998): **Exotic insects** – New (1994): Australian Academy of Science (1996): **Hawaiian insects and spiders** – Mlot (1995): Slater, Slater & Slater (1986): ***Mammals of Australia*** – Strahan (1995): **Slider** – Griffiths (1997): **CSIRO handbook** – Lazarides, Cowley & Hohnen (1997): **Recent Incursions** – Groves (1998): **Wet Tropics weeds** – list drawn up by Steve Goosem, Wet Tropics Management Authority: **Hosking** – of NSW Agriculture, Tamworth, pers. comm.: **Fungi** – Tom May, National Herbarium of Victoria, pers. comm.: **Mosses** – David Meagher, Zymurgy Consulting, Victoria, pers. comm.: **Lichens** – Gintris Kantvilas, Tasmanian Herbarium, pers. comm.: **Rosy barbs** – Arthington, Milton & McKay (1983): **Bureau of Fauna and Flora** – Dyne & Walton (1987): **Ostriches** – Slater, Slater & Slater (1986): **Sheep** – Lever (1985): **Dingoes** – Corbett (1995): **Squirrels** – Watts & Aslin (1981): **Elephants in Borneo** – Payne & Phillipps (1985): **Hawaii** – 'Alien Species in Hawaii' available at http://www.hear.org/AlienSpeciesInHawaii/index.html: **Florida** – Frank & McCoy (1995): **Exotic mammals in Africa and Europe** – Lever (1985)

25. Seizing the Advantage

Honeybee impact – Pyke (1990), Wood & Wallis (1998). Also, *Victorian Naturalist*, 1997, 114(1), has several bee articles: **Salvinia** – Parsons & Cuthbertson (1992): **Disturbance essential to invasion** – Hobbs (1991): **Tiwi** – Fensham & Cowie (1998): **Mosses** – David

Meagher, Zymurgy Consulting, Victoria, pers. comm.: Kirkpatrick (1994): **Amazonian earthworms** – Keith McDonald, Queensland Department of Environment, pers. comm.: Flannery (1994): **Beaches suffer enormously** – see Batianoff & Franks (1998) & chapter 34: Leichhardt (1847): **Bullrush seed** – Parsons & Cuthbertson (1992): **Paperbark seed** – Laroche (1994): **Dung beetles** – Williams (1986): **Annual weeds** – Fox (1991): **Mosquito fish** – Arthington & Lloyd (1989), Dove (1998): **Toad** – Covacevich & Archer (1975). Also see citations for chapter 7.

26. A Bad Rap

Stephen I. wren – Halliday (1978): **Macquarie I.** – Brothers (1984): **More birds in leafy suburbs** – Low (1994): Dickman (1993): **Cat predation** – a series of important papers are collected in Potter (1991): Bomford, Newsome & O'Brien (1995). See also May & Norton (1996) and various other papers appearing in *Wildlife Research*: **State of the Environment** – Alexander (1996): **Western Shield** – Morris et al. (1998): **1080 plants** – Oliver, King & Mead (1977): **Crocker** – Richards & Short (1996): **Lord Howe** – Hindwood (1940), Hutton (1990): **Maclear's rat** – Pickering & Norris (1996): **Rabbits** – Alexander (1996): **Honeybees** – Pyke (1990), Wood & Wallis (1998): **Green crab** – Grosholz & Ruiz (1995), Environment & Natural Resources Committee (1997): **Big-headed ant** – Hoffman & Hohenhaus (1998). See also Hoffmann (1998). This species is also called the coastal brown ant.

27. Where the Deer and the Antelope Roam

Blackbuck, eland, zebras – Long (1988), Jenkins (1977): **McCoy wanted eland** – McCoy (1862), Rolls (1969): **Blackbuck in 1930s** – notes by Klaus Weiss, held by Royal National Park: **Banteng** – Strahan (1995), Bowman (1992): **Camels** – Strahan (1995), Duncan-Kemp (1934, p. 92): Leichhardt (1847, p. 522): Grey (1841): **Donkeys** – Strahan (1995): **Ruminant densities** – Freeland (1990): **Buffalo hides exported** – Freeland (1992): **Five species in Gurig** – Freeland (1990): Deer selection – Strahan (1995): **Deer and kangaroos in Dandenongs** – Wheelwright (1861): Rolls (1969): **Royal National Park** – Mahood (1981): **Mammal weights** – from Strahan (1995)

28. The Ultimate Pest

Hudson – from an article posted on the Internet at http://www.s-p-h.com/grower/natives_vs_exotics.html: **Mammal book** – Lekagul & McNeely (1977): **Few predators** – Lions in Africa (but not India) readily eat people, but tigers, contrary to Flannery (1994), rarely do (see Lekagul & McNeely 1977): Mollison (1994): **Elephants smash woodlands** – Laws (1969): Harlan (1975): **Livestock** – Many papers address impacts of livestock – e.g., Cheal (1993), Bennett (1995), Chesterfield & Parsons (1985).

29. Expanding and Infilling

Asian house gecko – Arrival in Brisbane documented by Queensland Museum specimens: **Mynas in Brisbane** – Sightings listed in Queensland Ornithological Society Inc. newsletters. 1982 sighting by Ric Nattrass, Department of Environment: **Small black ants** – Geoff Monteith, Queensland Museum, & Ric Nattrass, Department of Environment, pers. comms.:

Mynas in Canberra - Phillipps (1994): **Canberra weeds** - Jeff Butler, weeds officer for the Conservation Council of the South East Region and Canberra, pers. comm.: **Canberra fish** - Lintermans, Rutzou & Kukolic (1990): **Toads at Cape York** - Burnett (1997): **Toads south of Sydney** - Sutherst, Floyd & Maywald (1995): **Invasion speeds** - Grosholz (1996): **Fly agaric** - Tom May, National Herbarium of Victoria, pers. comm.: **Asparagus fern** - Batianoff & Franks (1998): **Pasture grasses** - Walker et al. (1997): **African lovegrass** - Johnston, Aveyard & Legge (1984): **Willows** - *AQIS Bulletin*, May 1996: **Pampas grass** - Rawling (1994), *Pampas*, 1988 leaflet by the Forestry Commission, Tasmania: **Twining legumes** - Low (1997c): **Ten pines** - Richardson (1998): **Alligator weed** - O'Neill (1994): **Port Phillip Bay** - Environment & Natural Resources Committee (1997)

30. Sleepers Wake

Daugherty (1993): **Miconia** - S. Csurhes & P. Stanton, pers. comm., Meyer & Florence (1996): ***Recent Incursions*** - Groves (1998): **Athel pine** - Griffin et al. (1989): **Mimosa** - Lonsdale, Miller & Forno (1989): **Pampas grass** - Rawling (1994), *Pampas*, 1988 leaflet by the Forestry Commission, Tasmania: **Willows** - *AQIS Bulletin*, May 1996: Csurhes & Edwards (1998): **Sweet thorn** - Scott (1991): ***Noxious Weeds*** - Parsons & Cuthbertson (1992): **Mimosa** - Miller & Lonsdale (1987): **Cutch** - Csurhes & Edwards (1998): Keighery (1996): **Madras thorn etc.** - Csurhes & Edwards (1998): **Perth zoo** - Csurhes & Edwards (1998), G. Keighery, pers. comm.: **CSIRO in Brisbane** - Hacker (1997): **AGRC** - Snowball, Foster & Collins (1992): **Listing of Potential Crops** - Fletcher (1993): **250 foreign birds** - The more unusual birds are registered under the National Exotic Bird Registration Scheme (list available from Environment Australia): **Yabbies** - Kibria et al. (1996). Some South Africans want to import yabbies and advice has been sought here: **Silver perch** - Kibria et al. (1996): **Edwards** - *Tropical Fish Hobbyist*, July 1992

31. Knocking at the Door

Golden apple snails - Naylor (1996), Baker (1998): **Giant African snail** - Colman (1977): **Asian honeybee** - David Banks, AQIS, pers. comm.: **Bees in Adelaide** - *AQIS Bulletin*, July 1996: **Japanese encephalitis** - Hanna et al. (1996): **QUT scientist** - 'New lab scare raises alarm', *Courier-Mail*, 2 Aug. 1997: Newcastle disease causing extinctions - Carson Creagh, AQIS, pers. comm.: **Woman fined** - *AQIS Bulletin*, Apr. 1998; AQIS media release, 2 Mar. 1998: **AustVetPlan** - at http://www.brs.gov.au/aphb/aha/ausvet.htm: **Avian influenza** - *AQIS Bulletin*, Sept. 1992: **Screw worm on neck** - *AQIS Bulletin*, May 1992: **Crayfish disease** - *AQIS Bulletin*, Oct. 1996: **Sponge disease** - AQIS 1994: **Guava rust** - Australian Academy of Science (1996): **Gypsy moth** - *AQIS Bulletin*, July 1992: **NAQS Weed Target List** - at http://dpie.gov.au/aqis/homepage/quarantine/nweedtl.html: **Mile-a-minute** - *NAQS News*, Sept. 1998: **Torres Strait Treaty** - Lawrence (1994): **NZ gave us nodding thistle (in pasture seed)** - Parsons & Cuthbertson (1992): **Chilean mite** - Walter (1998): **DSIR wants exotic bees** - Donovan (1990): **Australia imports new plants** - The quick-growing timber tree *Paulownia*, imported recently and promoted heavily by entrepreneurs, is worrying weed experts.

32. Whither the Wet Tropics?

Cassowary – Steve Goosem, pers. comm.: **Pig study** – Pavlov, Chrome & Moore (1992). Pigs, believe it or not, are also a major threat to crocodiles, whose eggs they eat: **Phytophthora** – Brown (1998), Gadek (1998): **Palm leaf beetle** – Foulis & Halfpapp (no date): **25, 53 weeds** – Steve Goosem, unpublished lists. Goosem lists 52 but I have added mile-a-minute (*Mikania micrantha*), a recent invader: **Worst invaders** – Humphries & Stanton (1992): **Harungana** – Swarbrick (1993a): Humphries & Stanton (1992): **Sweet prayer plant** – Humphries & Stanton (1992): **Pond apple** – Swarbrick (1993a), Swarbrick (1993b), Swarbrick & Skarratt (1994): **Rare plants suffer** – Queensland has rare rainforest trees (in family Annonaceae) related to pond apple and likely to be attacked by any biocontrol agents: **Big-headed ant** – Dunn, Kitching & Dexter (1994): **Hawaii** – 'Alien Species in Hawaii' available at http://www.hear.org/AlienSpeciesInHawaii/index.html: **Florida** – Frank & McCoy (1995): **Internet appeal** – From Piers Barrow, Kakadu National Park, to ENVIROWEEDS@majordomo.nre.vic.gov.au, dated 2 Apr. 1998: **Lakefield** – Peter James, Department of Natural Resources, Townsville

33. The Homogocene

Bright (1998): **Supercontinent supports half the animals** – Vitousek et al. (1997): Lövei (1997): Stevens (1996): **Port Phillip Bay** – Environment & Natural Resources Committee (1997): Woolls (1884): **Lantana** – Citations will appear in my next book, *The New Nature*. Wild tobacco (*Solanum mauritianum*) is another weed well integrated into the Australian environment; many rainforest animals eat its fruits and pademelons browse the leaves: Bird dietary studies cited in Blakers, Davies & Reilly (1984). See also Long (1984): **Rare species need weeds** – Brown et al. (1991). Further citations will appear in my next book, *The New Nature*: **Mildura diets** – Baker-Gabb (1983): **Wedgetails** – see Brooker & Ridpath (1980), Blakers, Davies & Reilly (1984): **Barn owl diet** – Morton & Martin (1979): **Sparrow** – Parkin & Cole (1984): **Parma wallabies** – King (1990): **Wasps** – Davidson (1986–87): **Mallard hybrids** – Frith (1967): **Club-rush hybrids** – Hussey et al. (1997): **Lantana** – Swarbrick (1986): **Spartina** – Carr (1993): **Honeybee** – Pyke (1990)

34. The New Architects

Marram – Heathcote & Maroske (1996): **Recent Parliamentary inquiry** – Environment and Natural Resources Committee (1998): **Dune changes** – Heyligers (1985), Bell (1987–88): **Spartina** – Rawling (1994), Adair (1995): **Buffalo** – Freeland (1992): **Forty Mile Scrub** – Fensham, Fairfax & Cannell (1994): **Phytophthora** – Kennedy & Weste (1986): **Rabbits** – Rolls (1969), Stead (1935): **Asian mussel** – Willan (1987): **Athel pine** – Griffin et al. (1989), Beckmann (1990): **Willows** – Ladson & Gerrish (1996): **Cape Town** – Le Maitre et al. (1996), Low (1997b): **Iceplant** – Kloot (1983): **Creepers destroy forests** – Stockard & Hoye (1990)

35. Cryptogenic World

Stradbroke – Low (1997a) and unpublished notes held by the author: Other weeds used in Aboriginal medicine are listed by Low (1990): **Rabbits eaten** – Burbidge et al. (1988). Rabbits were also used in Aboriginal medicine (Low 1990): Cat meat is said to be 'rather sweet, tender and juicy white meat' but the only cat meat I have eaten was greasy and dark: **Tree tobacco** – Latz (1995): **Fox, camel names, cats considered native** – Burbidge et al. (1988): **Jerusalem thorn** – Latz (1995): **Rabbits, deer, apes** – Lever (1985): Hooker (1860): **Von Mueller** – quoted by Hamilton (1892): ***Noxious Weeds*** – Parsons & Cuthbertson (1992): Carlton (1996): **Hewitt** – pers. comm. in 1998: **Cinnamon fungus** – Newhook & Podger (1972): **Dove** – at the University of Queensland: Forster & Forster (1973): **Oxalis** – Conn & Richards (1994): Corbett (1995–96): **Cattle egret** – Long (1981): **Wanderer** – Tenison-Woods (1881): Freeland (1992): Bowman (1992): ***Mammals of Australia*** – Strahan (1995): Stein & Moxley (1992): **Coconut** – Low (1994–95), Buckley & Harries (1984): **Rowell** – quoted in the *Herbert River Express*, 8 Oct. 1996

36. It Happens Naturally

Brown – in Flinders (1814): **Monsoon plants** – Brock (1988): Hooker (1860): Unger (1866). 'The Sunken Island of Atlantis' appeared in the *Journal of Botany* the previous year: **Hops, fanflowers, koa** – Low (1998): **Gorteria** – Parsons & Cuthbertson (1992): **Hawaii invasion rate** – Vitousek et al. (1997)

37. Seeking Magic Bullets

Prickly pear story – Dodd (1940), Wright (1971): **In Florida** – Bennett & Halbeck (1995): **In South Africa** – Hoffman, Moran & Zeller (1998): **Percentages** – Briese (1993): **Lantana** – D. Panetta, pers. comm.: **St John's wort** – Parsons & Cuthbertson (1992): **Cats** – Dickman (1993): **Fertility control** – Taylor (1998): **Patterson's curse** – Parsons & Cuthbertson (1992): **Arum lily** – Scott & Neser (1996): **Snails** – Carter & Baker (1997): **Blackberry** – Stahle (1997), Bruzzese & Hasan (1986): **Trichopoda** – Sands & Coombs (1999): **Mimosa moth** – Davis et al. (1991). It now feeds on native sensitive plant (*Neptunia gracilis*): **Euclasta** – McFadyen & Marohasy (1990): **Rust** – Julien & White 1997: **Moth disappointing** – Marie Vitelli, Alan P. Dodd Research Centre, pers. comm.: **Euglandina** – Murray et al. (1988): **Hawaiian moth** – Simberloff & Stiling (1996a): **American weevil** – Simberloff & Stiling (1996b): **Australian ladybird (*Cryptolaemus montrouzieri*) eats scale** – Wilson (1963): **WA Ag.** – Wilson (1960): **Sugar industry** – see chapter 7: **Skunks etc.** – Stead (1935): Simberloff & Stiling (1996a, 1996b): **123 agents** – Julien & White (1997), McFadyen (1997): **Salvinia** – Parsons & Cuthbertson (1992): **Dung beetles** – Williams (1986): **Goats** – Allen & Lee (1994–95): **Fruit-piercing moths** – Thwaites (1996): **Fungus on grasshoppers** – Lockwood (1993): **Marine biocontrol** – Viney (1996)

38. The Quarantine Quandary

7 million people, 160 million letters etc. – *AQIS in Review* – a report to clients, available at the AQIS web site: http://www.aqis.gov.au/pubs/index.htm: **Nairn review** – Nairn et al. (1996), reviewed by Tyndale-Biscoe (1997a): **A third of passengers** – Nairn et al. (1996, p. 141): **Nothing but chocolate** – *AQIS Bulletin*, Nov. 1996: **Papaya fruit fly** – Drew (1997), *AQIS Bulletin*, Apr. 1997, Nov. 1996: **Recommended traps** – Australian Academy of Science (1996): **Mail poorly screened** – Nairn et al. (1996, p. 152): **At Melbourne's Mail Centre** – *AQIS Bulletin*, Mar. 1999: **Birds' nests** – *AQIS Bulletin*, Sept. 1997: **Human hand** – *AQIS*

Bulletin, Aug. 1997: **Cats** – *AQIS Bulletin*, Oct. 1996: **Items such as** – Nairn et al. (1996, p. 144): **Government rejects statutory authority** – Lovett (1997), *AQIS Bulletin*, Aug. 1997: **Funding for AQIS** – Tyndale-Biscoe (1997a), Nairn et al. (1996, p. 202): **Labor cut budget** – *AQIS Bulletin*, Aug. 1993: **Charge for services** – Nairn et al. (1996, p. 125): ***AQIS Bulletin*** – available at the AQIS web site: http://www.aqis.gov.au/pubs/index.htm: **Fish bait & meal** – Department of Primary Industries & Energy (1996): **Task Force** – Department of Primary Industries & Energy (1996): **Aquatic Animal Quarantine** – Humphrey (1995): **Second pilchard die-off** – see media releases by the Department of Primary Industries & Resources, South Australia, at http://www.pir.sa.gov.au/MediaReleases/index.html: **'Australia's Weed Scandal'** – Low (1995b): **20 million flowers** – *Australasian Flowers* magazine, Autumn 1998: **Methyl bromide** – *AQIS Bulletin*, Sept. 1992: **Fumigation procedures** – Ian Barass, AQIS, pers. comm.: **Plant-pest incursions report inaccurate** – Carson Creagh, AQIS, pers. comm.: **National Farmer's Federation** – Nairn et al. (1996, p. 85)

39. Are We Blind?

Mollison (1994): **Mesquite etc. noxious** – Parsons & Cuthbertson (1992): **Honey locust** – Csurhes & Kriticos (1994) Leucaena persistent – Jones & Jones (1995): **Fix up the land first** – *Permaculture International Journal* 61, Dec.–Feb. 1997: **Holmgren** – article in *Permaculture International Journal* 61, Dec.–Feb. 1997), reprinted in *Indigenotes* 10(2): 8–10, 1997, under the heading 'Weeds or Wild Nature': **Need good reasons** – *Permaculture International Journal* 5: 30–31: ***Permaculture International Journal*** admits – 61, Dec.–Feb. 1997, reprinted in *Indigenotes* 10(2): 8–10, 1997: **Thunbergia declared noxious** – Parsons & Cuthbertson (1992). *Burke's Backyard* features it under the horticultural name 'blue trumpet vine (*Thunbergia grandiflora*)': **Lavender and bluebell creeper** – *Burke's Backyard*, Jan. 1999: **Lavender noxious** – Parsons & Cuthbertson (1992), Rawling (1994): **Bluebell creeper in Arthurs Seat** – Wayne Hill, ranger, Arthurs Seat State Park: **Gloriosa invasive in southern Queensland** – Batianoff & Franks (1997). One of the 12 worst species: **A major rainforest weed in New South Wales** – The Big Scrub Landcare Group (1998): **Papaya fruit fly cost** – Drew (1997): Carr (1993): **Seed inspector** – V. Mungomery: Lövei (1997). Vitousek et al. (1997) make the same point: **Portuguese millipede** – Baker (1985): **Biological pollution ranked second** – Human overpopulation could be ranked the greatest problem of all, the root source of all others, but it is not usually listed directly as a threat to conservation: **You Yangs, NSW foreshores** – drowning under boneseed/bitou bush (Carr 1993, Humphries, Groves & Mitchell 1991): **NSW garden weeds** – see The Big Scrub Landcare Group (1998)

40. Wild Organisms

Belair – Cordes (1983), Department of Environment & Planning (1989). Only one person protested the planting of exotics – J.G. Tepper of the Adelaide Museum in 1904: Groves (1998): ***Western Weeds*** – Hussey et al. (1997): **Presume guilt** – Rawling (1994) is one who argues for this: WRA – Walton & Ellis (1997); at http://www.dpie.gov.au/aquis/homepage: Ewart (1909, p. 7): **260 species** – Walton, pers. comm.: **Parliamentary inquiry** – Environment & Natural Resources Committee (1998): Carr (1993, p. 264): **Bumblebees**

efficient, bumblebee flowers - Free (1982): **Introductions to Australia** - Buttermore (1997): **Bumblebees** - Hingston & McQuillan (1998), Semmens, Turner & Buttermore (1993), Semmens (1996), Donovan (1980), Buttermore (1997): None raised the point - Chris Mobbs, Environment Australia, pers. comm.: **Leafcutting bee** - Woodward (1996), Donovan (1980): **New NZ bees** - Donovan (1990): **Pollinate existing weeds** - Chris Mobbs (Environment Australia) remembered someone arguing that a weedy thistle in Victoria would benefit from bumblebees but could find no such submission in his files.

41. What to Do?

Ten weeds - Groves (1998): **Tributyl tin** - Environment & Natural Resources Committee (1997), Alexander (1996): **Bradley sisters** - Bradley (1988): **Pioneers used weeds** - Low (1985, 1989, 1990): **Weed recipes appear in my earlier book** - Low (1985)

42. Life Goes On

Glycine proclaimed wonderful - Allen (1961): **Singapore daisy in 1970s** - Batianoff (1992): **Chinese elm probably Brisbane's worst weed** - Low (1993): **Parrot's feather** - see Parsons & Cuthbertson (1992): **Indian mynas** - see chapter 29: **No platies** - Arthington, Milton & McKay (1983) recorded no platies in Brisbane in 1977–8, while warning of their likely appearance in future, nor were platies recorded from Moggill Creek by a later survey.

BIBLIOGRAPHY

A

ACIL Economics and Policy Pty Ltd. (1994) Australian Quarantine and Inspection Service Ballast Water Research Series Report No. 6.: Bio-economic risk assessment of the potential introductions of exotic organisms through ship's ballast water. Australian Government Publishing Service, Canberra.

Adair, R.J. (1995) The threat of environmental weeds to biodiversity in Australia: a search for solutions. In Bradstock, R.A., Auld, T.D., Keith, D.A., Kingsford, R.T., Lunney, D., & Sivertson, D.P. (eds). *Conserving Biodiversity: Threats and Solutions.* Surrey, Beatty & Sons, Sydney.

Alexander, N. (ed.). (1996) *Australia: State of the Environment 1996.* CSIRO Publishing, Melbourne.

Allen, F.E. (1953) Distribution of marine invertebrates by ships. *Australian Journal of Marine and Freshwater Research* 4: 260–316.

Allen, G.H. (1961) Glycine proves its value. *Queensland Agricultural Journal* 87(2): 96–7.

Allen, G.R. (1989) *Freshwater Fishes of Australia.* T.F.H. Publications, Neptune City (US).

Allen, J., Gosden, C., & White, J.P. (1989) Human Pleistocene adaptations in the tropical island Pacific: recent evidence from New Ireland, a Greater Australian outlier. *Antiquity* 63: 548–61.

Allen, L., & Lee, J. (1994-95) Technical Highlights 1994/5. Report on weed and pest animal control research conducted by the Land Protection Branch. Queensland Department of Lands Queensland Government, Brisbane.

Allmon, W.D., Jones, D.S., Ajello, R.L., Gowlett-Holmes, K., & Probert, P.K. (1994) Observations on the biology of *Maoricolpus roseus* (Quoy & Gaimard) (Prosobranchia: Turritellidae) from New Zealand and Tasmania. *The Veliger* 37(3): 267–79.

Anonymous. (1914) The algaroba or mesquit bean (*Prosopis juliflora*). *Queensland Agricultural Journal* 1: 351.

Anonymous. (1920) Birds and Insects. *Emu* 19: 248-51.

Anonymous. (1963) Parakeet City – a tourist attraction. *Florida Naturalist* 36: 1B.

Arthington, A.H., & Lloyd, L.N. (1989) Introduced Poeciliids in Australia and New Zealand. In Meffe, G.K., & Snelson, F.F. (eds), *Ecology and Evolution of Livebearing Fishes (Poecilidae).* Prentice-Hall, Englewood Cliffs, NJ.

Arthington, A.H., McKay, R.J., Russell, D.J., & Milton, D.A. (1984) Occurrence of the introduced cichlid *Oreochromis mossambicus* (Peters) in Queensland. *Australian Journal of Marine and Freshwater Research* 35: 267–72.

Arthington, A.H., Milton, D.A., & McKay, R.J. (1983) Effects of urban development and habitat alterations on the distribution and abundance of native and exotic freshwater fish in the Brisbane Region, Queensland. *Australian Journal of Ecology* 8: 87–101.

Ashburner, L.D. (1976) Fish diseases and potential fish diseases in Australia. *Animal Quarantine* 5(1): 1–7.

Attenborough, D. (1987) *The First Eden: The Mediterranean World and Man.* Collins, London.

Austin, D.F. (1978) Exotic plants and their effects in southeastern Florida. *Environmental Conservation* 5(1): 25–34.

Australian Academy of Science. (1996) Submission to the Review of the Australian Quarantine Inspection Service, March 1996. Available on-line at http://www.science.org.au/policy/statemen/aqis4.htm.

Australian Quarantine and Inspection Service. (1994) *Ballast Water Research Series Report No. 3: An epidemiological review of possible introductions of fish diseases, Northern Pacific seastar and Japanese kelp through ship's ballast water.* AGPS, Canberra.

Australian Quarantine and Inspection Service. (1994) *Ballast Water Research Series Report No. 5: Ballast water ports and shipping study.* AGPS, Canberra.

B

Bailey, F.M. (1900) Contributions to the flora of Queensland. *Queensland Agricultural Journal* 6: 498–9.

Baker, G.H. (1985) The distribution and abundance of the Portuguese millipede *Ommatoiulus moreletii* (Diplopoda: Iulidae) in Australia. *Australian Journal of Ecology.* 10: 249–59.

Baker, G.H. (1998) The golden apple snail, *Pomacea canaliculata* (Lamarck) (Mollusca: Ampullariidae), a potential invader of fresh water habitats in Australia. Pest Management: Future Challenges. In Sixth Australian Applied Entomological Research Conference. 2: 21–6.

Baker, G.H., Thumlert, T.A., Meisel, L.S., Carter, P.J., & Kilpin, G.P. (1997) 'Earthworms Downunder': a survey of the earthworm fauna of urban and agricultural soils in Australia. *Soil Biology and Biochemistry* 29(3–4): 589–97.

Baker-Gabb, D.J. (1983) The breeding ecology of twelve species of diurnal raptor in north-western Victoria. *Australian Wildlife Research* 10: 145–60.

Ballast Water Symposium (1994: Canberra, ACT). (1994) Ballast Water Symposium: EPIC Conference Centre, Canberra, 11–13 May 1994. AGPS, Canberra.

Barlow, N.D. (1991) Control of endemic bovine TB in New Zealand possum populations: results from a simple model. *Journal of Applied Ecology* 28: 794–809.

Batianoff, G.N. (1998) Coral cay terrestrial flora changes at Lady Elliot Island, Great Barrier Reef, Australia. *Proceedings of the Royal Society of Queensland.* 107: 5–14.

Batianoff, G.N. (1992) Notes on Wedelia trilobata (Singapore daisy), an environmental problem as a bushland weed in tropical and subtropical areas of Queensland. *Queensland Naturalist* 31(3–4):91–3.

Batianoff, G.N., & Franks, A.J. (1997) Invasion of sandy beachfronts by ornamental plant species in Queensland. *Plant Protection Quarterly* 12(4): 180–86.

Batianoff, G.N., & Franks, A.J. (1998) Environmental weed invasions on south-east Queensland foredunes. *Proceedings of the Royal Society of Queensland.* 107: 15–34.

Beckmann, R. (1990) Athels inland. *Ecos* 63: 18–20.

Bell, A. (1987–88) Alien dune plants reshape our beaches. *Ecos* 54: 3–6.

Bennett, B. (1995) Herefords or herbfields? *Ecos* 83: 29–34.

Bennett, B. (1996) Preparing for battle with *Bufo marinus*. *Ecos* 89: 28–30.

Bennett, F.D., & Habeck, D.H. (1995) *Cactoblastis Cactorum*: A successful weed control agent in the Caribbean, now a pest in Florida? In Delfose, E.S., & Scott, R.R. (eds), *Proceedings of the Eighth International Symposium on Biological Control of Weeds*, DSIR/CSIRO, Melbourne.

Bennett, G. (1862) *Acclimatisation: its eminent adaptation to Australia.* A lecture delivered in Sydney, reprinted by the Acclimatisation Society of Victoria. W.M. Goodhugh & Co., Melbourne.

Bennett, G. (1864) On the physiology, and also the utility, and importance of the acclimatisation or naturalisation of animals and plants in Australia. *Third Annual Report of the Acclimatisation Society of New South Wales.* Joseph Cook and Co., Sydney.

Berger, L., Speare, R., & Hyatt, A. (In press.) Chytrid fungi and amphibian declines: Overview, implications and future directions. In *Proceedings of the National Threatened Frog Workshop*, Canberra, November 1997. AGPS, Canberra.

Beveridge, I., & Spratt, D.M. (1996) The helminth fauna of Australian marsupials. *Advances in Parasitology* 37: 136–254.

The Big Scrub Rainforest Landcare Group. (1998) *Common Weeds of Northern NSW Rainforest.* The Big Scrub Rainforest Landcare Group, Mullumbimby.

Birkhill, A. (1997) PIJAC Submission to Department of Natural Resources Re: Aquatic Weeds Discussion Paper. Pet Industry Joint Advisory Council, Brisbane.

Bishop, R.A. (1993) Coff's Harbour Water Supply Augmentation EIS: Aquatic Studies – Freshwater Fishes.

Blackshaw, R.P., & Stewart, V.I. (1992) *Artioposthia triangulata* (Dendy, 1894), a predatory terrestrial planarian and its potential impact on Lumbricid earthworms. *Agricultural Zoology Reviews* 5: 201–18.

Blakers, M., Davies, S.J.J.F., & Reilly, P.N. (1984) *The Atlas of Australian Birds.* Melbourne University.

Bligh, W. (1936) *Bligh and the Bounty.* Methuen, London.

Blomfield, C.E. (1978) *Reminiscences of early New England.* Southern Publishers, Bega.

Boag, B., Evans, K.A., Yeates, G.W., Johns, P.M., & Neilson, R. (1995) Assessment of the global potential distribution of the predatory land planarian *Artioposthia triangulata* (Dendy) (Tricladida: Terricola) from ecoclimatic data. *New Zealand Journal of Zoology* 22: 311–8.

Boles, W. (1999) The big picture. *Wingspan.* 9(1): 16–21.

Bomford, M., Newsome, A., & O'Brien, P. (1995) Solutions to feral animal problems: ecological and economic principles. In Bradstock, R.A., Auld, T.D., Keith, D.A., Kingsford, R.T., Lunney, D., & Sivertson, D.P. (eds), *Conserving Biodiversity: Threats and Solutions*, Surrey Beatty & Sons, Sydney.

Bonny, M. (1994) Ballast water: The scourge of the oceans. *Search* 25(3): 72–4.

Bonny, M. (1995) Preventing the invasion of marine immigrants. *Search* 26(3): 81–3.

Borthwick, T. (1923) An anti-mosquito campaign in Adelaide. *Health* 1(9): 259–65.

Bowman, D. (1992) Banteng. *Australian Natural History* 24(3): 16.

Boyd, A.J. (1902) Extermination of the cane grub. *Queensland Agricultural Journal* 10: 469.

Bradley, J. (1988) *Bringing Back the Bush*. Ure Smith Press, Sydney.

Briese, D.T. (1993) The contribution of plant biology and ecology to the biological control of weeds. *Proceedings of the 10th Australian and 14th Asian-Pacific Weed Conference.* 1: 10–18.

Bright, C. (1998) *Life Out of Bounds: Bioinvasion in a Borderless World.* W.W. Norton & Co., New York.

Brock, J. (1988) *Top End Native Plants.* Self-published, Darwin.

Brooker, M.G., & Ridpath, M.G. (1980) The diet of the wedge-tailed eagle, *Aquila audax*, in Western Australia. *Australian Wildlife Research.* 7(3): 433–52.

Brothers, N.P. (1984) Breeding, distribution and status of burrow-nesting petrels at Macquarie Island. *Australian Wildlife Research* 11(1): 113–31.

Brown, B. (1998) Occurrence and impact of *Phytophthora cinnamomi* and other *Phytophthora* species in rainforests of the Wet Tropics World Heritage Area, and of the Mackay Region, Queensland. In Gadek (1998).

Brown, P.R., Wallis, R.L., Simmons, D., & Adams, R. (1991) Weeds and wildlife. *Plant Protection Quarterly* 6(3): 150–53.

Brown, R. (1814) General remarks, geographical and systematical, on the botany of Terra Australis. In Flinders, M., *A Voyage to Terra Australis*, G. & W. Nichol, London.

Brown, K., Innes, J., & Shorten, R. (1993) Evidence that possums prey on and scavenge birds' eggs, birds and mammals. *Notornis* 40(3): 169–77.

Brumley, A.R. (1991) Cyprinids of Australiasia. In Winfield, I.J., & Nelson, J.S. (eds), *Cyprinid Fishes: Systematics, Biology and Exploitation.* Chapman & Hall, London.

Bruzzese, E., & Hasan, S. (1986) Host specificity of the rust *Phragmidium violaceum*, a potential biological control agent of European blackberry. *Annals of Applied Biology* 108: 585–96.

Buchanan, R.A. (1981) *Common Weeds of Sydney Bushland.* Inkata Press, Melbourne.

Buckley, R. (1981) Alien plants in central Australia. *Botanical Journal of the Linnean Society* 82: 369–79.

Buckley, R., & Harries, R. (1984) Self-sown wild-type coconuts from Australia. *Biotropica* 16(2): 148–51.

Bunce, D. (1979) *Travels with Dr Leichhardt.* Oxford, Melbourne.

Burbidge, A.A., Johnson, K.A., Fuller, P.J., & Southgate, R.I. (1988) Aboriginal knowledge of the mammals of the central deserts of Australia. *Australian Wildlife Research* 15: 9–39.

Burnett, S. (1997) Colonizing cane toads cause population declines in native predators: reliable anecdotal information and management implications. *Pacific Conservation Biology* 3: 65–72.

Burrows, W.H. (1991) Sustaining productive pastures in the tropics 10. Forage and feeding systems for cattle. *Tropical Grasslands* 25: 145–52.

Butlin, N. (1983) *Our Original Aggression.* George Allen & Unwin, Sydney.

Buttermore, R.E. (1997) Observations of successful Bombus terestris (L.) (Hymenoptera:

Apidae) colonies in southern Tasmania. *Australian Journal of Entomology* 36: 251–4.

Buttermore, R.E., Turner, E., & Morrice, M.G. (1994) The introduced northern Pacific seastar *Asterias amurensis* in Tasmania. *Memoirs of the Queensland Museum* 36(1): 21–5.

C

Cadwallader, P.L. (1996) *Overview of the Impacts of Introduced Salmonids on Australian Native Fauna.* Australian Nature Conservation Agency, Canberra.

Campbell, D.J. (1990) Change in structure and composition of a New Zealand lowland forest inhabited by brushtail possums. *Pacific Science* 44(3): 277–96.

Campbell, M.H., Holtkamp, R.H., McCormick, L.H., Wykes, P.J., Donaldson, J.F., Gullan, P.J., & Gillespie, P.S. (1994) Biological control of the native shrubs *Cassinia* spp. using the native scale insects *Austrotachardia* sp. and *Paratachardina* sp. (Hemiptera: Keriidae) in New South Wales. *Plant Protection Quarterly* 9(2): 64–8.

Carlton, J.T. (1996) Biological invasions and cryptogenic species. *Ecology* 77(6): 1653–5.

Carlton, J.T., & Geller, J.B. (1993) Ecological roulette: the global transport of nonindigenous marine organisms. *Science* 261: 78–82.

Carr, G.W. (1993) Exotic flora of Victoria and its impact on indigenous biota. In Foreman, D.B., & Walsh, N.G. (eds), *Flora of Victoria. Vol. 1: Introductions.* Inkata Press, Melbourne.

Carter, P.J., & Baker, G.H. (1997) Biological control of white and conical snails. Farming Systems Developments 1997. Workshop papers, University of Adelaide, Adelaide.

Cato, M.P. (1934) *On Agriculture.* Heinemann, London.

Cavaye, J.M. Graham, T.W.G., & Robbins, G.B. (1989) Sown grass productivity: run down but not run out. *Queensland Agricultural Journal* 105: 281–3.

Center, T.D., Frank, J.H., & Dray, F.A. (1994) Biological invasions: stemming the tide in Florida. *Florida Entomologist* 78(1): 45–55.

Chaloupka, M.Y., & Domm, S.B. (1986) Role of anthropochory in the invasion of coral cays by alien flora. *Ecology* 67: 1536–47.

Cheal, D. (1986) A park with a kangaroo problem. *Oryx* 20: 95–9.

Cheal, D.C. (1993) Effects of stock grazing on the plants of semi-arid woodlands and grasslands. *Proceedings of the Royal Society of Victoria* 105(1): 57–65.

Chesterfield, C.J., & Parsons, R.F. (1985) Regeneration of three tree species in arid south-eastern Australia. *Australian Journal of Botany* 33: 715–32.

Chisholm, A.H. (1919) Introduced birds in Queensland. *Emu* 19: 60–62.

Christidis, L., & Boles, W. (1994) *The Taxonomy and Species of Birds of Australia and its Territories.* Royal Australasian Ornithologists Union Monograph 2. RAOU, Melbourne.

Clements, A. (1983) Suburban development and resultant changes in the vegetation of the bushland of the northern Sydney region. *Australian Journal of Ecology* 8: 307–19.

Coates, B.J. (1997) *A Guide to the Birds of Wallacea.* Dove Publications, Brisbane.

Coates, D.J., Yen, D.E., & Gaffey, P.M. (1988) Chromosome variation in Taro, *Colocasia esculenta*: implications for origin in the Pacific. *Cytologia* 53: 551–60.

Cobley, J. (1980) *Sydney Cove 1788.* Angus & Robertson, Sydney.

Cogger, H.G. (1996) *Reptiles and Amphibians of Australia.* Fifth edition. Collins, Sydney.

Cohen, A.N., & Carlton, J.T. (1995) *Nonindigenous Aquatic Species in the San Francisco Bay and Delta.* A report for the United States Fish and Wildlife Service, Washington, DC.

Colman, P.H. (1997) An introduction of Achatina fulica to Australia. *Malacological Review* 10: 77–8.

Commonwealth of Australia. (1997) *The National Weeds Strategy.* Commonwealth of Australia, Canberra.

Conn, B.J., & Richards, P.G. (1994) A new species of Oxalis Section *Corniculatae* (Oxalidaceae) from Australasia. *Australian Systematic Botany.* 7: 171–81.

Cooling, L.E. (1923) Mosquito-larvivorous fishes in relation to mosquito reduction work in Australia. *Health* 1(4): 94–8.

Corbet, G.B., & Harris, S. (1991) (eds). *The Handbook of British Mammals.* Blackwell Scientific Publications, Oxford.

Corbett, L. (1995) *The Dingo in Australia and Asia.* University of New South Wales Press, Sydney.

Corbett, L. (1995–96) Dingoes: expatriate wolves or native dogs? *Nature Australia* 25(3): 46–55.

Cordes, D.D. (1983) *The Park at Belair.* Self-published, Adelaide.

Covacevich, J., & Archer, M. (1975) The distribution of the cane toad, *Bufo marinus,* in Australia and its effects on indigenous vertebrates. *Memoirs of the Queensland Museum* 17(2): 305–10.

Craw, J. (1996) *The Good Plant Guide.* Northland Regional Council, Whangarei (NZ).

Creel, S., N.M. Creel, N., Matovelo, J.A., Mtambo, M.N.A., Batamuzi, E.K., & Cooper, J.E. (1995) The effects of anthrax on endangered African wild dogs (*Lycaon pictus*). *Journal of Zoology* 236: 199–209.

Crisp, D.J., & Chipperfield, P.N.J. (1948) Occurrence of *Elminius modestus* (Darwin) in British Waters. *Nature* 161: 64.

Cronk, Q.C.B., & Fuller, J.L. (1995) *Plant Invaders.* Chapman & Hall, London.

Crook, D., & Sanger, A. (1997) *Recovery Plan for the Pedder, Swan, Clarence, swamp and saddled galaxias.* Environment Australia, Canberra.

Crosby, A.W. (1986) *Ecological Imperialism: The Biological Expansion of Europe,* 900–1900. Cambridge University Press, Cambridge.

Csurhes, S., & Edwards, R. (1998) *Potential Environmental Weeds in Australia.* Environment Australia, Canberra.

Csurhes, S.M., & Kriticos, D. (1994) *Gleditsea triacanthos* L. (Caesalpiniaceae), another thorny, exotic fodder tree gone wild. *Plant Protection Quarterly* 9(3): 101–5.

Cumpston, J.H.L. (1923) Plague in Australia. *Health* 1(2): 34–5.

D

Dartnall, A.J. (1969) New Zealand sea stars in Tasmania. *Papers and Proceedings of the Royal Society of Tasmania* 103: 53–5.

Darwin, C. (1881) *The Formation of Vegetable Mould, through the Action of Worms: with observations on the habits.* Murray, London.

Daugherty, C.H. (1993) Introduced species: an overview. In Moritz, C., & Kikkawa, J. (eds), *Conservation Biology in Australia and Oceania*, Surrey Beatty & Sons, Sydney.

Davidson, S. (1984–85) Weeds: a legacy of drought. *Rural Research* 125: 4–6.

Davidson, S. (1986–87) The European wasp – here to stay? *Ecos* 50: 14–17.

Davis, D.R., Kassulke, R.C., Harley, K.L.S., & Gillet, J.D. (1991) Systematics, morphology, biology, and host specificity of *Neurostrota gunniella* (Busck) (Lepidoptera: Gracillariidae), and agent for the biological control of *Mimosa pigra* L. *Proceedings of the Entomological Society of Washington* 93(1): 16–44.

Davis, S.M., & Ogden, J.C. (1994) *Everglades: the Ecosystem and its Restoration.* St Lucie Press, Delray Beach, FL.

Dawson, F.H., & Warman, E.A. (1987) *Crassula helmsii* (T. Kirk) Cockayne: Is it an aggressive alien aquatic plant in Britain? *Biological Conservation* 42: 247–72.

DeBach, P. (1964) *Biological Control of Insect Pests and Weeds.* Chapman and Hall, London.

Department of Conservation & Environment (Victoria). (1991) *Dandenong Ranges National Park Management Plan.* Department of Conservation & Environment (Victoria).

Department of Environment & Planning. (1989) *Belair Recreation Park Management Plan.* Department of Environment & Planning, Adelaide.

Department of Natural Resources (Queensland). (1997) Discussion Paper: Sale of potentially invasive aquatic plants – notes from an aquatic plants industry workshop held on 14 March 1997. Department of Natural Resources (Queensland).

Department of Primary Industries & Energy. (1996) *Report of the National Task Force on Imported Fish and Fish Products.* DPIE, Canberra.

Dickman, C.R. (1993) Raiders of the last ark: cats in island Australia. *Australian Natural History* 24(5): 44–52.

Dixon, S. (1892) The effects of settlement and pastoral occupation in Australia upon the indigenous vegetation. *Transactions and Proceedings of the Royal Society of South Australia* 15: 195–206.

Dodd, A.P. (1940) *The Biological Campaign against Prickly Pear.* Government Printer, Brisbane.

Dodd, J. (1996) Comparison of the eradication programs for kochia (*Kochia scoparia* [L.] Schrad.) and skeleton weed (*Chondrilla juncea* L.) in Western Australia. In Shepherd, R.C.H. (ed.), *Proceedings of the Eleventh Australian Weeds Conference*, Weed Science Society of Victoria, Frankston.

Donovan, B.J. (1980) Interactions between native and introduced bees in New Zealand. *New Zealand Journal of Ecology* 3: 104–16.

Donovan, B.J. (1990) Selection and importation of new pollinators to New Zealand. *New Zealand Entomologist* 13: 26–32.

Dove, A.D.M. (1998) A silent tragedy: parasites and the exotic fishes of Australia. *Proceedings of the Royal Society of Queensland.* 107: 109–113.

Dove, A.D.M., Cribb, T.H., Mockler, S.P., & Lintermans, M. (1997) The Asian fish tapeworm, *Bothriocephalus acheilognathi*, in Australian freshwater fishes. *Marine and Freshwater Research* 48: 181–3.

Drew, R.A.I. (1997) The economic and social impact of the *Bactrocera papayae* Drew and

Hancock (Asian papaya fruit fly) outbreak in Australia. In Allwood, A.J., & Dunn, R.A.I. (eds), *Management of Fruit Flies in the Pacific*, Australian Centre for International Agricultural Research, Canberra.

Ducker, S.C. (ed.). (1988) *The Contented Botanist: Letters of W.H. Harvey about Australia and the Pacific*. Melbourne University Press, Melbourne.

Duncan-Kemp, A.M. (1934) *Our Sandhill Country*. Second edition. Angus & Robertson, Sydney.

Dunn, K.L., Kitching, R.L., & Dexter, E.M. (1994) *The Conservation Status of Australian Butterflies*. A report to Australian National Parks and Wildlife Service, Canberra.

Dyne, G.R., & Wallace, C.C. (1994) Biodiversity and conservation of the earthworm fauna of the Wet Tropics of Queensland's World Heritage Area. *Memoirs of the Queensland Museum* 36: 59–66.

Dyne, G.R., & Walton, D.W. (eds). (1985) *Fauna of Australia*. Vol. 1A. General Articles. AGPS, Canberra.

E

Easteal, S. (1981) The history of introductions of *Bufo marinus* (Amphibia: Anura); a natural experiment in evolution. *Biological Journal of the Linnean Society* 16: 95–113.

Elton, C. (1958) *The Ecology of Invasions of Animals and Plants*. Methuen, London.

Emery, K.O., & Bryan, W.B. (1989) Ballast from H.M.S. *Endeavour* left at Great Barrier Reef, Australia, in 1770. *Proceedings of the Royal Society of Queensland* 100: 71–8.

Emmery, P. (1997) Priorities of beef producers in northern Australia for new forage cultivars. *Tropical Grasslands* 31: 260–65.

Environment & Natural Resources Committee. (1997) *Report on Ballast Water and Hull Fouling in Victoria*. Victorian Government Printer, Melbourne.

Environment & Natural Resources Committee. (1998) *Report on Weeds in Victoria*. Environment & Government Printer, Melbourne.

Ericson, P.G.P, Tyreberg, T, Kjellberg, A.S., Jonsson, L., & Ullen, I. (1997) The earliest record of house sparrows (*Passer domesticus*) in northern Europe. *Journal of Archaeological Science* 24: 183–90.

Evans, L.V. (1981) Marine algae and fouling: a review, with particular reference to ship-fouling. *Botanica Marina* 24: 167–71.

Ewart, A.J. (1909) *The Weeds, Poison Plants, and Naturalized Aliens of Victoria*. Government Printer, Adelaide.

F

Fensham, R.J., & Cowie, I.D. (1998) Alien plant invasions on the Tiwi Islands; entent, implication and priorities for control. *Biological Conservation* 83: 55–68.

Fensham, R.J., Fairfax, R.J., & Cannell, R.J. (1994) The invasion of *Lantana camara* L. in Forty Mile Scrub National Park, north Queensland. *Australian Journal of Ecology* 19: 297–305.

Fisher, H.I. (1948) The question of avian introductions in Hawaii. *Pacific Science* 2: 59–64.

Flannery, T.F. (1994) *The Future Eaters.* Reed, Melbourne.

Flannery, T.F. (1995) *Mammals of the South-West Pacific & Moluccan Islands.* Reed, Sydney.

Flannery, T.F., & White, J.P. (1991) Animal translocation. *National Geographic Research & Exploration* 7(1): 96–113.

Fletcher, R.J. (1993) *Listing of Potential New Crops for Australia.* (Second edition, 1997.) The University of Queensland Gatton College, Lawes.

Flinders. M. (1814) *A Voyage to Terra Australis.* G. & W. Nichol, London.

Forster, R.R., & Forster, L.M. (1973) *New Zealand Spiders: An Introduction.* Collins, Auckland.

Foulis, S.L., & Halfpapp. (no date) Biological control of palm leaf beetle (*Brontispa longissima* Gestro) in far north Queensland. Report by the Department of Primary Industries, Queensland.

Fox, M.D. (1991) Developing control strategies for environmental weeds. *Plant Protection Quarterly* 6: 109–10.

Francis, G.W. (1862) The acclimatisation of plants and animals. Paper read before the Philosophical Society of Adelaide, 1862.

Frank, J.H., & McCoy, E.D. (1995) Invasive adventive insects and other organisms in Florida. *Florida Entomologist* 78(1): 1–15.

Free, J.B. (1982) *Bees and Mankind.* Allen & Unwin, Boston.

Freeland, W.J. (1985) The need to control cane toads. *Search* 16(7–8): 211–5.

Freeland, W.J. (1986) Invasion north. *Australian Natural History* 22(2): 69–72.

Freeland, W.J. (1990) Large herbivorous mammals: exotic species in northern Australia. *Journal of Biogeography* 17: 445–9.

Freeland, W.J. (1992) Water buffalo of the top end. *Australian Natural History* 24(1): 52–9.

Freeland, W.J. (1993) Parasites, pathogens and the impacts of introduced organisms on the balance of nature in Australia. In Moritz, C., & Kikkawa, J. (eds), *Conservation Biology in Australia and Oceania*, Surry Beatty & Sons, Sydney.

Frith, H.J. (1967) *Waterfowl in Australia.* Angus & Robertson, Sydney.

G

Gadek, P. (ed.). (1998) *Phytophthora cinnamomi* Workshop, 28–29 April 1998. Attendees' Manual. James Cook University, Cairns.

Garnett, S. (ed.). (1993) *Threatened and Extinct Birds of Australia.* Second Edition. RAOU Report 82. RAOU, Melbourne.

Gerard, P.J. (1994) Adult development and reproduction in *Anthrenocerus australis* Hope (Coleoptera: Dermestidae). *Journal of Stored Product Research* 30(2): 139–47.

Giles, E. (1875) *Geographic Travels in Central Australia, from 1872 to 1874.* Printed for the author by McCarron, Bird, Melbourne.

Gillbank (1986) The origins of the Acclimatisation Society of Victoria: Practical science in the wake of the Gold Rush. *Historical Records of Australian Science* 6(3): 359–73.

Gledhill, M. & McGrath, P. (1997) Call for a spin doctor. *New Scientist.* 2106: 4–5.

Godwin, H. (1975) *The History of the British Flora.* Cambridge, New York.

Goldsmid, J.M. (1984) The introduction of vectors and disease in Australia: an historical perspective and present-day threat. In Laird, M. (ed.), *Commerce and the Spread of Pests and Disease Vectors*, Praeger, New York.

Goodacre, W.A. (1947) The giant toad (*Bufo marinus*): an enemy of bees. *The Agricultural Gazette* 58: 374–5.

Gough, D. (1994) Palms turn up in Joondalup. *Landscope* 10(1): 6.

Gray, M., & Michael, P.W. (1986) List of plants collected at the old Flemington Saleyards, Sydney, New South Wales. *Plant Protection Quarterly* 1(4): 135–43.

Grey, G. (1841) *Journals of Two Expeditions of Discovery in North-west and Western Australia during the years 1837, '38, and '39*. T. & W. Boone, London.

Griffin, D.A., Thompson, P.A., Bax, N.J., Bradford, R.W., & Hallegraeff, G.M. (1997) The 1995 mass mortality of pilchards: no role found for physical or biological oceanographic factors in Australia. *Marine and Freshwater Research* 48: 27–42.

Griffin, G.F., Stafford Smith, D.M., Morton, S.R., Allan, G.E., Masters, K.A., & Preece, N. (1989) Status and implications of the invasion of tamarisk (*Tamarix aphylla*) on the Finke River, Northern Territory, Australia. *Journal of Environmental Management* 29(4): 297–316.

Griffiths, K. (1997) *Frogs and Reptiles of the Sydney Region*. University of New South Wales Press, Sydney.

Grosholz, E.D. (1996) Contrasting rates of spread for introduced species in terrestrial and marine systems. *Ecology* 77(6): 1680–86.

Grosholz, E.D., & Ruiz, G.M. (1995) Spread and potential impact of the recently introduced European green crab, *Carcinus maenus*, in central California. *Marine Biology* 122: 239–47.

Groves, R.H. (1998) *Recent Incursions of Weeds to Australia 1971–1995*. CRC for Weed Management Systems, Adelaide.

Groves, R.H., & Burdon, J.J. (eds). (1986) *The Ecology of Biological Invasions*. Australian Academy of Science and Cambridge University Press, Canberra.

Guild, E. (1938) Tahitian aviculture: acclimatization of foreign birds. *Aviculture Magazine* 3: 8–11.

Guild, E. (1940) Western bluebirds in Tahiti. *Aviculture Magazine* 5: 284–5.

H

Hacker, J.B. (1997) Priorities and activities of the Australian tropical forages genetic resource base. *Tropical Grasslands* 31: 243–50.

Hacker, J.B., Date, R.A., & Pengelly, B.C. (1997) Conclusions from the workshop: Forage Genetic Resources: Meeting the Requirements of Industry. *Tropical Grasslands* 31: 370–75.

Hall, AV., & Boucher, C.H. (1977) The threat posed by alien weeds to the Cape Flora. In (anon.), *Proceedings of the Second National Weeds Conference of South Africa*. Balkema, Cape Town. Hallegraeff, G.M., & Bolch, C.J. (1991) Transport of toxic dinoglagellate cysts via ships' ballast water. *Marine Pollution Bulletin* 22(1): 27–30.

Hallegraeff, G.M., & Bolch, C.J. (1992) Transport of diatom and dinoflagellate resting spores in ships' ballast water: implications for plankton biogeography and aquaculture. *Journal of Plankton Research* 14(8): 1067–84.

Hallegraeff, G.M., Bolch, C.J., Blackburn, S.I., & Oshima, Y. (1991) Species of the toxigenic dinoflagellate genus *Alexandrium* in southeastern Australian waters. *Botanica Marina* 34: 575–87.

Halliday, T. (1978) *Vanishing Birds.* Hutchinson, Australia.

Hamilton, A.G. (1892) On the effect which settlement in Australia has produced upon indigenous vegetation. *Journal of the Royal Society of New South Wales* 26: 178–239.

Hamilton, A.G. (1937) *Bush Rambles.* Angus & Robertson, Sydney.

Hamr, P. (1992) The Pedder galaxias. *Australian Natural History* 23(12): 904.

Hamr, P. (1994) Threatened fishes of the world: *Galaxias pedderensis* Frankenberg, 1968 (Galaxiidae). *Environmental Biology of Fishes* 43: 406.

Hanna, J.N., Ritchie, S.A., Phillips, D.A., Shield, J., Bailey, M.C., Mackenzie, J.S., Poidinger, M., McCall, B.J., & Mills, P.J. (1996) An outbreak of Japanese encephalitis in the Torres Strait, Australia, 1995. *Medical Journal of Australia* 165: 256–60.

Hardy, A. (1989) *The Nature of Cleland.* State Publishing, South Australia, Adelaide.

Hardy, G.E., & Sivasithamparam, K. (1988) *Phytophthora* spp. associated with container-grown plants in nurseries in Western Australia. *Plant Disease* 72: 435–7.

Haria, A.H. (1995) Hydrological and environmental impact of earthworm depletion by the New Zealand flatworm (*Artioposthia triangulata*). *Journal of Hydrology* 171: 1–3.

Harlan, J.R (1975) *Crops and Man.* American Society of Agronomy, Madison, WI.

Hayward, I.M., & Druce, G.C. (1919) *The Adventive Flora of Tweedside.* T. Bungle & Co., Arbroath.

Heathcote, J., & Maroske, S. (1996) Drifting sand and marram grass on the south-west coast of Victoria in the last century. *Victorian Naturalist* 113(1): 10–15.

Heinsohn, T. (1998) Captive ecology. *Nature Australia* 26(2): 36–43.

Helbaek, H. (1964) The Isca grain, a Roman plant introduction in Britain. *The New Phytologist* 63: 158–64.

Helms, R. (1898) Notes on *Emex australis.* Report of the Seventh Meeting of the Australasian Association for the Advancement of Science.

Henning, M.W. (1956) *Animal Diseases in South Africa.* Central News Agency, Johannesburg.

Heyligers, P.C. (1985) The impact of introduced plants on foredune formation in south-eastern Australia. *Proceedings of the Ecological Society of Australia* 14: 23–41.

Hickman, J.C. (ed.). (1993) *The Jepson Manual: Higher Plants of California.* University of California Press, Berkeley.

Hill, T.C.J., Tippett, J.T., & Shearer, B.L. (1994) Invasion of Bassendean dune Banksia woodland by *Phytophthora cinnamomi. Australian Journal of Botany* 42: 725–38.

Hindwood, K.A. (1940) Birds of Lord Howe Island. *Emu* 40: 1–86.

Hingston, A.B., & McQuillan, P.B. (1998) Does the recently introduced bumblebee *Bombus terrestris* (Apidae) threaten Australia ecosystems? *Australian Journal of Ecology* 23: 539–49.

Hobbs, R.J. (1991) Disturbance a precursor to weed invasion in native vegetation. *Plant Protection Quarterly* 6(3): 99–104.

Hoffman, B.D. (1998) The big-headed ant *Pheidole megacephala*: a new threat to monsoonal northwestern Australia. *Pacific Conservation Biology* 4: 250–55.

Hoffman, B.D., & Hohenhaus, R.N. (1998) *Wildlife Australia* 35(2): 13–16.

Hoffmann, J.H., Moran, V.C., & Zeller, D.A. (1998) Evaluation of *Cactoblastis cactorum* (Lepidoptera: Phycitidae) as a biological control agent of *Opuntia stricta* (Cactaceae) in the Kruger National Park, South Africa. *Biological Control* 12: 20–24.

Holmes, B. (1998) Day of the sparrow. *New Scientist* 158(2140): 32–5.

Holmes, M.J. (1927) Rat control on vessels. *Health* 5(2): 38–45.

Hooker, J.D. (1860) *The Botany of the Antarctic Voyage of H.M. Discovery ships Erebus and Terror, in the years 1839–1845*. Lovell Reeve, London.

Hopkins, G.H.E. (1949) Host-associations of the lice of mammals. *Proceedings of the Zoological Society of London* 119: 387–604.

Hort, A. (1968) *Theophrastus: Enquiry into Plants and Minor Works on Odours and Weather Signs*, William Heinemann.

Howarth, F.G., & Moll, W.P. (1992) *Insects and Their Kin*. University of Hawaii Press, Honolulu.

Humphrey, J.D. (1995) *Australian Quarantine Policies and Practices for Aquatic Animals and their Products: a Review for the Scientific Working Party on Aquatic Animal Quarantine*. Bureau of Resource Sciences, Canberra.

Humphrey, J.D., & Ashburner, L.D. (1993) Spread of the bacterial fish pathogen *Aeromonas salmonicida* after importation of infected goldfish, *Carassius auratus*, into Australia. *Australian Veterinary Journal* 70(12) 453–4.

Humphries, S.E., Groves, R.H., & Mitchell, D.S. (1991) *Plant Invasions of Australian Ecosystems: A status review and management directions*. Kowari 2, Australian National Parks and Wildlife Service, Canberra.

Humphries, S.E., & Stanton, J.P. (1992) *Weed Assessment in the Wet Tropics World Heritage Area of north Queensland*. Report to the Wet Tropics Management Agency.

Hussey, B.M.J., Keighery, G.J., Cousens, R.D., Dodd, J., & Lloyd, S.G. (1997) *Western Weeds: A guide to the weeds of Western Australia*. Plant Protection Society of Western Australia, Perth.

Hutton, I. (1990) *Birds of Lord Howe Island: Past and Present*. Self-published, Coffs Harbour.

Hyatt, A.D., Hine, P.M., Jones, J.B., Whittington, R.J., Kearns, C., Wise, T.G., Crane, M.S., & Williams, L.M. (1997) Epizootic mortality in the pilchard *Sardinops sagax neopilchardus* in Australia and New Zealand in 1995. II. Identification of a herpesvirus within the gill epithelium. *Diseases of Aquatic Organisms*. 28: 17–29.

Hylton, C.G. (1927) Colombo crows reach Australia. *Emu* 27: 44.

I

Illingworth, J.F. (1921) Natural enemies of sugar-cane beetles in Queensland. Division of Entomology *Bulletin No. 13*. Government Printer, Brisbane.

J

Jenkins, C.F.H. (1946) Biological control in Western Australia. *Journal of the Royal Sociey of Western Australia* 32: 1–17.

Jenkins, C.F.H. (1959) Introduced birds in Western Australia. *Emu* 59(3): 201–7.

Jenkins, C.F.H. (1977) *The Noah's Ark Syndrome*. Zoological Gardens Board, Perth.

Jenkins, P.T. (1996) Free trade and exotic species introductions. *Conservation Biology* 10(1): 303–4.

Johnson, P.M., Speare, R., & Beveridge, I. (1998) Mortality in wild and captive rock-wallabies and nailtail wallabies due to hydatid disease caused by *Echinococcus granulosus*. *Australian Mammalogy* 20: 419–23.

Johnston, W.H., Aveyard, J.M., & Legge, K. (1984) Selection and testing of consol lovegrass for soil conservation and pastoral use. *Journal of Soil Conservation* 40(1): 38–45.

Johnstone, R.A. (1984) *Reminiscences of Pioneering in North Queensland*. J.W. Johnstone-Need, Cairns.

Jones, H.D., & Boag, B. (1996) The distribution of New Zealand and Australian terrestrial flatworms (Platyhelminthes: Turbellaria: Tricladida: Terricola) in the British Isles – Scottish survey and Megalab Worms. *Journal of Natural History* 30: 955–75.

Jones, M.M. (1991) Marine Organisms Transported in Ballast Water *Bulletin No. 11*, Bureau of Rural Resources. AGPS, Canberra.

Jones, R., & Jones, D. (1995) Can Leucaena thicken up? *Tropical Grassland Society of Australia Newsletter* 11(4): 8–9.

Julien, M., & White, G. (1997) *Biological Control of Weeds: theory and application*. Australian Centre for International Agricultural Research, Canberra.

K

Kailola, P.J. (1990) Translocated and exotic fishes: towards a cooperative role for industry and government. In Pollard (1990).

Keighery, G.J. (1991) Environmental weeds of Western Australia. List published with Humphries, Groves & Mitchell (1991).

Keighery, G.J. (1996) *Succowia balearica* (Brassicaceae): a new and potentially serious weed in Western Australia. *Nuytsia* 11(1):139–40.

Keighery, G.J., Coates, D.J., & Gibson, N. (1994) Future ecosystems – ecological balance (ecological impact of disease causing fungi in south-western Australia). *Journal of the Royal Society of Western Australia* 77: 181–4.

Kennedy, J., & Weste, G. (1986) Vegetation changes associated with invasion by *Phytophthora cinnamomi* on monitored sites in the Grampians, western Victoria. *Australian Journal of Botany* 34:251–79.

Kibria, G., Nugegoda, D., Fairclough, R., & Lam, P. (1996) Australian native species in aquaculture. *Victorian Naturalist* 113(5): 264–7.

King, C.M. (ed.). (1990) *The Handbook of New Zealand Mammals*. Oxford University Press, Auckland.

Kingsmill, W. (1918) Acclimatisation. *Proceedings of the Royal Society of Western Australia* 5: 33–8.

Kirkpatrick, J.B. (1974) Plant invasion and extinction in a suburban coastal reserve. *Australian Geographic Studies* 12: 107–18.

Kirkpatrick, J.B. (1994) *A Continent Transformed.* Oxford University Press, Melbourne.

Kleinschmidt, H.E., & Johnson, R.W. (1977) Weeds of Queensland. Department of Primary Industries, Brisbane.

Kloot, P.M. (1983) The role of common iceplant (*Mesembryanthemum crystallinum*) in the deterioration of medic pastures. *Australian Journal of Ecology* 8: 301–6.

Kloot, P.M. (1985a) Spread of native Australian plants as weeds in South Australia and in other Mediterranean regions. *Journal of the Adelaide Botanic Gardens* 7(2): 145–57.

Kloot, P.M. (1985b) Plant introductions to South Australia prior to 1840. Journal of the Adelaide Botanic Gardens 7(3): 217–231.

Kloot, P.M. (1987a) The naturalised flora of South Australia 2. Its development through time. *Journal of the Adelaide Botanic Gardens* 10(1): 91–8.

Kloot, P.M. (1987b) The naturalised flora of South Australia 3. Its origin, introduction, distribution, growth forms and significance. *Journal of the Adelaide Botanic Gardens* 10(1): 99–111.

Kluge, R.L., & Neser, S. (1991) Biological control of *Hakea sericea* (Proteaceae) in South Africa. *Agriculture, Ecosystems and Environment* 37: 91–113.

L

La Billardière, J.-J. de. (1800) *Voyage in Search of La Perouse.* Stockdale, London.

Ladson, A.R., & Gerrish, G. (1996) Managing willows along Victorian waterways. In Shephard, R.C.H. (ed.), *Proceedings of the Eleventh Annual Weeds Conference.* Weed Science Society, Melbourne.

Langdon, J.S. (1990) Disease risks of fish introductions and translocations. In Pollard (1990).

Laroche, F.B. (ed.). (1994) *Melaleuca Management Plan for Florida.* Exotic Pest Plant Council.

Last, P.R., & Bruce, B. (1996-97) Spotted Handfish. *Nature Australia* 25(7): 20.

Latz, P. (1995) *Bushfires & Bushtucker: Aboriginal Plant Use in Central Australia.* IAD Press, Alice Springs.

Laurance, W.F., McDonald, K.R., & Speare, R. (1996) Epidemic disease and the catastrophic decline of Australian rain forest frogs. *Conservation Biology* 10(2): 406–13.

Lawrence, D. (1994) Customary exchange across Torres Strait. *Memoirs of the Queensland Museum* 34(2)l: 241–426.

Laws, R.M. (1969) Elephants as agents of habitat and landscape change. *Oikos* 21: 1–15.

Lazarides, M., Cowley, K., & Hohnen, P. (1997) *CSIRO Handbook of Australian Weeds.* CSIRO, Canberra.

Lazo-Wasem, E.A. (1983) Additional records of the terrestrial amphipod *Arcitalitrus sylvaticus* (Haswell, 1880) in California, US. *Crustaceana* 45(2): 213–4.

Leichhardt, L. (1847) *Journal of an Overland Expedition in Australia, from Moreton Bay to Port Essington, a Distance of Upwards of 3000 miles, during the Years 1844–1845.* T. & W. Boone, London.

Leigh, J., Boden, R., & Briggs, J. (1984) *Extinct and Endangered Plants of Australia.* Macmillan, Melbourne.

Lekagul, B., & McNeely, J.A. (1977) *Mammals of Thailand.* Association for the Conservation of Wildlife, Bangkok.

Le Maitre, D.C., Van Wilgen, B.W., Chapman, R.A., & McKelly, D.H. (1996) Invasive plants and water resources in the Western Cape Province, South Africa: modelling the consequences of a lack of management. *Journal of Applied Ecology* 33: 161–72.

Le Souef, A.S. (1923) The Australian native animals. How they stand today, and the cause of the scarcity of certain species. *Australian Zoologist* 3(3): 108–11.

Leutert, A. (1988) Mortality, foliage loss, and possum browsing in southern rata (*Metrosideros umbellata*) in Westland, New Zealand. *New Zealand Journal of Botany* 26: 7–20.

Lever, C. (1985) *Naturalized Mammals of the World.* Longman, London.

Lewis Smith, R.I. (1996) Introduced plants in Antarctica: potential impacts and conservation issues. *Biological Conservation* 74: 135–46.

Liddle, M.J. (1986) Noogoora burr – a successful suite of weeds. In Groves & Burdon (eds) (1986).

Lintermans, M., Rutzou, T., & Kukolic, K. (1990) Introduced fish of the Canberra region – recent range expansions. In Pollard (1990).

Lockwood, J.A. (1993) Environmental issues involved in biological control of rangeland grasshoppers (Orthoptera: Acrididae) with exotic agents. *Environmental Entomology* 22(3): 503–18.

Long, J.L. (1981) *Introduced Birds of the World.* Reed, Sydney.

Long, J.L. (1984) The diets of three species of parrots in the south of Western Australia. *Australian Wildlife Research* 11: 357–71.

Long, J.L. (1988) *Introduced Birds and Mammals in Western Australia.* Technical Series 1. Agriculture Protection Board of Western Australia, Perth.

Longman, H.A. (1923) The rat menace. *Health* 1(2): 35–6.

Lonsdale, W.M. (1994) Inviting trouble: introduced pasture species in northern Australia. *Australian Journal of Ecology* 19: 345–54.

Lonsdale, W.M., & Lane, A.M. (1990) Vehicles as vectors of weed seeds in Kakadu National Park. *Proceedings of the 9th Australian Weeds Conference*, Adelaide 1990. Crop Science Society of South Australia.

Lonsdale, W.M., Miller, I.L., & Forno, I.W. (1989) The biology of Australian weeds 20. *Mimosa pigra* L. *Plant Protection Quarterly* 4(3): 121–30.

Lövei, G.L. (1997) Global Change through Invasion. *Nature* 388: 627–8.

Lovett, J. (1997) Quarantine efforts receive a boost. *Search* 28(9): 280–81.

Low, T. (1985) *Wild Herbs of Australia & New Zealand.* Angus & Robertson, Sydney.

Low, T. (1989) *Bush Tucker.* Angus & Robertson, Sydney.

Low, T. (1990) *Bush Medicine.* Angus & Robertson, Sydney.

Low, T. (1993) *Dinkum Gardening.* Greening Australia, Brisbane.

Low, T. (1994) Invasion of the savage honeyeaters. *Australian Natural History* 24(10): 27–33.

Low, T. (1994-95) Palms: plants with hearts. *Australian Natural History* 24(11): 22–3.

Low, T. (1995a) Mushrooms: sample with care. *Nature Australia* 25(2): 24–5.

Low, T. (1995b) Australia's weed scandal. *Nature Australia* 25(2): 80.

Low, T. (1996–97) Trees of the Future. *Nature Australia* 25(7):46–53.

Low, T. (1997a) Strolls on Stradbroke. *Nature Australia* 25(9): 22–3.

Low, T. (1997b) Invading Africa. *Nature Australia* 25(10): 22–3.

Low, T. (1997c) Tropical Pasture Plants as Weeds. *Tropical Grasslands* 31: 337–43.

Low, T. (1998) Made in Australia. *Nature Australia* 25(12): 22–3.

Low, T. (1999) Superior creations. *Nature Australia* 26(4): 20–21.

Lowery, R. (1996) Should redclaw crayfish be introduced to Fiji? *Pacific Conservation Biology* 2: 312.

Lowry, B. (1991) Integrated production from *Albizia lebbeck* trees and tropical pastures. *Agricultural Science* 4(4): 36–8.

Ludyanskiy, M.L., McDonald, D., & MacNeill, D. (1993) Impact of the zebra mussel, a bivalve invader. *Bioscience* 43(8): 533–44.

M

Macinnis, P. (1996) When learned men schemed to create a land of alien species. *Geo Australasia* 18(3): 68–74.

McCoid, M.J. (1991) Brown tree snake (*Boiga irregularis*) on Guam: a worst case scenario of an introduced predator. *Micronesica* Supplement 3: 63–9.

McCoid, M.J. (1993) The 'new' herpetofauna of Guam, Mariana Islands. *Herpetological Review* 24(1): 16–17.

McCoy, F. (1862) *Acclimatisation, its nature & applicability to Victoria.* Anniversary address delivered at the 1st Annual Meeting of the Acclimatisation Society of Victoria. Wilson & Mackinnon, Melbourne.

McCoy, F. (1878–90) *Natural History of Victoria: Prodromus of the Zoology of Victoria.* Government Printer, Melbourne.

Macdonald, D.W. (1996) Dangerous liaisons and disease. *Nature* 379: 400–401.

Macdonald, I.A.W. (1985) The Australian contribution to southern Africa's invasive alien flora: an ecological analysis. *Proceedings of the Ecological Society of Australia* 14: 225–36.

McDowall, R.M. (1994) *Gamekeepers for the Nation.* Canterbury University Press, Christchurch.

McDowall, R. (ed.). (1996) *Freshwater Fishes of South-Eastern Australia.* Reed, Sydney.

McFadyen, R.E.C. (1998) Biological control of weeds. *Annual Review of Entomology* 43: 369–93.

McFadyen, R.E., & Marohasy, J.J. (1990) A leaf feeding moth *Euclasta whalleyi* [Lep. Pyralidae] for the biological control of *Cryptostegia grandiflora* [Asclepiadaceae] in Queensland, Australia. *Entomophaga* 35(3): 431–5.

McLoughlin, R., & Thresher, R. (1994) The Northern Pacific Seastar: Australia's most damaging marine pest? *Search* 25(3): 69–71.

McIvor, J.G., & McIntyre, S. (1997) Responsible use of exotic tropical pasture cultivars – an ecological framework. *Tropical Grasslands* 31: 332–6.

McKay, R.J. (1984) Introductions of exotic fishes in Australia. In Courtenay, W.R., &

Stauffer, J.R. (eds), *Distribution, Biology and Management of Exotic Fishes*. Johns Hopkins University Press, Baltimore.

McKell, K. (1924?) *Old Days and Gold Days in Victoria*. Edward A. Vidler, Melbourne.

McKeown, S. (1978) *Hawaiian reptiles and amphibians*. Oriental Publishing Company, Honolulu.

Mackey, A.P. (1996) *Cabomba* (Cabomba *spp.*) *in Queensland*. Department of Natural Resources, Brisbane.

Macknight, C.C. (1972) Macassans and Aborigines. *Oceania* 62(4): 283–321.

Macwhirter, P., & Ambrose, S. (1997) Introduction of exotic avian diseases through legal importation. *Ecclectus* 3: 50–52.

Mahood, I.T. (1981) Rusa of Royal National Park. *Australian Deer* 6: 15–21.

Main, H.B.Y. (1984) *Spiders*. Collins, Sydney.

Marshall, J.A. (1973) The British Orthoptera since 1800. In Hawksworth, D.L. (ed.), *The Changing Flora and Fauna of Britain*, Academic Press, London.

Martin, R.W., Handasyde, K.A., & Skerratt, L.F. (1998) Current distribution of sarcoptic mange in wombats. *Australian Veterinary Journal* 76(6): 411–4.

May, S.A., & Norton, T.W. (1996) Influence of fragmentation and disturbance on the potential impact of feral predators on native fauna in Australian forest ecosystems. *Wildlife Research* 23: 387–400.

Medeiros, A.C., Loope, L.L., & Anderson, S.J. (1993) Differential colonization by epiphytes on native (Cibotium spp.) and alien (Cyathea cooperi) tree ferns in a Hawaiian rain forest. *Selbyana* 14: 71–4.

Medeiros, A.C., Loope, L.L., Flynn, T., Anderson, S.J., Cuddihy, L.W., & Wilson, K.A. (1992) Notes on the status of an invasive Australian tree fern (*Cyathea cooperi*) in Hawaiian rain forests. *American Fern Journal* 82(1): 27–33.

Meredith, C.W., Specht, J.R., & Rich, P.V. (1985) A minimum date for Polynesian visitation to Norfolk Island, southwest Pacific, from faunal evidence. *Search* 16: 304–6.

Metro, A. (1976) *Eucalypts for planting*. Food and Agricultural Organisation, Rome.

Meyer, J., & Florence, J. (1996) Tahiti's native flora endangered by the invasion of *Miconia calvescens* DC (Melastomataceae). *Journal of Biogeography* 23: 775–81.

Milburn, C. (1996) Melbourne the weed capital, research finds. *The Age*. 5 May 1996, p. A9.

Miles, J.A.R. (1984) On the Spread of Ross River Virus in the Islands of the South Pacific. In Laird, M. (ed.), *Commerce and the Spread of Pests and Disease Vectors*, Praeger, New York.

Miller, I.L., & Lonsdale, W.M. (1987) Early records of *Mimosa pigra* in the Northern Territory. *Plant Protection Quarterly* 2(3): 140–42.

Mitchell, T.L. (1848) *Journal of an Expedition into the Interior of Tropical Australia, in search of a route from Sydney to the Gulf of Carpentaria*. Longman, Brown, Green, and Longmans, London.

Mlot, C. (1995) In Hawaii, taking inventory of a biological hot spot. *Science* 269: 322–3.

Moldenke, H.N., & Moldenke, B.A. (1952) *Plants of the Bible*. Chronica Botanica Company, Waltham, MT.

Mollison, B. (1994) *Introduction to Permaculture*, Second edition. Tagari Publications, Tyalgum.

Morgan, D.L, Gill, H.S., & Potter, I.C. (1998) Distribution, identification and biology of freshwater fishes in south-western Australia. *Records of the Western Australian Museum*. Supplement No. 56.

Morgan, L.A., & Buttemer, W.A. (1996) Predation by the non-native fish *Gambusia holbrooki* on small *Litoria aurea* and *L. dentata* tadpoles. *Australian Zoologist* 30(2): 143–9.

Morison, A., & Hume, D. (1990) Carp (*Cyprinus carpio* L.) in Australia. In Pollard (1990).

Morris, K., Armstrong, R., Orell, P., & Vance, M. (1998) Bouncing back. *Landscope* 14(1): 28–35.

Morton, S.R., & Martin, A.A. (1979) Feeding ecology of the barn owl, *Tyto alba*, in arid southern Australia. *Australian Wildlife Research* 6: 191–204.

Morton, W.L. (1860) Suggestions for the introduction of animals and agricultural seeds into Victoria. *Transactions of the Royal Society of Victoria* 5: 153–7.

Mott, J.J. (1986) Planned invasions of Australian tropical savannas. In Groves, R.H., & Burdon, J.J. (eds), *Ecology of Biological Invasions: an Australian Perspective*. Australian Academy of Science, Canberra.

Mulvaney, D.J. (1975) *The Prehistory of Australia*. Penguin, Melbourne.

Mungomery, R.W. (1935a) The giant American toad (*Bufo marinus*). *Cane Growers' Quarterly Bulletin* 3(1): 21–27.

Mungomery, R.W. (1935b) A short note on the breeding of *Bufo marinus* in captivity. *Proceedings of the 5th Congress of the International Society of Sugar Cane Technologists*, 589–91.

Mungomery, R.W. (1936) A survey of the feeding habits of the giant toad, (*Bufo marinus* L), and notes on its progress since its introduction into Queensland. *Proceedings of the Annual Conference of the Queensland Society of Sugar Cane Technologists*, 63–74.

Mungomery, R.W. (1937) The present situation regarding the giant American toad in Queensland. *Cane Growers' Quarterly Bulletin* 5(1): 12.

Mungomery, R.W. (1949) Control of the 'greyback' cane grub pest, *Dermolepida albohirtum* Water., by means of 'Gammexane' (Benzene Hexachloride). *The Queensland Journal of Agricultural Science* 6(4): 205–26.

Mungomery, R.W., & Buzacott, J.H. (1935) Control of the 'greyback cane beetle' (Lepidoderma albohirtum Waterh.) in north Queensland. *Proceedings of the 5th Congress of the International Society of Sugar Cane Technologists*, 456–64.

Murray, J., Murray, E., Johnson, M.S., & Clarke, B. (1988) The extinction of *Partula* on Moorea. *Pacific Science* 42(3–4): 150–53.

Myers, R.J.K, & Robbins, G.B. (1991) Sustaining productive pastures in the tropics 5. Maintaining productive sown grass pastures. *Tropical Grasslands* 25: 104–10.

N

Nairn, M.E., Allen, P.G. Inlis, A.R., & Tanner, C. (1996) *Australian Quarantine: A Shared Responsibility*. (Often called the 'Nairn review'.) Department of Primary Industries & Energy, Canberra.

Naylor, R. (1996) Invasions in agriculture: assessing the cost of the golden apple snail in Asia. *Ambio* 25(7): 443–8.

New, T.R. (1994) *Exotic Insects in Australia.* Gleneagles Publishing, Adelaide.

Newhook, F.J., & Podger, F.D. (1972) The role of *Phytophthora cinnamomi* in Australian and New Zealand forests. *Annual Review of Phytopathology* 10: 299–326.

Nicols, A. (1882) *The Acclimatisation of the Salmonidae at the Antipodes: Its History and Results.* Sampson Low, Marston, Searle and Rivington, London.

North, M. (1980) *A Vision of Eden: The Life and Work of Marianne North.* Holt, Rinehart and Winston, New York.

Nunn, M.J. (1995) *Aquatic Animal Quarantine in Australia: Report of the Scientific Working Party on Aquatic Animal Quarantine.* Bureau of Resource Sciences, Canberra.

O

Oakwood, M. & Pritchard, D. (1999) Little evidence of toxoplasmosis in a declining species, the northern quoll (*Dasyurus hallucatus*). *Wildlife Research.* 26: 329–333.

O'Hare, N.K., & Dalrymple, G.H. (1997) Wildlife in southern Everglades wetlands invaded by melaleuca (*Melaleuca quinquenervia*). *Bulletin of the Florida Museum of Natural History* 41(1): 1–68.

Ohmart, C.P., & Edwards, P.B. (1991) Insect herbivory on eucalypts. *Annual Review of Entomology* 36: 637–57.

Oliver, A.J., King, D.R., & Mead, R.J. (1977) The evolution of resistance to fluoroacetate intoxication in mammals. *Search* 8(4): 130–32.

O'Neill, G. (1994) Alligator weed demands guerilla tactics. *Ecos* 81: 10–15.

Ormsby, A.I. (1955) Notes on the giant toad. *Proceedings of the Royal Zoological Society of New South Wales,* 54–5.

Osborne, P.J. (1971) An insect fauna from a Roman site at Alcester, Warwickshire. *Britannia* 2: 156–65.

P

Pain, S. (1987) Australian invader threatens Britain's waterways. *New Scientist* 115(1570): 26.

Panetta, F.D. (1993) A system of assessing proposed plant introductions for weed potential. *Plant Protection Quarterly* 8(1): 10–14.

Panton, W.J. (1993) Changes in post World War II distribution and status of monsoon rainforests in the Darwin area. *Australian Geographer* 24: 50–59.

Parkin, D.T., & Cole, S.R. (1984) Genetic differentiation and rates of evolution in some introduced populations of the house sparrow, *Passer domesticus*, in Australia and New Zealand. *Heredity.* 54: 15–23.

Parsons, J.J. (1972) Spread of African pasture grasses to the American tropics. *Journal of Range Management.* 25: 12–17.

Parsons, W.T., & Cuthbertson, E.G. (1992) *Noxious Weeds of Australia.* Inkata Press, Melbourne.

Paterson, D., & Colgan, K. (1998) *Invasive Marine Species.* AQIS, Canberra.

Pavlov, P.M., Chrome, F.K.J., & Moore, L.A. (1992) Feral pigs, rainforest conservation and exotic disease in north Queensland. *Wildlife Research* 19: 179–93.

Payne, J., & Phillipps, K. (1985) *A Field Guide to the Mammals of Borneo.* The Sabah Society with World Wildlife Fund Malaysia, Sabah.

Pemberton, C.E. (1921) The fern weevil parasite. *The Hawaiian Planters' Record* 25(5): 196–201.

Pernetta, J.C., & Watling, D. (1978) The introduced and native terrestrial vertebrates of Fiji. *Pacific Science* 32(3): 223–43.

Petrie, T. (1904) *Tom Petrie's Reminiscences of Early Queensland.* Watson, Ferguson & Co., Brisbane.

Phillipps, H. (1994) Cane toads with wings. *Wingspan* 13: 11–12.

Pickering, J., & Norris, C.A. (1996) New evidence concerning the extinction of the endemic murid *Rattus macleari* from Christmas Island, Indian Ocean. *Australian Mammalogy* 19: 19–25.

Pigott, P., & Armstrong, R. (1997) Garden plants gone wild. *Landscope* 13(1): 23–7.

Pliny the elder (1940–63) *Natural History.* William Heinemann, London.

Pollard, D.A. (ed.). (1990) Introduced and Translocated Fishes and their Ecological Effects. *Bureau of Rural Resources Proceedings* 8: 98–107.

Pollard, D.A., & Hutchings, P.A. (1990a) A review of exotic marine organisms introduced to the Australian region. I. Fishes. *Asian Fisheries Science* 3: 205–21.

Pollard, D.A., & Hutchings, P.A. (1990b) A review of exotic marine organisms introduced to the Australian region. II. Invertebrates and algae. *Asian Fisheries Science* 3: 223–50.

Ponder, W.F. (1975) The occurrence of *Lymnaea (Pseudosuccinea) columella*, an intermediate host of *Fasciola hepatica* in Australia. *Australian Veterinary Journal* 51: 494–5.

Ponder, W.F. (1986) Mound springs of the Great Artesian Basin. In DeDeckker, P., & Williams, W.D. (eds), *Limnology in Australia*, CSIRO, Melbourne.

Ponder, W.F. (1988) *Potamopyrgus antipodarum* – a molluscan coloniser of Europe and Australia. *Journal of Molluscan Studies* 54: 271–85.

Potter, C. (ed.). (1991) The impact of cats on native wildlife. Proceedings of a workshop, 8–9 May 1991. Australian National Parks and Wildlife Service.

Priddel, D., & Carlile, N. (1995) Mortality of adult Gould's petrels *Petrodroma leucoptera leucoptera* at the nesting site on Cabbage Tree Island, New South Wales. *Emu* 95(4): 259–64.

Pyke, G.H. (1990) Apiarists versus scientists: a bittersweet case. *Australian Natural History.* 23(5): 386–92.

Q

Quammen, D. (1996) *The Song of the Dodo.* Pimlico, London.

R

Raff, J.W. (1956) Australian rain-frog in England. *Victorian Naturalist* 72: 158.

Randall, J.M., & Marinelli, J. (eds). (1996) *Invasive Plants: Weeds of the Global Garden.* Brooklyn Botanic Garden, New York.

Rawling, J. (1994) Australia's environmental weeds - whose responsibility? *Landscape Australia* 16(1): 36–40, 58.

Reed, (1969) *Captain Cook in Australia.* Reed, Sydney.

Richards, J.D., & Short, J. (1996) History of the disappearance of native fauna from the Nullarbor Plain through the eyes of long time resident Amy Crocker. *The Western Australian Naturalist* 21(2): 89–95.

Richardson, A.M.M. (1980) Notes on the occurrence of *Talitris dorrieni* Hunt (Crustacea Amphipoda: Talitridae) in south-west England. *Journal of Natural History* 14: 751–7.

Richardson, D.M. (1998) *Ecology and Biogeography of* Pinus. Cambridge University Press, Cambridge.

Robb, J. (1986) *New Zealand Amphibians & Reptiles.* Collins, Auckland.

Roberts, M. (1991) Origin, dispersal routes, and geographic distribution of *Rattus exulans*, with special reference to New Zealand. *Pacific Science* 45(2): 123–30.

Rodda, G.H., Fritts, T.H., & Conry, P.J. (1992) Origin and population growth of the brown tree snake, *Boiga irregularis*, on Guam. *Pacific Science* 46(1): 46–57.

Roelke-Parker, M.E., Munson, L., Packer, C., Kock, R., Cleaveland, S., Carpenter, M., O'Brien, S.J., Pospischil, A., Hofmann-Lehmann, R., Lutz, H., Mwamengele, G.L.M., Mgasa, M.N., Machange, G.A., Summers, B.A., & Appel, M.J.G. (1996) A canine distemper virus epidemic in Serengeti lions (*Panthera leo*). *Nature* 379: 441–5.

Rolls, E.C. (1969) *They All Ran Wild.* Angus & Robertson, Sydney.

Rostramel, G. (1972) *Iais californica and Sphaeroma quoyanum*, two symbiotic isopods introduced to California (Isopoda, Janiridae and Sphaeromatidae). *Crustaceana.* Supplement II: 193–7.

Runge, C.F. (1994) *Freer Trade, Protected Environment.* Council on Foreign Relations Press, New York.

Russell, R.C., Rajapaksa, N., Whelan, P.I., & Langsford, W.A. (1984). In Laird, M. (ed.), *Commerce and the Spread of Pests and Disease Vectors*, Praeger, New York.

S

Sainty, G., Hosking, J. & Jacobs, S. (1998) *Alps Invaders: Weeds of the Australian High Country.* Australian Alps Liaison Committee.

Sanderson, J.C. (1990) A preliminary survey of the distribution of the introduced macroalga, *Undaria pinnatifida* (Harvey) Suringer on the east coast of Tasmania, Australia. *Botanica Marina* 33: 153–7.

Sands, D., & Scott, S. (1996–97) Richmond birdwing. *Nature Australia* 25(7): 24–9.

Sands, D.P.A. & Coombs, M.T. (1999) Evaluation of the Argentinian parasitoid; *Trichopoda giacomellii* (Diptera: Tachinidae), for biological control of *Nezara viridula* (Hemiptera: Pentatomidae) in Australia. *Biological Control.* 15: 19–24.

Sankaran, K.V., Sutton, B.C., & Balasundaran, M. (1995) *Cryptosporiopsis eucalypti* sp. nov., causing leaf spots of eucalypts in Australia, India and U.S.A. *Mycological Research* 99(7): 827–30. Sainty, G., Hosking, J., & Jacobs, S. (eds). (1998) *Alps Invaders: Weeds of the Australian High Country.* Australian Alps Liaison Committee.

Scott, J.K. (1991) *Acacia karoo* Hayne (Mimosaceae), a potentially serious weed in Australia. *Plant Protection Quarterly* 6(1): 16–18.

Scott, J.K., & Neser, S. (1996) Prospects for the biological control of the environmental weed, *Zantedeschia aethiopica* (arum lily). In Shepherd, R.C.H. (ed.), *Proceedings of the Eleventh Australian Weeds Conference*, Weed Science Society of Victoria, Frankston.

Seabrook, W.A., & Dettmann, E.B. (1996) Roads as activity corridors for cane toads in Australia. *Journal of Wildlife Management* 60(2): 363–8.

Selkirk, P.M., Seppelt, R.D., & Selkirk, D.R. (1990) *Subantarctic Macquarie Island: Environment and Biology*. Cambridge University Press, Cambridge.

Semmens, T.D. (1996) Flower visitation by the bumble bee *Bombus terrestris* (L. Hymenoptera: Apidae) in Tasmania. *Australian Entomologist* 23(2): 33–5.

Semmens, T.D., Turner, E., & Buttermore, R. (1993) *Bombus terrestris* (L.) (Hymenoptera: Apidae) now established in Tasmania. *Journal of the Australian Entomological Society* 32: 346.

Shaner, D.L. (1996) Introduction of transgenic crops and their potential as future weeds. In Shepherd, R.C.H. (ed) *Proceedings of the Eleventh Australian Weeds Conference*. Weed Science Society of Victoria, Frankston.

Sharrock, J.T.R. (1978) Free-winged budgerigars in the Isles of Scilly. *British Birds* 71: 82–3.

Shearer, K.D., & Mulley, J.C. (1978) The introduction and distribution of the carp, *Cyprinus carpio* Linnaeus, in Australia. *Australian Journal of Marine and Freshwater Research* 29: 551–63.

Silcock, R.G., & Johnston, P.W. (1993) Tropical pasture establishment. 9. Establishing new pastures in difficult tropical environments – do we expect too much? *Tropical Grasslands* 27: 349–58.

Simberloff, D., & Stiling, P. (1996a) How risky is biological control? *Ecology* 77(7): 1965–74.

Simberloff, D., & Stiling, P. (1996b) Risk of species introduced for biological control. *Biological Conservation* 78: 185–92.

Simon, N. (1962) *Between the Sunlight and the Thunder: The Wild Life of Kenya*. Collins, London.

Sims, R.W., & Gerard, B.M. (1985) *Earthworms*. E.J. Brill/Dr. W. Blackhuys, London.

Skerratt, L.F., Martin, R.W., & Handasyde, K.A. (1998) Sarcoptic mange in wombats. *Australian Veterinary Journal* 76(6): 408–10.

Slater, P., Slater, P., & Slater, R. (1986) *The Slater Field Guide to Australian Birds*. Rigby Publishers, Sydney.

Sluys, R., Joffe, B., & Cannon, L.R.G. (1995) An alien flatworm in Australian waters. *Memoirs of the Queensland Museum* 38(2): 642.

Smith, A., & Carter, I.D. (1984) International transportation of mosquitoes of public health importance. In Laird, M. (ed.), *Commerce and the Spread of Pests and Disease Vectors*, Praeger, New York.

Smith, N.M. (1995) *Weeds of Natural Ecosystems: A Field Guide to Environmental Weeds of the Northern Territory*, Australia. Environment Centre NT Inc., Darwin.

Smith, R.I.L. (1996) Introduced plants in Antarctica: potential impacts and conservation issues. *Biological Conservation* 76: 135–46.

Snowball, R., Foster, K., & Collins, B. (1992) Australian genetic resources of *Trifolium* and *Ornithopus* species. *Western Australian Journal of Agriculture* 33: 103–109.

Solem, A. (1964) New records of New Caledonian nonmarine mollusks and an analysis of the introduced mollusks. *Pacific Science* 18: 130–7.

Somerfield, K.G. (1977) Insects of economic significance recently recorded in New Zealand. *New Zealand Journal of Agricultural Research* 20: 421–8.

Spencer, D.M. (ed.). (1981) *The Downy Mildews*. Academic Press, London.

Spicer, W.W. (1878) Alien plants. *Papers and Proceedings and Report of the Royal Society of Tasmania for 1877*. 62–9.

Stace, C. (1991) *New Flora of the British Isles*. Cambridge University Press, Cambridge.

Stace, C.A. (1975) (ed.). *Hybridization and the Flora of the British Isles*. Academic Press, London.

Stahle, P. (1997) *Evaluation of the Efficacy of Selected Biological Control Agents for Environmental Weeds. Blackberry and Boneseed/Bitou Bush: 2 Case Studies*. Environment Australia, Canberra.

Stead, D.G. (1935) *The Rabbit in Australia*. Sydney.

Stein, A.B., & Moxley, J.C. (1992) In Defense of the Nonnative: The Case of the Eucalyptus. *Landscape Journal* 11(1): 35–50.

Stevens, W.K. (1996) Bay inhabitants are mostly aliens. *The New York Times*. 20 August 1996. Page C1.

Stockard, J., & Hoye, G. (1990) Wingham Brush: resuscitation of a rainforest. *Australian Natural History* 23(5): 402–9.

Strahan, R. (1995) *The Mammals of Australia*. Reed Books, Sydney.

Summers-Smith (1988) *The Sparrow: A Study of the Genus* Passer. T. & A.D. Poyser, Calton (UK). Sutherst, R.W., Floyd, R.B., & Maywald, G.F. (1995) The potential geographic distribution of the cane toad, *Bufo marinus* L. in Australia. *Conservation Biology* 9(6): 294–99.

Swarbrick, J.T. (1986) History of the lantanas in Australia and origins of the weedy biotypes. *Plant Protection Quarterly* 1(3): 115–21.

Swarbrick, J.T. (1993a) The biology, distribution, impact and control of five weeds of the Wet Tropics World Heritage Area. A report to the Wet Tropics Management Agency.

Swarbrick, J.T. (1993b) The commercial uses of pond apple (*Annona glabra*) and their implications for its declaration in Queensland. A note for the Wet Tropics Management Agency.

Swarbrick, J.T., & Skarratt, D.B. (1994) The ecological requirements and potential Australian distribution of pond apple (*Annona glabra*). A report to the Wet Tropics Management Agency.

Swarbrick, J.T., & Timmins, S.M. (1997) *Annotated Bibliography of Environmental Weeds in Australia and New Zealand*. Environment Australia, Canberra.

Swezey, O.H. (1926) Recent introductions of beneficial insects in Hawaii. *Journal of Economic Entomology* 19: 714–20.

Swezey, O.H. (1928) Present status of certain insect pests under biological control in Hawaii. *Journal of Economic Entomology* 21: 669–76.

T

Taylor, R. (1998) Cunning contraceptions. *Ecos* 95: 20–24.

Telford, I.R. (1986) Dioscoreaceae. *Flora of Australia* 46: 1–66.

Temple, S.A. (1977) Plant-animal mutualism: coevolution with dodo leads to near extinction of plant. *Science* 197: 885–6.

Tenison-Woods, J.E. (1881) On Some Introduced Plants of Australia and Tasmania. *Papers & Proceedings, & Report, of the Royal Society of Tasmania for 1880*, 44–5.

Tesh, R.B., McLean, R.G., Shroyer, D.A., Calisher, C.H., & Rosen, R. (1981) Ross River virus (Togaviridae: *Alphavirus*) infection (epidemic polyarthritis) in American Samoa. *Transactions of the Royal Society of Tropical Medicine and Hygiene* 75(3): 426–431.

Thomson, G.M. (1922) *The Naturalisation of Animals & Plants in New Zealand*. Cambridge University Press, London.

Thwaites, T. (1996) Sweet tooth. *Ecos* 87: 6–11.

Tidemann, S.C., McOrist, S., Woinarski, J.C.Z., & Freeland, W.J. (1992) Parasitism of wild Gouldian finches (*Erythrura gouldiae*) by the air-sac mite *Sternostoma tracheacolum*. *Journal of Wildlife Diseases* 28(1): 80–84.

Tindale, B. (1959) Baron von Mueller gave us blackberries! *Victorian Naturalist* 76: 33.

Tomich, P.Q. (1986) *Mammals in Hawai'i*. Second Edition. Bishop Museum Press, Honolulu.

Travis, J. (1993) Invader threatens Black, Azov Seas. *Science* 262: 1366–7.

Tribe, G.D. (1991) *Drosophila flavohirta* Malloch (Diptera: Drosophilidae) in *Eucalyptus* flowers: Occurrence and parasites in eastern Australia and potential for biological control on *Eucalyptus grandis* in South Africa. *Journal of the Australian Entomological Society* 30: 257–62.

Troughton, E. (1941) *Furred Animals of Australia*. Angus & Robertson, Sydney.

Tyler, M.J. (1976) *Frogs*. Collins, Sydney.

Tyndale-Biscoe, H. (1997a) A fresh approach to quarantine. *Search* 28(2): 54–7.

Tyndale-Biscoe, H. (1997b) Culling koalas with kindness. *Search* 28(8): 250–51.

U

Unger, F. (1866) New Holland in Europe. *Journal of Botany* 3: 39–70.

V

van Riper, S.G., & van Riper, C. (1982) *A Field Guide to the Mammals in Hawaii*. The Oriental Publishing Company, Honolulu.

Varro, M.T. (1934) *On Agriculture*. Heinemann, London.

Viney, C. (1996) Pest control in the deep. *Ecos* 89: 26–7.

Vitousek, P.M. (1994) Beyond global warming: ecology and global change. *Ecology* 75(7): 1861–6.

Vitousek, P.M., D'Antonio, C.M., Loope, L.L., Reimanek, M., & Westbrooks, R. (1997) Introduced species: a significant component of human-caused global change. *New Zealand Journal of Ecology* 21(1): 1–16.

Von Mueller (1870) On the application of phytology to the industrial purposes of life. A popular discourse delivered at the Industrial Museum of Melbourne, on 3 November

1870. Reprinted in Cooper, E. (ed.), *Forest Culture and Eucalyptus Trees.* Cubery and Company, San Francisco.

Von Mueller, F. (1885) *Select Extra-Tropical Plants Readily Eligible for Industrial Culture.* Government Printer, Melbourne.

W

Wace, N.M. (1968) Australian red-backed spiders on Tristan da Cunha. *Australian Journal of Science* 31(5): 189–90.

Wace, N. (1977) Assessment of dispersal of plant species – the car-borne flora in Canberra. *Proceedings of the Ecological Society of Australia* 10: 168–86.

Walker, B., & Weston, E.J. (1990) Pasture development in Queensland – a success story. *Tropical Grasslands* 24: 257–68.

Walker, B., Baker, J., Becker, M., Brunckhorst, R., Heatley, D., Simms, J., Skerman, D.S., & Walsh, S. (1997) Sown pasture priorities for the subtropical and tropical beef industry. *Tropical Grasslands* 31: 266–72.

Walter, D.E., Azam, G.N., Waite, G., & Hargreaves, J. (1998) Risk assessment of an exotic biocontrol agent: *Phytoseiulus persimilis* (Acari: Phytoseiidae) does not establish in rainforest in southeast Queensland. *Australian Journal of Ecology* 23: 587–92.

Walton, C., & Ellis, N. (1997) A manual for using the Weed Risk Assessment system (WRA) to assess new plants. Australian Quarantine and Inspection Service, Canberra.

Watts, C.H.S., & Aslin, H.J. (1981) *The Rodents of Australia.* Angus & Robertson, Sydney.

Webb, C., & Joss, J. (1997) Does predation by the fish *Gambusia holbrooki* (Atheriniformes: Poeciliidae) contribute to declining frog populations? *Australian Zoologist* 30(3): 316–24.

Webb, C.J., Sykes, W.R., & Garnock-Jones, P.J. (1988) *The Flora of New Zealand, Volume IV, Naturalized Pteridophytes, Gymnosperms, and Dicotyledons.* Manaaki Whenua Press, Wellington.

Webb, C.J., Sykes, W.R., Garnock-Jones, P.J., & Brownsey, P.J. (1995) Checklist of dicotyledons, gymnosperms, and pteridophytes naturalised or casual in New Zealand: additional records 1988–1993. *New Zealand Journal of Botany* 33: 151–82.

Weir, A., McLeod, J., & Adams, C.E. (1995) The winter diet and parasitic fauna of a population of Red-necked Wallabies *Macropus rufogriseus* recently introduced to Scotland. *Mammal Review* 25(3): 111–6.

Weste, G. (1993) The cinnamon fungus. Is it a threat to Australian native plants? *Victorian Naturalist.* 110(2): 78–84.

Weste, G., & Ashton, D.H. (1994) Regeneration and survival of indigenous dry sclerophyll species in the Brisbane Ranges, Victoria, after a *Phytophthora cinnamomi* epidemic. *Australian Journal of Botany* 42: 239–53.

Weste, G., & Marks, G.C. (1987) The biology of *Phytophthora cinnamomi* in Australian forests. *Annual Review of Phytopathology* 25: 207–29.

Wheelwright, H.W. (1861) *Bush Wanderings of a Naturalist, or Notes on the Field Sports and Fauna of Australia.* Routledge, Warne & Routledge, London.

White, N.H. (1981) A history of plant pathology in Australia. In Carr, D.J., & Carr, S.G.M. (eds), *Plants and Man in Australia,* Academic Press, Sydney.

White, P., & Anderson, A. (1999) A first for Norfolk. *Nature Australia* 26(6): 26–29.

Whittington, R.J., & Cullis, B. (1988) The susceptibility of salmonid fish to an atypical strain of *Aeromonas salmonicida* that infects goldfish, *Carassius auratus* (L.), in Australia. *Journal of Fish Diseases* 11: 461–71.

Whittington, R.J., Jones, J.B., Hine, P.M., & Hyatt, A.D. (1997) Epizootic mortality in the pilchard *Sardinops sagax neopilchardus* in Australia and New Zealand in 1995. L. Pathology and epizootiology. *Diseases of Aquatic Organisms* 28: 1–16.

Willan, R.C. (1987) The mussel *Musculista senhousia* in Australasia; another aggressive alien highlights the need for quarantine at ports. *Bulletin of Marine Science* 41(2): 475–89.

Williams, G. (1986) Dung beetles. *Australian Natural History* 22(2): 62–5.

Williams, P.A. (1992) *Hakea salicifolia*: biology and role in succession in Abel Tasman National Park, New Zealand. *Journal of the Royal Society of New Zealand* 22(1): 1–18.

Williams, R.J., van der Wal, E.J., & Story, J. (1978) Draft inventory of introduced marine organisms. *Australian Marine Science Bulletin* 61: 12.

Williams, R.J., Griffiths, F.B., Van der Wal, E.J., & Kelly, J. (1988) Cargo vessel ballast water as a vector for the transport of non-indigenous marine species. *Estuarine, Coastal and Shelf Science* 26: 409–20.

Wills, R.T. (1992) The ecological impact of *Phytophthora cinnamomi* in the Stirling Range National Park, Western Australia. *Australian Journal of Ecology* 17: 145–59.

Wilson, B.A., Newell, G., Laidlaw, W.S., & Friend, G. (1994) Impact of plant diseases on faunal communities. *Journal of the Royal Society of Western Australia* 77: 139–43.

Wilson, E. (1858) On the introduction of the British song bird. *Transactions of the Philosophical Institute of Victoria* 11: 77–88.

Wilson, F. (1960) *A Review of the Biological Control of Insects and Weeds in Australia and Australian New Guinea.* Commonwealth Agricultural Bureaux, Bucks, England.

Wilson, F. (1963) *Australia as a Source of Beneficial Insects for Biological Control.* Technical Communication No. 3. Commonwealth Agricultural Bureaux, Bucks.

Wilson, I.M., Walshaw, D.F., & Walker, J. (1965) The new groundsel rust in Britain and its relationship to certain Australasian rusts. *Transactions of the British Mycological Society* 48(4): 501–11.

Wodzicki, K., & Flux, J.E.C. (1967) Re-discovery of the white-throated wallaby, *Macropus parma* Waterhouse 1846, on Kawau Island, New Zealand. *Australian Journal of Science* 29: 429–30.

Wood, M. (1997) Aussie weevil opens attack on rampant melaleuca. *Agricultural Research* 45(12): 4–7.

Wood, M.S., & Wallis, R.L. (1998) Potential competition for nest sites between feral European honeybees (*Apis mellifera*) and common brushtail possums (*Trichosurus vulpecula*). *Australian Mammalogy* 20: 377–81.

Woodward, D.R. (1996) Monitoring for impact of the introduced leafcutting bee, *Megachile rotundata* (F.) (Hymenoptera: Megachilidae), near release sites in South Australia. *Australian Journal of Entomology* 35: 187–91.

Woolls, W. (1884) Plants which have become naturalized in N.S. Wales. *Proceedings of the Linnean Society of New South Wales* 9: 185–205.

Wriggley, J.W., & Fagg, M. (1989) *Banksias, Waratahs & Grevilleas.* Collins, Sydney.

Wright, P.A. (1971) *Memories of a Bushwacker.* University of New England, Armidale.

Y

Yalden, D.W. (1988) Feral wallabies in the Peak District, 1971–1985. *Journal of Zoology,* London 215: 369–74.

Yu, D. (1994) Free trade is green, protectionism is not. *Conservation Biology* 8(4): 989–96.

Yu, D. (1996) New factor in free trade: reply to Jenkins. *Conservation Biology* 10(1): 303–4.

Z

Ziesing, P. (1996) Phillip Island – before and after. *Conservation Australia.* 2: 32–34.

INDEX

Abrus precatorius. See bean, jequirity
acacia, prickly, 33, 86, 181, 218, 293, 302, 318, 319, 329
Acacia arabica, 219
A. auriculiformis. See wattle, earpod
A. baileyana. See wattle, cootamundra
A. catechu. See tree, cutch
A. cyclops, 167
A. giraffae, 219
A. karoo. See thorn, sweet
A. koa, 264
A. longifolia, 167
A. mearnsii, 171
A. melanoxylon. See wattle, blackwood
A. nilotica. See acacia, prickly
A. podalyriifolia. See wattle, Brisbane golden
A. saligna, 167
A. xanthophloea, 219
Acanthaster planci. See crown-of-thorns
Acanthogobius flavimanus, 321
Acanthophis antarcticus, 52
acara, blue, 67, 68, 321
acclimatisation societies, 31–8, 55, 57, 154–5, 195, 314
acclimatisers , 30–8, 56, 126, 150–51, 153–4, 158–9, 173, 195, 293
Acentrogobius flaumi. See goby
Acer pseudoplatanus. See sycamore
Acetosa vesicaria. See dock, bladder
Achatina fulica. See snail, giant African
Achillea millefolium. See yarrow
Acmena smithii, 155
Acridotheres tristis. See myna, Indian
Adansonia gregorii, 264
Aedes albopictus, 130
Aequidens pulcher. See acara, blue
A. rivulatus. See green terror
Aeromonas salmonicida, 70
agaric, fly, 212
agave, 77, 329
Agave sisalana. See sisal
Ageratina adenophora, 42
A. riparia. See mistflower
Ageratum houstonianum. See weed, billygoat
Agrotis infusa, 28
Alauda arvensis. See skylark
albizia, 89
Albizia lebbeck, 89
Alexandrium catavella, 112
A. minutum, 112
A. tamarense, 110, 112
allamanda, 233, 329
Allamanda cathartica. See allamanda
Allen, G.H., 81
alligator weed, 110, 209, 213, 319, 330
Allium triquetrum, 329
Alocasia macrorrhizos. See taro
A rivulatus, 322
alstroemeria, 78
Alstroemeria aurea. See alstroemeria
Alternathera philoxeroides. See weed, alligator
A. sessilis, 209
Amanita muscaria. See fungus, fly agaric
A. phalloides. See fungus, 105
amaranth, 8, 312, 329
Amaranthus spp. *See* amaranth
Ambassis agassizii. See perchlet, olive
Ammophila arenaria. See grass, marram
Amphibromus neesii, 146
Amphilophus citrinellum, 321
amphipod, 148, 161, 324, 325
amulla, 255
Anagallis arvensis, 25
Anas platyrhynchos. See duck, mallard
Andropogon gayanus, 84–5
Annona glabra. See apple, pond
Anopholepis gracilipes, 99
Anredera cordifolia. See vine, Madeira
ant, 4, 51, 101, 154, 194, 237, 325
  big-headed, 186, 194, 235, 236, 238, 329
  crazy, 99
  Iridomyrmex, 208
  South American army, 275
antelope, 32, 33, 84, 125, 183, 195–6, 260, 316
anthracnose, 87
Antilope cervicapra. See blackbuck
Anthrenocerus australis, 146
Aonidiella aurantii, 275
Aphanomyces astaci, 128, 225
aphid, 101, 180, 287
Apis cerana. See bee, Asian honeybee

A. dorsata. See bee, giant honeybee
A. mellifera. See bee, honeybee
apple, 17, 18, 19, 23*n*, 181, 285
  balsam, 260*n*
  Chinese, 22
  kangaroo, 121, 175
  monkey, 155
  pond, xv, 170, 181, 187, 211, 234–5, 236, 240, 319, 329
Aranea pustulos, 255
Araujia hortorum, 45
arawana, 68
Archontophoenix cunninghamiana, 155
Arcitalitrus dorrieni, 141, 148, 325
A. sylvaticus, 161, 325
Arctotheca calendulacea. See capeweed
Argemone ochroleuca, 42
Aristolochia elegans, 77
Artemesia, 10*n*
Artioposthia triangulata, 103–4, 133
Asclepias curassavica. See cotton bush, red-headed
Asimov, Isaac, 238
Asparagus aethiopicus. See fern, asparagus
A. asparagoides. See creeper, bridal
Asterias amurensis. See seastar, Northern Pacific
Astronotus ocellatus, 321
Atriplex, 146
Attenborough, David, 124
aurochs, 201, 204
Australian Academy of Science, 131, 132, 180
Australian Conservation Foundation, 131, 296
Australian Quarantine and Inspection Service (AQIS), 62–5, 67–71, 96–7, 98, 106–7, 109, 112–5, 128, 134, 214, 223–7, 273, 279–89, 300, 305
Avena fatua. See oats, wild
Axis axis. See deer, chital
A. porcinus. See deer, hog
Azadirachta indica, 220

Baccharis halimifolia. See bush, groundsel
Bactrocera cucurbitae, 327
B. frauenfeldi, 326
B. papayae. See fruit fly, papaya
B. philippinensis, 327
B. trivialis, 327
B. tryoni. See fruit fly, Queensland
Bailey, Frederick, 15
ballast, dry, 100, 110
ballast pests, xvi, 97, 108–15, 130–32, 161, 227, 284, 295
  economic impact of, 132
ballast water, 109–15, 284, 311
banana, 14, 15, 23*n*, 201
bandicoot, 125, 126, 151, 193, 194, 240, 251
  eastern barred, 192
  southern barred, 241
  southern brown, 241
Banks, David, 224
Banks, Joseph, 262
banksia, 91, 117, 118–9, 120, 121, 167, 218, 305, 329
  coast, 155
  heath, 167, 220
Banksia ericifolia. See banksia, heath
B. integrifolia, 155
banteng, (Bali), 196–9, 203, 242, 251, 258–9, 260, 320
Barkly, Sir Henry, 30–31
barley, 6, 24, 334
barnacle, xiv, 13, 95, 111, 146, 164, 210, 276, 322, 324
barramundi, 61*n*, 220, 265, 329
Barrass, Ian, 288
basil, sacred, 262
bat, 22, 75, 150
  fruit, 264
Batianoff, George, 76
Batrachochytrium, 124
Baudin, Nicolas, 20, 128
bean, 18
  jequirity, 15, 334
  mung, 262
  phasey, 87
beaver, 183, 202
bêche-de-mer, 14
bee, 101, 194
  honeybee, xv, xvii, 26, 49, 119, 184–5, 187, 189, 194, 203, 224, 243, 281, 305, 330
  Africanised honeybee, 217, 225
  Asian honeybee, 189, 223–4, 226, 227, 228, 232, 236, 243, 306
  bumblebee, 44, 194, 227–8, 243, 304–6
  giant honeybee, 225, 330
  leafcutting, 44, 305–6
  Mexican carpenter, 306
  reed, 240
beech, 104, 212
  antarctic, 212
Beers, Peter, 134
beetle, xiv, 4–5, 51, 95, 99, 101, 154, 163, 179, 227, 325, 328
  'Australian carpet beetle', 146
  common eucalypt longicorn, 164
  dung, 52, 188, 276
  elm leaf, 277
  flour, 7
  greyback cane, 47–8, 50
  jewel, 117
  khapra, 188, 328
  palm leaf, 229, 230, 232, 234
  salvinia, 277
Begic, Zelco, 69–70
Bell, Arthur, 46
bellbird, 151, 175
Bemisia tabaci. See whitefly, poinsettia
Bennett, George, 30
bettong, brush-tailed, 192–3
Bidens pilosa. See cobblers pegs
Bidyanus bidyanus. See perch, silver
bilharzia, 63

billy buttons, 147
biocontrol, 44, 46–54, 121, 127, 143, 159, 161–4, 169, 185, 227, 228, 232, 234, 239, 269–78
Biomphalaria straminea, 63
Bipalium kewense, 106
blackberry, 22, 29, 36, 38, 192, 240, 241, 272, 273, 274, 312, 316, 319, 329
blackbird, 32, 37, 38, 173, 192, 240, 275, 321
blackbuck, 195, 220, 329
black-eyed susan, 233, 331
blackfish, 173, 329
blackwood, 142, 264
bleeding heart, 155
Bligh, Captain William, 18, 19, 259
blight, 4
  azalea petal, 287
  potato, 104, 121, 293
  See also Phytophthora spp.
Blomfield, Charles, 125
blowfly, 154
blue eye, 315, 329
Boiga irregularis, 160
Bombus terrestris. See bee, bumblebee
boneseed, 73, 298, 318, 319, 329
Boopiidae, 142
borer
  European corn, 226, 329
  sugar cane, 327
Borthwick, T., 59
Bos javanicus. See banteng
B. taurus. See cow
Bothriocephalus acheilognathi, 70
Bowman, David, 258
box, brush, 167, 329
boxthorn, African, 36–7, 141, 290, 329
Boyd, A.J., 53
Brachiaria mutica. See grass, para
Brachionichthys hirsutus, 111–12
Brachystegia boehmii, 304
B. spiciformes, 303–4
Brachythecium albicans, 181, 186
Bracteantha bracteata, 148
Bradley, Joan and Eileen, 313
Brassica napus, 23
breadfruit, 18, 20, 23*n*
Bright, Chris, vi, 237
Briza maxima. See grass, quaking
B. minor. See grass, shivery
brolga, 159, 192
Brontispa longissima. See beetle, palm leaf
bronzewing, 151, 159
broom
  English, 36, 75, 79, 209, 329
  Montpellier, 209
  Scotch, 75
Brown, Bruce, 231
Brown, Robert, 25, 262
brumby, 258, 316, 320
Bryophyllum tubiflorum. See mother-of-millions
Bubalus bubalis. See buffalo, water
Buchanan, Robin, 76
buddleia, 75, 298, 308, 329
Buddleja madagascariensis. See buddleia
budgerigar, 140, 149, 151, 167, 324
buffalo, 33, 125
  water, 6–7, 185–7, 197–9, 203, 236, 247, 249, 251, 258, 260, 265, 293, 320
Bufo bufo, 53
B. calamita, 53
B. marinus. See toad, cane
B. periglenes, 124
bug, xiv, 101, 142, 154, 157, 163, 180, 325
  green vegetable, 273
  psyllid, 169, 329
bulbul, red-whiskered, 38, 321
bullfinch, 34
bullfrog, 175
bulrush, 187, 260*n*
bumblebee. *See* bee
Bunce, Daniel, 21
Burke, Don, 292, 297
burr
  Bathurst, 27–8
  Chinese, 15
  noogoora, 276, 329
  urena, 15
bush,
  bitou, 73, 75, 91, 110, 141, 296, 298, 318, 319, 328
  groundsel, 42, 141, 187, 239, 329
  mimosa, xviii, 89, 187, 211, 217, 218, 235, 240, 242, 274, 277, 293, 303, 318, 319, 329
butcherbird, 175
buttercup, small-flowered, 155
butterfly,
  apollo jewel, 235
  banana skipper, 226–7
  big greasy, 77
  cabbage white, 156, 180, 210, 227, 329
  Cairns birdwing, 77
  monarch, 257
  painted lady, 28–9
  palmdart, 174
  Richmond birdwing, 77, 240
  wanderer, 180, 257
button-quail
  black-breasted, 151, 241
  painted, 159
Buzzacott, J.H., 48

cabbage, 36
cabomba, xix*n*, 64–5, 213, 230, 233, 235, 236, 295, 304, 319
Cabomba caroliniana. See cabomba
Cacatua galerita. See cockatoo, sulphur-crested
C. roseicapilla. See galah
cactoblastis, 163, 269–71, 329
Cactoblastis cactorum. See cactoblastis
cactus, 265, 270, 281
  harrisia, 276
  jumping prickly pear, 270
  pitaya, 220
  prickly pear, 36, 73, 82, 163, 188, 212, 250, 269–71, 274, 290, 293, 312, 330
  semaphore, 270

Caenoplana coerulea, 325
C. sulphurea, 325
Cakile edentula, 246
Cakile, 110
Ciedentula, 246
C. maritima, 246
calicivirus, 272
Callipepla californica, 320
calopo, 86–7, 233, 329
Calopogonium mucunoides. See calopo
Calotis hispidula, 147
calotrope, 45*n*
Calotropis procera, 45*n*
camel, one-humped, xvii, 33, 44, 81, 196–7, 199, 251, 316, 320
Camelus dromedarius. See camel, one-humped
Canis lupus. See wolf
C. lupus dingo. See dingo
canna, 300, 329
Canna indica. See canna
Cape York, v, 15, 210, 223, 226–7, 235, 296
caper, 262
capeweed, 28–9, 252
Capparis spinosa, 262
Capra hircus. See goat
Capsella bursa-pastoris, 8
Carassius auratus. See goldfish
Carcinus maenas. See crab, European green
Carduelis carduelis. See goldfinch, European
C. chloris. See greenfinch, European
C. flammea. See redpoll, common
Carduus nutans, 227
Carlton, James, 254
carp, 31, 38, 41, 55, 60–61, 67, 70, 71, 194, 208, 213, 217, 277, 321
Carpobrotus edulis, 252
C. glaucescens. See pigface
Carr, Geoff, 294, 296, 304
cassia, easter, 74
cassowary, 6, 128, 151, 230, 232, 234
Cassytha, 264
C. pubescens, 167
Castanospermum australe, 225
casuarina, 167, 171, 329
Casuarina cunninghamiana. See oak, river
C. equisetifolia. See oak, beach she-oak
cat, vi, xv, 6, 44, 124–5, 126, 139, 142, 190–4, 203, 272, 282, 320, 345
domestication of, 7, 190
feral, 7, 16, 139, 190–4, 251, 256–7, 265, 316
tiger, 125
catchfly, 25
caterpillar, 4, 28, 175
mopane, 302
red-banded mango, 327
catfish, 175
Amazonian zebra, 68
armour-plated, 68
Chinese walking, 167
Catharanthus roseus, 250
Cato, 4
Caulerpa taxifolia, 65–6
celery, wild, 22
Celtis sinensis. See elm, Chinese
Cenchrus ciliaris. See grass, buffel
C. echinatus, 100
Centaurium erythraea, 312
centaury, 312
centro, 84, 86, 90, 233, 329
Centrosema pubescens. See centro
Cephrenes augiades, 174
C. trichopepla, 174
Cerastium glomeratum, 25
Ceratitis capitata, 211
Cernuella virgata, 272
Cervus elaphus. See deer, red
C. timorensis. See deer, rusa
C. unicolor. See deer, sambar
cestrum, green, 45*n*
Cestrum parqui, 45*n*
chaffinch, common, 32, 151, 181, 321
Chamaecytisus palmensis. See tagasaste
Chamaelaucium uncinatum, 155
chamomile, 22, 27
charlock, 4
Chenopodium album. See fat hen
C. auricomiforme, 141, 147
C. murale, 7
Cherax destructor. See crayfish, yabby
C. quadricarinatus. See crayfish, redclaw
cherry, 17, 20
'buttercup brush', 167
cherry-plum, 79, 329
chestnut
horse, 298, 329
Moreton Bay, 255
water, 85
chicken, 17, 18, 20, 24–5, 201, 202, 225
feral, 18, 22
chickweed, mouse-eared, 25
Chilo terrenellus, 327
chilodinella, 70, 255
Chiloscyphus semiteres, 148
Chloris gayana. See grass, Rhodes
Chromolaena odorata. See weed, Siam
Chrysanthemoides monilifera. See bush, bitou
Chrysemys scripta. See slider
Chrysocephalum apiculatum, 147
Chuckrasia velutina, 233
Cichlasoma nigrofasciatum, 321
cichlid, 181
banded, 321
convict, 321
jewel, 321
midas, 321
Niger, 322
pearl, 321
Cinnamomum camphora. See laurel, camphor
Cissus hypoglauca, 121
Citrulus lanatus. See melon
Claridge, George, 144
Clements, Annemarie, 75
Clifford, Trevor, 148
clitoria, 230, 235
Clitoria laurifolia. See clitoria

clover, 21, 23*n*, 91, 219, 329
 red, 304
 subterranean, 91
club-rush, 243, 260*n*
cobblers pegs, 315, 330
Coccinia grandis, 14
Cochlospermum mopane, 302
cockatoo, sulphur-crested, 33, 151, 159, 173, 324
cockroach, 4, 8, 49, 97, 154, 188, 325, 333
coconut, 15, 21
 Also see palm
Cocos nucifera. *See* coconut
cod
 Murray, 33, 173, 183, 329
 trout, 58, 173, 329
Codium fragile tomentosoides, 96
Coffea arabica. *See* coffee
coffee, 17, 21–2, 23*n*, 38, 163, 232, 233, 234, 240, 329
Colocasia esculenta. *See* taro
Columba livia, 320
Commonwealth Scientific and Industrial Research Organisation (CSIRO), 38, 52, 82–91, 114, 180, 219, 221, 228, 272–3, 276, 294, 301, 303
Compere, George, 275
Congeria sallei, 228
Conium maculatum. *See* hemlock
conservation movement, 295–7, 310
Cook, Captain James, 14, 17–18, 110, 259, 314
Cooke, David, 79
Coprinus comatus, 255
Coprosma repens. *See* plant, mirror
coral, lace, 322
Corbett, Laurie, 12–13, 256–7
Corchorus cunninghamii, 241
corella, 173
 long-billed, 241
 western, 241
cormorant, 56–7
Coronopus didymus, 25
Cortaderia richardii, 156
C. selloana. *See* grass, pampas
cotoneaster, 78, 216, 329
Cotoneaster glaucophyllus. *See* cotoneaster
cotton bush, red-headed, 78, 250
Cotula, 146
C. coronopifolia. *See* waterbuttons
Coturnix ypsilophora. *See* quail, brown
cow, 18, 26, 124–5, 147, 154, 185–6, 188, 197, 198, 199, 201, 202, 203, 234, 243, 249, 257
 feral, v, 5, 22, 24–5, 180, 245, 320
cowbird, 175
coypu, 183
crab, 111, 151, 280, 322
 Cornwall, 33
 European green, xv, 110, 194, 252, 276, 329
 red, 99
Crassostrea gigas. *See* oyster, Pacific
Crassula helmsii, 145
Crataegus, 290
Craterocephalus stercusmuscarum. *See* hardyhead
crayfish, 55, 58
 redclaw, 161, 173, 220, 225, 325, 329
 yabby, 220, 331
Creagh, Carson, 106, 284
Crean, Simon, 283
creeper
 bluebell, 174, 292
 bridal, 73, 75, 249, 298, 318, 319, 329
 cats-claw, 77, 249, 277, 329
cress, 19, 21. *See also* watercress
Crocker, Amy, 125, 193
crocodile, 167, 168
 caiman, 66
 salt-water, 45
Crocodylus porosus, 45
Crosby, Alfred, 239
Crotalaria goreensis, 236
crow, 52, 264
 Asian house, 98
crowfoot, native, 147
crown-of-thorns, 175, 329
cryptogenic species, 250–60
Cryptolaemus montrouzieri, 274
Cryptolepis grayi, 273
Cryptosporiopsis eucalyptii, 141
Cryptostegia grandiflora. *See* vine, rubber
Csurhes, Steve, 217
Cucumis melo, 262
Cunningham, Allan, 20
Cupaniopsis anacardioides, 169
currawong, 159, 175, 191
cuscus, 5, 6, 7
 admiralty, 6
Cuscuta campestris. *See* dodder, golden
cut flower trade, 131, 133, 135, 287–8
Cyathea cooperi, 165–6
C. woollsiana, 166
Cyathospirura seurati, 13
Cydia pomonella, 211
Cygnus atratus. *See* swan, black
C. olor, 320
Cynara cardunculus. *See* thistle, artichoke
Cynodon dactylon. *See* grass, blue couch
Cyperus rotundus. *See* nutgrass
Cyprinus carpio. *See* carp
Cytisus scoparius. *See* broom, English

Dacelo novaeguineae. *See* kookaburra, laughing
daddy-long-legs. *See* spider
daffodil, 216
daisy, 146, 147, 305
 Bogan flea, 147
 paper, 148
 Singapore, xiv, 74, 209, 314, 329
Dama dama. *See* deer, fallow

Danaus plexippus. See butterfly, wanderer
dandelion, xvii, 8, 29, 36, 203, 329
Darling, Sir Charles, 31
darnel, 10*n*
Darwin, Charles, 103–4, 166, 263
Dasyurus, 124
D. viverrinus, 125
date, 22, 35
Datura ferox, 100
D. leichhardtii, 253
Deanolis sublimbalis, 327
deer, 33, 44, 183, 153, 197–9, 220, 247–8, 258, 260
  chital, 26, 198, 320
  fallow, 9, 183, 198, 252, 320
  hog, 38, 198, 320
  red, 38, 198, 320
  rusa, 198–9, 320
  sambar, 198, 320
D'Entrecasteaux, Joseph-Antoine, 19, 97
Desomyces albus, 141
devil's fig, 36
devil's twine, 167, 264
Dexter, Raquel, 46, 50
Dickman, Chris, 192, 272
Dicksonia antarctica, 148
dieback. *See* phytophthora
Digitalis purpurea. See foxglove
Digitaria milanjiana, 90
dingo, 7, 12, 44, 124, 126, 141, 174, 182, 191, 256–7, 276, 314, 320
dinoflagellates, toxic, 108, 110, 112, 326
Dioscorea bulbifera, 15
D. pentaphylla, 14–15
Dipsacus fullonum, 43
Dirofilaria immitis, 333
disease
  goldfish ulcer, 70
  internal papillamatous, 128
  Newcastle, 128, 225
  Panama, 328
diseases, xv, 42, 45, 122–8, 220, 224, 226, 230, 239, 253, 255, 277, 327, 328
  bird, 126–8, 135, 225, 326, 327, 328
  crustacean, 110, 128, 225
  fish, xvi, 62–3, 69–71, 110, 113, 122–3, 126, 127, 131, 133–5, 210, 225, 255, 286, 295
  frog, 123–4, 127, 225, 294
  human, 25–6, 60, 64, 101, 128, 141, 224, 327
  livestock and mammal, 63, 123–8, 154, 182, 194, 197, 224–5, 271–2, 288, 326
  plant, 101, 116–125, 120, 225, 287, 326, 327
distemper, canine, 125
Dixon, Samuel, 76
dock, 29, 312, 329
  bladder, 45*n*, 100
  curled, 312
  mud, 260*n*
  swamp, 147, 155
Dodd, Alan, 269, 270
dodder, 3–4, 329
  golden, 99, 327
dodo, 8
Dodonaea viscosa, 264
dog, 6, 7, 26, 51, 125, 159–60, 225, 230, 257
  feral, 12–13, 182
Donax deltoides. See pipi
donkey, 44, 194, 196–9, 203, 252, 320
dove
  bar-shouldered, 159
  diamond, 159
  peaceful, 140, 324
  rock, 320
  Senegal, 211
Dove, Alastair, 255
Dreissena polymorpha, See mussell, zebra
dromedary. *See* camel, one-humped
Druce, George, 147
dryandra, 117, 118
duck, 25, 225
  black, 243
  Burdekin, 247
  mallard, 243, 320
Duncan-Kemp, Alice, 196
Dutchman's pipe, 77
eagle, wedge-tailed, 241–2
earthball, 142
earthworm, xiv, 25, 103, 105, 149, 174, 188, 228, 322, 325
  Amazonian, xiii, 101, 105, 186, 194, 230, 231, 329
echidna, 6
Echinochloa polystachya. See grass, aleman
Echinococcus granulosus, 13, 126
Echium plantagineum. See Patterson's curse
ecotones, 202
ecotourism, 235
Edwards, Les, 221
Edwards, R., 217
Edwardsiella ictaluri, 63
eel, xiv, 33, 34, 55, 123
  electric, 67
egret, cattle, 257, 264
Ehrharta calycina, 85
Eichhornia crassipes. See hyacinth, water
eland, 195–6
elephant
  African, 84, 175, 202, 270, 302
  Indian, v, 183, 202
elm, 4
  Chinese, 314–5, 329
Elminius modestus, 146, 164, 210, 324
Elton, Charles, 146, 237, 242
Emberiza citrinella. See yellowhammer
Emerson, Ralph Waldo, 312
emex, spiny, 43, 141, 329
Emex australis. See emex, spiny
emu, 32, 151, 152, 173
encephalitis, Japanese, 224, 226
England, Australian pests in, 142, 144–9
Entomophaga, 276
Entwisle, Tim, 254
Epiphyas postvittana. See moth, light brown apple
Equus asinus. See donkey
E. caballus. See brumby, horse

Eragrostis curvula. See grass, lovegrass
Eremophila debile, 255
Erica lusitanica. See heath, Spanish
Erionota thrax, 226–7
Erodium crinitum, 146
E. cygnorum, 147
eucalypt, 140, 142–3, 148, 154–6, 157, 161, 163, 164, 167, 171, 175, 208, 220, 225, 240, 252, 259, 305
'Eucalyptopolis', 171
Eucalyptus globulus. See gum, blue
E. gomphocephala, 85
E. marginata. See jarrah
Euclasta gigantalis, 273
E. whalleyi, 273
Eudocima, 276
Euglandia rosea, 274
Euphorbia hirta, 312
E. paralias. See spurge, sea
E. peplus. See spurge, petty
Ewart, Alfred, 29, 31, 300

fairywren, 240, 241
falcon, 191, 242
  peregrine, 242, 264
fanflower, 264
Fasciola gigantea, 63
F. hepatica. See fluke, liver
fat hen, 8, 312, 330
Felis catus. See cat
fennel, 4, 22, 290, 329
fern, 47, 107, 163, 179
  asparagus, 74–5, 209, 212, 272, 329
  peacock, 234
  staghorn, 167
  See also salvinia, tree fern
ferret, 44, 275
fever, Ross River, 141
fever tree, 219
Ficopomatus enigmaticus, 255
Ficus rubiginosa, 155
fig, 4, 19, 22, 23*n*, 300
  African Hottentot, 252
  Port Jackson, 155
finch, 151, 173, 240
  double-barred, 158
  Gouldian, 126–7, 158
  long-tailed, 158
  mannikin, chestnut-breasted, 158–9, 324, 340
  nutmeg, 321
  plum-head, 158
  red-browed, 158, 324, 340
  star, 158
  zebra, 158
  See also firetail
firemouth, 66, 321
firetail
  beautiful, 151
  diamond, 158, 159
fish, aquarium, xvi, xviii, 44, 62–71, 134–5, 181, 220, 255, 284, 288, 295
  ballast arrivals, 110
  mosquito, 41, 44, 59–60, 66, 67, 85, 182, 189, 194, 202, 237, 275, 293, 315, 321
  sportfish, 38, 55–8
  See also trout, salmon
First Fleet, 8, 24–5, 95, 106
fish bait and meal, 97, 123, 134–5, 286–7, 295
flamingo, 31, 33
Flannery, Tim, 44–5, 187
flatworm, 66, 103–4, 105–6, 133, 148, 161, 174, 188, 325
flax, New Zealand, 156
Flindersia brayleyana, 167
Flinders, Matthew, 17, 19–20, 25, 259, 262
Florida, feral species in, xviii, 140–1, 149, 167–70, 173, 183, 188, 220, 235, 239, 270
fluke
  liver, 63–4, 159
  Asian liver, 63
fly, 101, 157, 180, 273, 325
  blowfly, 154
  bush, 276
  Trichopoda giacomelli, 273
  See also fruit fly
Foeniculum vulgare. See fennel
Forsterygion lapillum. See triplefin, variable
fox, vi, xiv, xv, 22, 38, 41, 44, 124–5, 139, 152, 179, 192–4, 203, 209, 241, 244, 251, 257, 265, 272, 312, 316, 320
foxglove, xvii, 329
foxtail, swamp, 253
Francis, George, 33, 37
Frankliniella occidentalis. See thrips, western flower
Franks, Andrew, 77
free trade
  cost estimates of trade-related pests, xv, 113, 132
  ecological implications, 129–35
Freeland, Bill, 124, 258
freesia, xvii, 300, 308, 329
Freesia leichtinii. See freesia
Fringilla coelebs. See chaffinch, common
frog, 48, 51, 52, 58, 71, 124, 127–8, 140, 160, 167, 174, 182, 194, 225, 239, 240, 255, 294, 327
  armoured mistfrog, 124
  brown tree, 149, 154, 324
  disappearance of species, 123–4, 230
  dwarf treefrog, 160, 324
  golden bell, 60, 154, 242, 324
  green swamp, 154, 242, 324
  green treefrog, 124, 127, 221
  mountain mistfrog, 124
  northern platypus, 123–4
  ornate burrowing, 317, 329
  platypus, 123
  poison arrow, 183
  southern day, 123
  white-lipped tree, 182–3, 331
frogmouth, tawny, 159
fruit fly, 226, 232, 326
  mango, 326
  Mediterranean, 211
  melon, 327
  papaya, 226, 230, 280, 294, 326, 327, 329
  Philippine, 327
  Queensland, 173, 276, 329
Fumaria officinalis, 99–100

fumitory, 99–100
fungus, xv, 42, 106, 141, 142, 179, 181, 200, 212, 220, 225, 253, 255, 271, 287
 canker, 121, 175
 chytrid, 71, 124, 194, 234, 255
 cinnamon. *See* phytophthora
 Cryptosporiopsis eucalypti, 141
 deathcap mushroom, 105
 Desomyces albus, 141
 earthballs, 142–3
 Entomophaga sp., 276
 fly agaric, 105, 212
 grape downy mildew, 121
 inkcap, 255
 parasol mushroom, 255
 Spongiophaga communis, 225
 tobacco blue mould, 141, 175, 330
 See also Phytophthora, rust, whitespot
Furneaux, Tobias, 17
fynbos, 141, 167

Gadopsis marmoratus. *See* blackfish
galah, 324
galaxias
 barred, 58
 Clarence, 58
 Pedder, 57
 saddled, 58
 swan, 57–8
Galaxias fontanus, 57–8
G. fuscus, 58
G. johnstoni, 58
G. pedderensis, 57
G. tanycephalus, 58
Gallirallus australis. *See* weka
Gallus gallus. *See* junglefowl, red
Gambusia affinis, 335
G. holbrooki. *See* fish, mosquito
gannet, 122
garden centres, xviii, 65, 78–9, 132, 166, 167, 252, 291, 302, 304, 310
gardens, botanic,
 Adelaide, 33, 219
 Brisbane, 216, 219, 303
 Castlemaine, 219
 Darwin, xviii, 91, 218
 Flecker (Cairns), 216
 Kings Park, 218
 Melbourne, 32, 34, 35, 181, 216, 219
 Perth, 303
 Sydney, 216, 219
garlic, three-cornered, 329
Garnett, Stephen, 127
Gastrolobium, 193
Gecarcoidia natalis, 99
gecko
 Asian house, 99, 207–8, 210, 321
 mourning, 7, 180, 188, 321
Genista monspessulana, 209
Geopelia striata. *See* dove, peaceful
Geophagus brasiliensis, 321
Geraldton wax, 155
geranium, garden, 76
Giles, Ernest, 21
ginger
 common, 15
 wild, 14–15
Girardia tigrina, 66
giraffe, 31, 33, 125
gladiolus, 300, 329
 pink, 242
Gladiolus. *See* gladiolus
 G. caryophyllaceus, 242
Gleditsea triacanthos, 290–1
globalisation of ecology, xiii, 102, 140, 171
Gloriosa superba. *See* lily, gloriosa
glow-worm, 34, 44
Glyceria maxima. *See* grass, reed sweetgrass
glycine, 86, 209, 233, 294, 314, 315, 329
goanna, 44–5, 52, 191, 210
goat, 3, 6, 7, 18, 19, 22, 23, 24, 194, 197–9, 203, 243, 247, 265, 276, 320
goby, 60, 321
 striped, 321
 yellowfin, 321
goldfinch, European, 32, 37, 38, 188, 321
goldfish, 61, 70, 71, 321
Gomphocarpus fruticosus, 25
Gonipterus scutellatus, 164
goose, 18, 25
 Cape barren, 150
 magpie, 85, 247
 snow, 175
gooseberry
 Cape, 25
 native, 253
goosefoot, nett leleaf, 7
Goosem, Steve, 230, 232–3
gorse, 36, 78, 241, 245, 319, 329
gorteria, 264
Gorteria personata, 264
Gould, John, 126
gourami, 71, 329
gourd, ivy, 14
grape, native, 121
grass, 84–91, 187, 212, 287
 African fountain, 253
 aleman, 86, 318
 blady, 239
 blue couch, 253
 buffel, 81–3, 85, 91, 272, 301, 318, 329
 canary, 43, 99, 329
 carpet, v
 Chilean needle, 319
 cord, 243, 247
 couch, 253, 262
 digitaria, 90
 gamba, 84–5
 green panic, 84, 330
 guinea, 85, 233, 314, 315, 316, 329
 hymenachne, xiv, xv, xix*n*, 84, 86, 230, 233, 235, 318, 319, 330
 kangaroo, 119, 261–2, 329
 lovegrass, 91, 213, 330
 marram, 91, 241, 245–6, 252, 329
 Mediterranean barley, 100
 mission, xvi, 84–5, 187, 318, 329
 molasses, 85, 233, 329
 Mossman River, 100
 pampas, 156, 213, 217, 272, 329

para, 60, 84, 85, 186, 233, 236, 315, 318, 330
paspalum, 91, 231, 330
prairie, 21, 23*n*
quaking, 45*n*
reed sweetgrass, 91, 318, 331
Rhodes, 84, 330
rye, 21
sea wheat-grass, 246
setaria, 84, 330
shivery, 25
swamp foxtail, 253
swamp wallaby, 146
tussock, serrated, 319, 327
veldt, 85
winter, 149
grasshopper, 244, 257, 276
grasstree, 117, 119, 330
grayling, Australian, 126
greenfinch, European, 38, 321
green terror, 66, 321
Grevillea robusta. See oak, silky
Grey, George, 17, 20–1, 152, 197
groundsel, 148
See also bush, groundsel
Groves, Richard, xvi, 300
grub, grass, 53
guava, 233, 250, 290, 330
cherry, 241
gudgeon, 315
firetail, 58–9
western carp, 70–71
Guild, Eastham, 158–9
Gulubia costata, 231
gum, 174
blue, 21, 23*n*, 166, 314, 330
river red, 73, 216
guppy, 59, 66, 67, 71, 322
Gutteridge, Tony, 38
Gymnocoronis spilanthoides, 209
Gymnodinium catenatum, 112
Gymnorhina tibicens. See magpie
Gymnothera oblonga, 273

Hacker, Bryan, 84, 87, 89, 91, 294
Haddon, William, 15
hakea, 117, 167
silky, 155, 167, 172
willow-leafed, 155
Hakea drupacea, 167
H. gibbosa, 167
H. salicifolia, 155
H. sericea. See hakea, silky
Hamilton, Alex, 75, 80
handfish, spotted, 111–2
haplochromid, Burton's, 66, 321
Haplochromus burtoni. See haplochromid, Burton's
Hardy, Anne, 240
hardyhead, 315, 330
hare, brown, 44, 320
Haria, A.H., 104
Harlan, Jack, 203
Harris, John, 26
Harrisia, 276
harungana, xv, 229, 230, 232, 293, 304, 330
Harungana madagascariensis. See harungana
Harvey, William, 27, 31
Hawaii, feral species in, 46–7, 49, 140, 142, 159–60, 161, 163, 165–6, 180, 183, 235, 265
hawthorn, 29, 290, 298
Hayward, Ida, 146–7
heath, 117, 120
Spanish, 74, 75, 330
Hedera helix. See ivy, English
hedgehog, 31, 53, 154, 275
Helipterum floribundum, 147
Helix aperta, 211
H. aspera, 9
Hemichromus guttatus, 321
Hemidactylus frenatus. See gecko, Asian house
hemigraphis, 230, 330
Hemigraphis colorata. See hemigraphis
hemlock, 4, 8, 42–3, 78, 330
Herichthys meeki. See firemouth
Heros severus, 321
Heterodoxus spiniger, 141–2
H. longitarsus, 142
Heteropsylla cubana, 329
Hewitt, Chad, 16, 182, 254
hibiscus, 175
Phillip Island, 22–3
swamp, 255
Hibiscus diversifolius, 225
H. insularis, 22–3
Hill, Walter, 21–2
holly, English, 75, 79, 330
Holmes, Genta Hawkins, 135
Holmgren, David, 291
Homogocene, 237–44
honeybee. *See* bee
honeyeater, 117, 119, 224, 305
Hooker, Joseph, 253, 262–3
hopbush, 264
hops, native, 252
Hordium hystrix, 100
horehound, 26, 27, 43
horse, 24–5, 44
feral, 5, 197–9, 245, 252
See also brumby
hull foulers, 95–6, 131
Humphrey, J.D., 69–70, 71, 286
Humphries, Stella, xvi, 233
Hunter, Captain John, 26
hyacinth, water, 47, 72, 73, 213, 275, 318, 330
Hydrocotyle ranunculoides, 65
hydroid, 154
Hylocereus undatus, 220
Hylton, C.G., 98
Hymenachne amplexicaulis. See grass, hymenachne
hypericum, tangled, 218, 221
Hypericum perforatum, 271
H. triquetrum. See hypericum, tangled
Hypochrysops apollo, 235
Hypseleotris galii, 58–9
H. klunzingen, 70–71
hyptis, v, 249, 251
Hyptis suaveolens. See hyptis

Iais californica, 141, 161, 255, 324
iceplant, 248–9, 290, 293, 330

Icerya purchasi. *See* scale, cottony cushion
Idioscopus nitidulus, 328
Ilex aquifolium. *See* holly, English
Imperata cylindrica. *See* grass, blady
Indigofera circinella, 90
influenza, avian, 225, 328
Ingram, Glen, 123
inkweed, 250
Inland Fisheries Commission, 38, 57–8
Invasive Species Management Plan (US), 310
Ischioganus syagrii, 163
Iselin, Margaret, 250
Isolepis cernua, 243
I. marginata, 260*n*
I. prolifera, 243
isopod, 160–4, 324
ivy, English, xv, 74, 75, 79, 249, 272, 290, 298, 330

jacaranda, v, 74, 329
Jacaranda mimosaefolia. *See* jacaranda
Jardine, Frank, 15
jarrah, 117
'jarrah dieback', 117
Jenkins, Peter, 130
Juncus effusus, 147
J. pallidus, 147
J. squarrosus, 105
junglefowl, red, 180, 201, 320
jute, native, 241

Kailola, Patricia, 62
kangaroo, 7, 13, 32, 119, 142, 151, 174, 176, 198
  grey, 199
  red, 199
  tree, 5, 13
kelp
  giant, 254
  Japanese, xvi, 96–7, 109, 110–1, 131, 211, 244, 318, 327, 330
Keighery, Greg, 218
Kennedy, Jill, 119
Khadira, 276
King, Lieutenant Phillip, 20, 25
kingfisher, paradise, v
Kingsmill, Walter, 56, 150
Kirkpatrick, Jamie, 76, 186
kite, 52, 170, 242
Kloot, P.M., 100
koala, 125, 157, 172, 174, 176
kochia, xiv, 326, 330
Kochia scoparia. *See* kochia
Koebele, Albert, 162
koel, 52
Komodo dragon, 44–5
kookaburra, laughing, 151, 152, 159, 173, 324
kudzu, 86
Kuranda Skyrail, 229

La Billardière, Jacques-Julien de, 13, 19, 97
Lachenalia mutabilis, 219
ladybird
  mealybug, 274
  vedalia, 162–3
lagarosiphon, 65, 304, 330
Lagarosiphon major. *See* lagarosiphon
Lampona cylindrata, 154
lamprey, 175
Lampropholis delicata. *See* skink, eastern grass
land reclamation, xiv, 43, 91, 149, 220, 228, 245
lantana, 73, 75, 78, 82, 203, 212, 230, 240–1, 243, 247, 271, 319, 330
  creeping, 74, 330
Lantana camara. *See* lantana
L. montevidensis. *See* lantana, creeping
Lapita voyagers, 7, 15
Lates calcarifer. *See* barramundi
L. niloticus, 61*n*
Latrodectus geometricus, 99
L. hasseltii. *See* spider, redback
Latz, Peter, 251
laurel, camphor, xvi-xvii, 73, 75, 79, 240, 241, 272, 312, 330
Lavandula stoechas. *See* lavender, Spanish
lavender, Spanish, 292, 308, 330
leech, 150
Leichhardt, Ludwig, 21, 187, 197
Leiolopisma hawaiiensis, 160
Le Souef, Albert, 125, 139
Leonotis leonurus, 235
Lepidium peregrinum. *See* peppercress
Lepidodactylus lugubris. *See* gecko, mourning
Lepidoderma albohirtum. *See* beetle, greyback
Leptospermum laevigatum. *See* tea-tree, coast
Lepus capensis. *See* hare, brown
lettuce, water, 309–310
leucaena, xviiii, 86, 91, 213, 237, 272, 290–1, 301, 303, 330
Leucaena leucocephala. *See* leucaena
lichen, 181, 262
Ligustrum lucidum. *See* privet, broadleaf
L. sinense. *See* privet, Chinese
lillypilly, 155
lily
  arum, xv, 74, 76, 79, 272, 300, 330
  flax, 264
  gloriosa, 77, 292–3, 309, 330
Limax maximus, 183
Limnodynastes ornatus. *See* frog, ornate burrowing
lion, 190, 342
  mountain, 44
lion's tail, 235
Litoria aurea. *See* frog, golden bell
L. caerulea. *See* frog, green treefrog
L. ewingii. *See* frog, brown tree
L. fallax. *See* frog, dwarf treefrog
L. infrafrenata, 182

L. *lorica*, 124
L. *nyakalensis*, 124
L. *raniformis*. *See* frog, green swamp
livebearer, one-spot, 322
liverwort, 104, 148, 181
lizard, 174, 191
anole, 169, 286
lobster, 149–50
locust, 4, 54, 173, 263
honey, 290–1
Lolium temulentum, 10*n*
Lonchura castaneothorax. *See* finch, chestnut-breasted mannikin
L. oryzivora, 321
L. punctulata, 321
Longman, Heber, 97–8
Lonsdale, Mark, 84, 88, 296, 301
loosestrife, purple, 262
Lophostemon confertus. *See* box, brush
lorikeet, rainbow, 151, 173–4, 324
lotus, sacred, 262
louse, 24, 141–2, 154
head, 13, 188
Lövei, Gábor, 238, 295
Lowe, Professor Ian, 222
Lowry, Brian, 89
lucerne, 305
Ludwigia peploides, 260*n*
lupin, 91, 252, 330
Lycium ferocissimum. *See* boxthorn, African
Lycodon aulicus, 321
Lygosoma bowringii, 321
Lymantria dispar. *See* moth, Asian gypsy
Lymnaea rubiginosa, 63
lyrebird, 75
Lythrum salicaria, 262

Macadamia tetraphylla. *See* macadamia
Macassan voyagers, 14, 15, 16
macaw, 33, 128
Maccullochella macquariensis, 58
Macfadyena unguis-cati. *See* creeper, cats-claw
Macinnis, Peter, 55
Macrocystis pyrifera, 254
Macrolepiota procera, 255
Macroptilium atropurpureum. *See* siratro
M. lathyroides, 87
Macropus eugenii. *See* wallaby, tammar
M. parma. *See* wallaby, parma
M. rufogriseus. *See* wallaby, red-necked
magpie, 151, 152, 159, 188, 324
magpie-lark, 159, 191
mahogany, East Indian, 233
Mahood, Ian, 199
mange, 128
mango, 22, 38, 163
mannikin
chestnut-breasted, 324
nutmeg, 321
Manorhina melanocephala. *See* miner, noisy
Maoricolpus roseus, 156
Maravalia cryptostegiae, 273
Martin, Roger, 172
Marrubium vulgare. *See* horehound
masasa, 303–4
May, Tom, 181
McCoy, Frederick, 28, 33–4, 55, 195–9, 260
McDonaldization
of world culture, xiv
of biosphere, 238
McKell, Katherine, 26–7
mealybug, citrophilous, 163
medic, 91, 330
Medicago sativa, 305
medicinal plants, Aboriginal, 250
Megachile rotundata. *See* bee, leafcutting
Melaleuca quinquenervia. *See* paperbark, Australian
Melanesian voyagers, 5–6
Melanotaenia, 220
M. duboulayi, 58–9
Meleagris gallopavo. *See* turkey, wild
Melinis minutiflora. *See* grass, molasses
melon, 20, 22, 23*n*, 36, 262
Melopsittacus undulatus. *See* budgerigar
Mesembryanthemum crystallinum. *See* iceplant
mesquite, 36, 41–2, 86, 102, 290, 292, 293, 318, 319, 330
methyl bromide, 289*n*
mfuti, 304
miconia, 159, 215–6, 230, 330
Miconia calvescens. *See* miconia
midge, 163
Mikania micrantha. *See* mile-a-minute
mildew, 4
grape downy, 121
mile-a-minute, 226, 230, 232, 282–3, 303, 330
millipede, xvii, 194, 322,
Portuguese, 295, 330
mimosa, sensitive, 87
Mimosa pigra. *See* bush, mimosa
M. pudica, 87
miner, noisy, 151, 175, 191, 324
mintweed, 100
Misgurnus anguillicaudatus. *See* weatherloach, oriental
mistflower, 42, 187, 315
Mitchell, David, xvi
Mitchell, Major Thomas, 26
mite, v, 15, 25, 42, 104, 106, 179, 287, 322
air-sac, 127
Chilean, 227
European red, 211
mange, 128
varroa, 224, 226, 326
Mnemiopsis leidyi, 132
mole, 4, 53, 54, 104, 275
Mollison, Bill, 202, 290–2
molly, sailfin, 67, 68, 322
Momordica balsamina, 260*n*
M. charantia, 260*n*
mongoose, Indian, 31, 44, 54, 183, 275
monkey, 7, 8, 30–1, 152, 183
mopane, 302

morning glory, 72
Morus alba. See mulberry
mosquito, 58–9, 99, 101, 154, 175, 325
  Asian tiger, 130
moss, v, 104, 174, 181, 186, 252, 253
moth, 99, 101, 273, 276, 325
  Asian gypsy, 225–6, 328
  bogong, 28
  codling, 211
  light brown apple, 154, 163
mother-in-laws tongue, 77, 233, 330
mother-of-millions, 74, 330
mould, tobacco blue, 141, 175, 330
mouse
  field, 4
  house, xv, 7, 126, 182, 192, 241, 242, 320
mulberry, 22, 315, 330
mullein, 29, 78, 330
Mungomery, Reginald, 46–50, 53, 293, 307
Mus musculus. See mouse, house
Musculista senhousia. See mussel, Asian
mushroom, xiv, 140, 212
  See also fungi
mussel, 95, 111, 112
  Asian, 113, 248
  black-striped, 228
  edible, 254
  farming, threats to, 146
  zebra, 113, 130, 132, 228
mustard, hedge, 9
Myiopsitta monarchus. See parakeet, monk
Mylittus edulis planulatus, 254
myna, Indian, xviii, 38, 41, 44, 54, 208–9, 211, 275, 314, 321
Myriophyllum aquaticum. See parrot's feather
Myrmecodia beccarii, 235
Myrsiphyllum asparagoides. See creeper, bridal
myxomatosis, 127, 272
Nairn quarantine review, 107*n*, 132, 223, 261, 279, 281–3, 288, 311
Nannoperca variegata, 58
Nassella charruana, 327
N. neesiana, 319
N. trichotoma. See tussock, serrated
national parks
  Belair, 75, 298–9
  Blue Mountains, 75
  Brisbane Ranges, 118, 120
  Burleigh Heads, 77
  Carnarvon, 85
  Epping Forest, 81
  Expedition Range, 85
  Ferntree Gully, 75
  Forty Mile Scrub, 247
  Grampians, 118, 119
  Gurig, 196, 198, 251, 258
  Hattah-Kulkyne, 174
  Kakadu, xviii, 12, 51, 85, 100, 185, 186, 210, 211, 236, 251, 260, 296, 307
  Lakefield, 235
  Millstream–Chichester, 23
  Mount Glorious, 75
  Mt Kosciuszko, xiii, 149
  Mt Bellenden Ker, 232
  Palm Grove, 85
  Royal, 196, 198–9
  Simpson's Gap, 85
  South West, 120
  Stirling Range 118, 119
  Sundown, 198
  Wilsons Promontory, 118, 174, 246
native-hen, Tasmanian, 12
Naylor, Rosamond, 222
Nelumbo nucifera, 262
nematode, potato cyst, 188
Neochmia temporalis. See finch, red-browed
Neonotonia wightii. See glycine
nettle, 25, 29, 312, 330
Neurostrota gunniella, 273
New Guinea, 5, 6, 15, 160, 226–7, 277, 280, 326
New Ireland, 5, 6
New Zealand, xiv, 18, 33, 65, 80, 104, 111, 133, 140, 141, 142, 150–6, 160, 161, 163, 164, 167, 190–1, 199, 212, 227, 242–3, 304–5
Nezara viridula, 273
Nicols, Arthur, 55, 56
Nicotiniana glauca, 251
nightingale, 32, 150
nightshade, Brazilian, 74
Normanbya normanbyi, 231
North, Marianne, 24, 29
Northern Australian Quarantine Strategy, 226–7, 280, 284, 303
Nothofagus, 212
numbat, 192–3
Nunn, M.J., 69
nurseries. *See* garden centres
Nursery Industry Association of Australia, 78–9
nut, macadamia, 118, 155
nutgrass, 105, 253

oak, 18, 23*n*, 31, 35, 298, 330
  beach she-oak, 168, 171
  river, 168, 315, 330
  she-oak, 155, 243
  silky, 140, 155, 167, 171
oats, wild, 22, 202
Ocimum tenuifolium, 262
Oenothera, 78
O. drummondii, 252
Olea europaea. See olive
olive, xv, 22, 23, 192, 213, 229, 241, 272, 290, 298, 312, 330
olive knot, 272
Omalanthus populifolius, 155
Ommatoiulus moreleti. See millipede, Portuguese
Oncopera, 53
Oncorhynchus gairdneri. See trout, rainbow
Onopordium acanthium. See thistle, Scotch
Operation Western Shield, 192–3
Opuntia spinosissima, 270
O. stricta. See cactus, prickly pear
O. triacantha, 270
orchid, 117, 172, 179

Oreochromis mossambicus. *See* tilapia
Ornithoptera richmondia. *See* butterfly, Richmond birdwing
Oryctes rhinoceros, 328
Oryctolagus cuniculus. *See* rabbit
oscar, 321
Osteoglossum, 68
ostrich, 32, 33, 44, 179, 182, 220, 320
Ostrinia nubilalis. *See* borer, European corn
Othreis, 276
owl, 54, 151, 173, 191
  barn, 242, 264
Oxalis pes-caprae. *See* soursob
O. thompsoniae, 256
Oxylobium, 193
Oxyops vitiosa, 169
oyster, xvii, 44, 55, 96, 156, 179, 254
  Pacific, 108, 252
  pearl, 228
oyster farms, threats to, 108, 112, 146

pademelon, 5, 13
palm, 35, 174, 231, 232–3
  bangalow, 155
  coconut, 14, 20–1, 23*n*, 233, 258, 259–60
  date, 23, 330
Pandorea pandorana, 167
Panicum maximum. *See* grass, guinea
P. maximum var. *trichoglume*. *See* grass, green panic
Panonychus ulmi, 211
Papaver aculeatum, 253
paperbark, Australian (melaleuca), xviii, 140, 141, 167–70, 173, 188, 189, 220–1, 234, 239, 240, 247, 249, 330
parakeet
  Macquarie Island, 156, 191
  monk, xviii, 220, 330
Parakontikia ventrolineata, xiv, 148, 161, 325
Park, King's, 85
Parkinsonia aculeata. *See* thorn, Jerusalem
parrot, 33, 128, 173, 185, 208, 209, 224, 241
  Norfolk Island, 241
  orange-bellied, 241
  See also parakeet, lorikeet
parrot's feather, 65, 315, 330
Parthenium hysterophorus, 319
partridge, 38
Paspalum conjugatum, 231
Passer domesticus. *See* sparrow, house
P. luteus. *See* sparrow, 220
P. montanus, 320
Passiflora foetida. *See* passionfruit, stinking
P. subpeltata, 250
passionfruit, 201
  stinking, 187, 251
  white, 250
pasture plants, 81–91, 233
Patasson nitens, 164
Patterson, David, 254
Paterson's curse, 42, 252, 272
Paulownia, 343
Pavo cristatus. *See* peafowl, Indian
pawpaw (papaya), 201
pea, 18, 117
  Phillip Island glory, 23
  sweet, 216
pea-bush, 196, 197, 305
peacock, 182
peafowl, Indian, 180, 320
pear
  balsam, 260*n*
  prickly. *See* cactus
Pelargonium 'domesticum', 76
penguin, 122, 192
Pennisetum alopecuroides, 253
P. polystachion. *See* grass, mission
P. setaceum, 253
pennyroyal, 43
pennywort, water, 65
peppercress, 141, 147, 194
peppermint, 22
peppertree, Brazilian, 168
Perca fluviatilis. *See* perch, redfin
perch, 209
  Macquarie, 126
  Nile, 61*n*
  pygmy, 58
  redfin, 56, 126, 322
  silver, 70, 220, 330
perchlet, olive, 59, 330
periwinkle, 298
  blue, 73, 330
  Madagascar, 250
  pink, 77, 78, 330
Perkinsiella saccharicida, 163
permaculture, 23, 202, 233, 290–3, 297
Peronospora tabacina. *See* fungus, tobacco blue mould
Pet Industry Joint Advisory Council, 65
petrel, Gould's, 175
Petrogale penicillata. *See* wallaby rock
Phalanger orientalis, 5, 6
Phalaris minor. *See* grass, canary
Phalloceros caudimaculatus, 322
phascogale, 125, 193
Phasianus colchicus. *See* pheasant
pheasant, 9, 32, 33, 38, 180, 252, 320
Pheidole megacephala. *See* ant, big-headed
Pheloung, Paul, 301
Phoenix dactylifera. *See* palm, date
Pholcus phalagiodes. *See* spider, daddy-long-legs
Phoracantha semipunctata, 164
Phormium tenax, 156
phosphonate, 119, 120
Phragmidium violaceum. *See* rust, blackberry
Phycticiplex doddi, 270
Physalis minima, 253
P. peruviana, 25
Phytolacca octandra, 250

phytophthora (cinnamon fungus), 101, 104, 116–21, 194, 212, 225, 229, 230–2, 234, 244, 248, 277, 295, 310, 330
Phytophthora, 104–5, 117, 121, 121*n*, 231, 255
P. cinnamomi. *See* phytophthora
P. citricola, 121
P. cryptogea, 121
P. drescheri, 121
P. infestans. *See* blight, potato
P. megasperma, 121
P. nicotiniana, 121
Phyticiplex doddi, 270
Phytoseiulus persimilis, 227
Pieris rapae. *See* butterfly, cabbage white
pig, 6–7, 17, 18, 24–5, 32, 320
  feral, v, 5, 8, 22, 26, 41, 44, 105, 187, 194, 198–9, 202, 230–1, 234, 247, 265
pigeon, 33
  crested, 149, 151, 159, 191
  flock, 151
  fruit, 241
  spinifex, 159
  squatter, 151
  street, 242
  wonga, 33, 151
  See also bronzewing
pigface, 148, 252
pigweed, 18
pilchard, 122–3, 127, 131, 210, 330, 338
pilchard die-offs, 122–3, 127, 134, 210, 225, 286, 327
pine, 118, 189, 212, 213, 235, 240, 265, 298, 330
  athel, 73, 76, 216, 248, 293, 318, 319, 330
  'Australian pine', 168–70, 243
  Caribbean, 233, 235
  radiata, 209
  slash, 239, 316, 330
pineapple, 118, 231
Pinus caribaea. *See* pine, Caribbean
P. elliottii. *See* pine, slash
P. radiata, 209
pipi, 112, 330
piranha, 66, 67
Pistia stratioites. *See* lettuce, water
Pithecellobium dulce, 219
pittosporum, sweet, 75, 140, 142, 172, 173–4, 176, 330
Pittosporum undulatum. *See* pittosporum, sweet
plague, bubonic, 8, 9, 97–8, 210
plant,
  ant, 235
  asthma, 312
  castor oil, 36, 42, 43, 250, 315, 330
  giant sensitive, 333
  mirror, 155–6, 290, 330
  Senegal tea, 209
  swan, 25
  sweet prayer, 233–4
  white moth, 45
plantain, 8, 25, 186–7
planthopper
  green, 163
  sugarcane, 163
plants, aquarium, xvii, 62–71, 135, 220, 233, 288, 295
Plantago. *See* plantain
P. major, 25
Plasmopara viticola, 121
platy, 66, 67, 68, 208–9, 315, 322
Platycercus elegans. *See* rosella, crimson
P. eximius. *See* rosella, eastern
Platycerium superbum, 167
platypus, 192
Pliny the Elder, 4, 7, 9
Podger, Frank, 117
Poecilia latipinna. *See* molly, sailfin
P. reticulata. *See* guppy
poinciana, 216
Polistes humilis, 154
Pomacea bridgeii. *See* snail, mystery
P. canaliculata. *See* snail, golden apple
Pontoscolex corethrurus. *See* earthworm, Amazonian
poplar, 298, 330
poppy
  Mexican, 42
  native, 253
Populus spp. *See* poplar
Porcellionides pruinosus, 105–6
porcupine, 7, 183
Port Phillip Bay, xiii, 26, 96, 111, 112, 113, 115, 214, 239, 327
Portulaca oleracea, 18
possum, 125, 151–2, 170, 185, 199
  brushtail, 33, 141, 142, 151–4, 173, 188, 243, 324
  honey, 117, 119
  pygmy, 117
  ringtail, 154
pot plants, soil contaminants of, 4, 25, 99, 104, 105–6, 107, 117, 148, 211
Potamopyrgus antipodarum, 156, 188
potato, 18, 19, 23*n*
potoroo, 151
prawn, 33, 154
  farming, 286
Praxelis clematidea, 326
primrose
  beach, 252
  evening, 78
  water, 260*n*
privet, 73, 75, 272, 298
  broadleaf, 209, 330
  Chinese, 330
Prosopis juliflora. *See* mesquite
Protasparagus densiflorus. *See* fern, asparagus
Protroctes maraena, 126
Prunella vulgaris, 262
Prunus cerasifera. *See* cherry-plum
Pseudechis guttatus, 52
P. porphyriacus, 52
Pseudococcus calceolariae, 163

Pseudomonas syringae, 272
Pseudomugil signifer. *See* blue eye
Pseudopeas tuckeri, 324
Pseudoscleropodium purum, 181
Pseudosuccinea columella, 63
Psidium cattelianum, 241
P. guajava. *See* guava
psyllid, *See* bug, psyllid
Ptergygoplichthys, 68
Puccinia graminis, 25
P. lagenophorae, 148–9
P. psidii, 225
Pueraria lobata, 86
P. phaseoloides. *See* puero
puero, 233, 330
Puntius conchonius. *See* rosy barb
purple-top, 78
purslane, 18
Pycnonotus jocosus. *See* bulbul, red-whiskered
Pyrrhaltaluteola, 277
Python, amythestine, 210

quail
  brown, 151, 159, 324
  California, 320
  stubble, 159
Quammen, David, 238
quarantine laws, inconsistencies in, 62–3, 70–1, 134–5
Quercus spp. *See* oak
quoll, 52, 124–5, 151, 191, 193, 210

rabbit, xiii, xv, 4, 8, 9, 18, 22, 23, 25, 26, 38, 41, 124, 127, 139, 142, 147, 173, 179, 182, 185, 192, 193–4, 229, 240, 241–2, 248, 251, 256, 265, 272, 275, 312, 320, 344, 345
rabies, 125
raccoon, 170, 175, 183, 316
radish, 18
  wild, 4, 23
ragwort, 45*n*
rainbowfish, 58–9, 220
Ramphotyphlops braminus. *See* snake, flowerpot
Ranunculus sessiliflorus, 155
rape, oilseed, 23
Raphanus raphanistrum. *See* radish, wild
rat, 8, 97–8, 126, 154, 173, 183, 192, 203
  Asian house, 8, 180, 320
  black, 8, 9, 52, 193, 241, 320
  brown, 9, 320
  bush, 9, 330
  canefield, 9, 54, 142, 173, 330
  kangaroo, 125
  Maclear's, 193
  native bush, 9
  Pacific, 7, 9, 13–14, 44, 320, 334
  water, 52
  white-tailed, 259
Rattus, 9
R. exulans. *See* rat, Pacific
R. fuscipes. *See* rat, bush
R. macleari, 193
R. norvegicus. *See* rat, brown
R. rattus. *See* rat, black
R. sordidus. *See* rat, canefield
R. tanezumi. *See* rat, Asian house
redclaw, *See* crayfish, redclaw
redpoll, common, 182, 321
Retropinna semoni. *See* smelt
Rhadinocentrus ornatus, 60
Rheobatrachus silus, 123
R. vitellinus, 123–4
rice, 6–7, 64, 222
Ricinis communis. *See* castor oil plant
Riley, Charles, 162
rinderpest, 125
Rinodina pyrina, 181
roach, 38, 55, 322
rocket, sea, 110
  American sea, 246
  European sea, 246
Rodolia cardinalis, 162–3
Rolls, Eric, vi, 199, 295
Rorippa nasturtiumaquaticum. *See* watercress
Rosa rubiginosa. *See* sweetbriar
rosella
  crimson, 151, 324
  eastern, 151, 324
rosemary, 18, 23*n*
rosy barb, 67, 181, 329
Rowell, Marc, 259–60
Rubus fruticosus. *See* blackberry
Rumex bidens, 260*n*
R. brownii, 147, 155
R. crispus, 312
Runge, C. Ford, 129–30
rush, 147
  European heath, 105
Russo, Sal, 288
rust, 181, 273
  black stem, 25
  blackberry, 121, 272, 273–4
  groundsel, 148–9
  guava, 225
  rubber vine, 273–4, 278
  white, 288
Rutilis rutilis. *See* roach

Sabella spallazanii. *See* worm, giant sea
Sacculina carcini, 276
Saint-Hillaire, Isidore Geoffroy, 31
St John's wort, 271
Salix babylonica, 331
S. nigra, 78
Salmo trutta. *See* trout, brown
salmon, 37
  Altantic, 56–8, 70
  farms, 113, 286
  imported meat, 133–5
saltbushes, 141, 142, 146, 147, 174, 175
salvation jane, 72
Salvelinus fontinalis, 322
Salvia reflexa, 100
salvinia, 185, 233, 275, 318, 319, 330
Salvinia molesta. *See* salvinia

sanchezia, 229, 230, 232
Sanchezia parvibracteata. *See* sanchezia
San Francisco Bay, exotic pests in, 141, 160–1, 238
Sansevieria trifasciata. See mother-in-laws tongue
Sarcoptes scabiei, 128
Sardinops sagax. See pilchard
Saville-Kent, William, 126
sawfly, 273
scaevola, 264
scale
  cottony cushion, 161–2, 163, 173, 330
  red, 275
scarlet pimpernel, 25
Schefflera actinophylla. See tree, umbrella
Schinus terebinthifolia, 168
Schistosoma mansoni, 63
Scleroderma, 142
Scolymus hispanicus, 42
Scotland, exotic species in, 103–4, 142, 144–9
screwshell, New Zealand, 156
scrubwren, 240
seastar, Northern Pacific, 109, 110–2, 131, 187, 194, 276, 311
seasquirt, 322
seaweed, 13, 65–6, 96, 140, 179, 182
  codium, 96
  See also kelp
seed contaminants, 4, 25, 99–100, 227, 294
Selaginella willdenovii, 234
self-heal, 262
Senecio jacobaca, 45*n*
S. vulgaris, 148
Senna pendula, 74
sensitive plant, giant, 318
Serventy, Dom, 57
Setaria sphacelata. See grass, setaria
Setchelliogaster tenuipes, 141
shag, 56–7
sheep, 6, 18, 25, 26, 33, 44, 124, 152–3, 154, 182, 188, 199, 203
shepherd's purse, 8
shipworms, 95
shrew, 53, 54, 104, 275
  Asian, 226
shrimps, freshwater, 154, 179, 322
Sida rhombifolia. See sida-retusa
S. spinosa, 260*n*
sida, spiny, 260*n*
sida-retusa, 15, 27, 312, 331
Silene gallica, 25
silvereye, 158–9, 173, 324
Simberloff, Daniel, 275
Sinapis arvensis, 4
Sinoxylon conigerum, 327
Siphanta acuta, 163
siratro, 84, 86, 331
sisal, 21
Sisymbrium officinale, 9
skink, 321
  eastern grass, 160, 324
skipper, banana, 226–7
skylark, 32, 37, 38, 321
slater, 105, 174, 322
  delicate, 105–6
sleepers, 215–21, 299, 300, 302, 303, 305–6
slider, 44, 180, 321
slug, 99, 104–5, 275, 322
  great striped garden, 183
smallpox, 25–6, 188
smelt, 315, 331
Smith, Lockwood, 135
smuggling, 226–7, 279–86
snail, xiv, 15, 42, 53, 63, 95, 104–5, 111, 140, 156, 188, 239, 273, 274, 275, 280, 285, 286, 295, 309, 322, 324
  American ribbed fluke, 63
  common garden, 9
  giant African, 188, 223, 230, 234, 274, 331
  golden apple, 64, 222–3, 230, 235, 236, 331
  green, 211
  Mediterranean white, 272
  mystery, 64, 230
  pond, 235
  ribbed fluke, 63
  rosy wolfsnail, 274
snake, 47, 51, 52, 179, 191, 243, 280
  amythestine python, 210
  boa-constrictor, 31
  brown tree, 160
  death adder, 52
  flowerpot, 104, 106, 188, 211, 321, 331
  keelback, 52, 264, 331
  red-bellied black, 52
  spotted black, 52
  wolf, 321
Solanum aviculare. See apple, kangaroo
S. mauritianum. See tobacco, wild
S. seaforthianum, 74
S. torvum, 36
Sollya heterophylla. See creeper, bluebell
Sonchus asper, 4, 331
S. oleraceus. See sowthistle
soursob, 42, 141, 331
South Africa, exotic species in, 140–2, 161, 166, 167, 171, 173, 270, 302
sowthistle, 8, 253, 262, 312, 331
  prickly, 4
sparrow
  Eurasian tree, 320
  house, xiv, 6, 7, 8, 31, 32, 34–5, 37, 38, 41, 44, 98, 151, 188, 203, 211, 220, 242, 251, 320
  Java, 321
  Sudan golden-headed, 220
Spartina maritima, 247
S. townsendii, 243, 247
Spathodea campanulata. See tulip, African
Spenceriella minor, 149, 325
Sphaeroma quoyanum, 160–1, 164, 324, 339
Spicer, Rev. W., 99
spider, 95, 101, 140, 180, 191, 255
  brown widow, 99
  daddy-long-legs, 99, 252, 331
  huntsman, 154
  redback, xiii, 107, 140,

154, 161, 164, 173, 174, 325, 331
whitetailed, 154
Spilocuscus kraemeri, 6
spinebill, eastern, 305
spinifex, beach, 246
Spinifex sericeus, 246
SPS Agreement, 133
sponge, xvii, 95, 96, 225, 323
Spongiophaga communis, 225
spurge
petty, 25, 315, 331
sea, 110, 252, 331
squid, 122
squill, medicinal, 34
squirrel, 34, 44, 170, 183
American grey, 182
Indian palm, 182
Stanley, Trevor, 277
Stanton, Peter, 233
starfish, 322. *See also* seastar, Northern Pacific
starling, xv, 7, 9, 22, 31, 32, 37, 38, 41, 149, 173, 208, 209, 210, 211, 275, 316, 321
Sternostoma tracheacolum, 127
stevia, 217, 221, 303, 331
Stevia eupatoria. *See* stevia
stoat, 150, 154, 275
stonecrop, swamp, 145
storksbill, 146
Streblorrhiza speciosa, 23
Streptopelia chinensis, 320
S. senegalensis, 320
Struthio camelus. *See* ostrich
Sturnus vulgaris. *See* starling
stylo, Townsville, 87, 91
Stylosanthes humilis. *See* stylo, Townsville
succowia, 218, 303
Succowia balearica. *See* succowia
sugar cane, 9, 46–50, 53–4, 84, 163, 223, 326
Suncus murinus, 226
sunfish, ornate, 60
Sus scrofa. *See* pig
swallow, welcome, 151
swan, xvii, 33
black, 151, 158, 324
mute, 320
sweetbriar, 27, 29, 78, 331
swift, 264
swinecress, 25
swordtail, 66, 67, 85, 315, 322, 331
Syagrius fulvitarsus, 163
sycamore, 75, 78, 331
Syzygium paniculatum, 167

tagasaste, 91, 264, 290, 331
Tahiti, introductions, 158–9
tamarind, 14
Tamarindus indica, 14
Tamarix aphylla. *See* pine, athel
tapeworm, 13, 70, 126
Taraxacum officinale. *See* dandelion
tares, 10*n*
taro, 15, 47, 315, 331
Taronga Park Zoo, 182, 196
Tasmanian devil, 12
Tasmanian tiger. *See* thylacine
Taudactylus diurnis, 123
taupata. *See* plant, mirror
Taylor, Alan, 114
teasel, wild, 43
tea-tree, 141, 167, 175, 305
coast, 167, 172–3, 331
paperbark. *See* paperbark
tench, 38, 55, 322
Tench, Watkin, 24
Tenison-Woods, Rev. J.E., 27–8, 31, 78, 99
termite, 51, 327
Thaumastochloa danielii, 233–4
Thayer, Dan, 168
Theba pisana, 272
Thellung, Albert, 147
Themeda triandra. *See* grass, kangaroo
Theophrastus, 3
Thinopyrum junceum, 246
thistle, 3, 203, 239, 274
artichoke, 42, 202
golden, 42
nodding, 227
Scotch, 4, 8, 26–7, 42–3, 78, 251
Thompson, Barbara, 15
thorn, 3, 10
buffalo, 219
Jerusalem, 73, 239–40, 251, 302, 304, 318, 331
Madras, 219
sweet, 217, 218, 331
thornapple, 181
fierce, 100
native, 253
thrips, 101, 287
melon, 326
western flower, 131, 287–8, 326
Thrips palmi, 326
thrush, song, 32, 38, 321
thunbergia, blue, 73, 229, 232, 249, 292, 293, 318, 331
Thunbergia alata. *See* black-eyed susan
T. grandiflora. *See* thunbergia, blue
thylacine, 11–12, 124, 256
Thylogale brownii, 5
tibouchea, 252
tilapia, xv, 41, 66–7, 208–9, 213, 235, 236, 277, 322, 331
Tilapia mariae, 322
Timor, 6, 7
Tinca tinca. *See* tench
tipuana, 315, 331
Tipuana tipu. *See* tipuana
toad, xiv, xv, 51, 150, 327
cane, 41, 44, 46–54, 179, 182, 187, 189, 194, 203, 210, 235, 236, 242, 243, 264, 265, 275, 277, 307, 314, 316–7, 321, 331
European common, 53
golden, 124
natterjack, 53
tobacco
native, 15, 141
South American tree, 251
wild, 187, 344
toetoe, 156
Toxocara canis, 333
Toxoplasma gondii, 126
toxoplasmosis, 124

trade, travel and tourism, significance of growth in, xv, 16, 101–2, 129–35, 156, 235, 288
Tradescantia albiflora. See wandering jew
tree, boab, 264
  cutch, xviii, 218, 331
  fever, 219
  koa, 264
  neem, 220
  umbrella, v, 74–5, 166, 167, 169, 172–3, 209, 313, 331
tree fern, xiv, 35, 165–7, 175
  soft, 148
Trent Richard, 159
tributyl tin (TBT), 96, 311
Trichogaster trichopterus. See gourami
Trichoglossus haematodus. See lorikeet, rainbow
Trichopoda giacomelli, 273
Trichosurus vulpecula. See possum, brushtail
Tridentiger trigonocephalus, 321
Trifolium pratense, 304
triplefin, variable, 254, 321
Triumfetta rhomboidea, 15
Trogoderma granarium. See beetle, khapra
Tropidonophis mairii. See snake, keelback
Troughton, Ellis, 152
trout, xv, 41, 44, 154, 175, 187, 188, 193, 194, 295
  brook, 322
  brown, 38, 56–8, 60, 183, 322
  rainbow, 38, 321
Tryon, Henry, 54
tuart, 85
tuberculosis, bovine, 125
tuckeroo, 169
tulip, African, 234, 331
tuna
  bluefin, 123
  farming, 123, 131, 286
turbina, 229, 230, 232
Turbina corymbosa. See turbina
Turdus merula. See blackbird
T. philomelos. See thrush, song
turkey, 25, 70,
  brush, 149, 151, 173
  wild, 180, 320
turnip, 18
  wild, 100
turtledove
  laughing, 320
  spotted, 320
tussock, serrated, 319, 327
Typha, 187
T. latifolia, 260*n*

Ulex europaeus. See gorse
Uluru, xiii, 100
Undaria pinnatifida. See kelp, Japanese
Unger, F., 263
urchin, sea, 112
Urena lobata, 15
Urtica. See nettle
U. urens, 25

Vancouver, George, 17
Vanessa kershawi, 28–9
Varro, 4
vedalia, 162–3
Verbascum thapsus. See mullein
Verbena bonariensis, 78
V. incompta, 256
V. officinalis, 7
vervain, 7
Vespula germanica. See wasp, European
Vicia sativa, 10*n*
Vigna marina, 262
Vinca major. See periwinkle, blue
vine
  blue trumpet, 292
  Dutchman's pipe, 77
  Madeira, 77, 264, 277, 331
  rubber, 73, 79, 211, 249, 264, 273–4, 277–8, 293, 318, 319
  wonga, 167
Von Mueller, Baron Ferdinand, 35–6, 43, 245, 246, 253, 312
Vulpes vulpes. See fox
vulture, 31, 170

Wachendorfia thyrsiflora, 327
Wagner, Paul, 16
wagtail, Willie, 159, 191
Wallabia bicolor. See wallaby, swamp
wallaby, xiv, 7, 13, 33, 126, 151–2, 183, 193–4, 240, 251
  bridled nailtail, 13, 126
  brush-tailed rock, 152, 159–60
  parma, 152, 242–3, 324
  Proserpine rock, 13, 126
  red-necked (Bennett's), 144–5, 152–3, 324
  rock, 140, 152–3, 192, 242, 324
  swamp, 152, 243, 324
  tammar, 152–3, 324
Walton, Craig, 300–302, 303
wandering jew, 75, 77, 234, 249, 330
wasp, 101, 157, 163, 164, 174, 180, 232, 270, 274, 276, 281, 325
  European, 99, 156, 211, 227, 243, 306, 331
  paperwasp, 154
  sirex, 156, 306
waterbuttons, 253
waterlily, 47, 247, 252, 331
Watersipora arcuata, 96
watercress, 23*n*, 315, 331
waterweeds, 59, 63–7, 110, 209, 213, 275, 309–310
watsonia, 141, 252, 331
Watsonia bulbifera. See watsonia
wattle, 17, 21, 23*n*, 140–41, 154–5, 165, 166, 167, 174, 189, 220, 308, 331
  black, 171
  blackwood, 142, 264
  Brisbane golden, 173, 331
  cootamundra, xiv, 172, 209, 331
  earpod, 166, 169
  golden, 173
weasel, 150, 275

weatherloach, oriental, 67, 209, 213, 321
Webb, Alan, 66
Wedelia trilobata. See daisy, Singapore
weed bans, public reactions to, 79, 169–70
Weed Risk Assessment, 65, 287, 301
weed,
alligator, 110, 209, 213, 319, 331
billygoat, 78, 250
crofton, 42
parthenium, 319
Siam, 99, 226, 227, 230, 232, 235, 303, 326, 328, 331
weeds
biologists' ranking of, 180–81
definition of, xvi
'ecological engineers', xvii, 181
Weeds Cooperative Research Centre, 79
weevil, 24, 97, 157, 169, 185, 274
eucalyptus, 164
large fern, 163
weka, 156, 331
Weste, Gretna, 119
Wet Tropics, 17, 42, 73, 170, 180, 229–35, 259, 296, 304, 326
Wet Tropics Management Authority, 22, 180, 230–3
whale, humpback, 113
wheat, 6, 23*n*, 24
whitefly, poinsettia, 132, 287, 326
whitespot, 70, 255
wildebeest, 152, 261
willow, xix*n*, 79, 213, 217, 248, 298, 300, 319, 331
black, 78
weeping, 330
Wilson, Edward, 30–31, 32
Wilson, Frank, 164
Wingham Scrub, 77
wolf, 6, 13
wombat, 33, 75, 128, 241
northern hairy-nosed, 81, 241
Wonsborough, D., 43
woodchip trade, 109, 111, 294
wool aliens, 146–8
Woolls, Rev. William, 240
World Trade Organization, 133–5
worm, 42, 104, 106, 140, 149, 322
giant sea, 109, 113, 131
marine, 95, 96, 110, 140
New World screw, 225, 331
Old World screw, 225, 327, 331
parasitic, 7, 322
See also earthworm
wormwood, 10*n*
wren
Stephen Island, 190–1, 331
superb blue, 151
Wright, Phillip, 269

Xanthium spinosum, 27–8
X. strumarium. See burr, noogoora
X. pungens. See burr, noogoora
Xanthomonas fragariae, 326
Xanthoria parietina, 181
Xanthorrhoea spp. *See* grasstree
Xenicus lyalli. See wren, Stephen Island
Xiphophorus helleri. See swordtail
X. maculatus. See platy
Xylothrips flavipes, 327

yabby, 173, 220, 331
yam, 13, 202, 227
Asian five-leaf, 14–15
round, 15
yarrow, 78, 331
yellowhammer, 182, 321
Yorkshire fog, 36
Yu, Douglas, 130

Zantedeschia aethiopica. See lily, arum
zebra, 152, 196
Zingiber officinale, 15
Z. zerumbet, 14–15
Zizyphus mauritania, 22
Z. mucronata, 219
Zosterops lateralis. See silvereye